Evolution and Genetics
for Psychology

Evolution and Genetics
for **Psychology**

DANIEL NETTLE

OXFORD
UNIVERSITY PRESS

Great Clarendon Street, Oxford OX2 6DP

Oxford University Press is a department of the University of Oxford.
It furthers the University's objective of excellence in research, scholarship,
and education by publishing worldwide in

Oxford New York

Auckland Cape Town Dar es Salaam Hong Kong Karachi
Kuala Lumpur Madrid Melbourne Mexico City Nairobi
New Delhi Shanghai Taipei Toronto

With offices in

Argentina Austria Brazil Chile Czech Republic France Greece
Guatemala Hungary Italy Japan Poland Portugal Singapore
South Korea Switzerland Thailand Turkey Ukraine Vietnam

Oxford is a registered trade mark of Oxford University Press
in the UK and in certain other countries

Published in the United States
by Oxford University Press Inc., New York

British Library Cataloguing in Publication Data
Data available

Library of Congress Cataloging in Publication Data
Data available

Typeset by Graphicraft Limited, Hong Kong
Printed in Italy by L.E.G.O. S.p.A

ISBN: 978–0–19–923151–5

1 3 5 7 9 10 8 6 4 2

'How extremely stupid not to have thought of that'

T.H. Huxley, on hearing of Darwin's theory of evolution by natural selection

Preface

Those of us who are interested in evolution find ourselves living in exhilarating times. Twenty or 30 years ago, to bring evolutionary or genetic ideas to bear on problems in the human sciences was to risk being reviled or dismissed as a crank. Now, 150 years after the publication of *The Origin of Species*, and 200 years after Darwin's birth, an efflorescence of evolutionary human science is underway. There is serious research in evolutionary psychology (Buss 2005), evolutionary anthropology (Holden & Mace 2003), evolutionary medicine (Nesse & Williams 1995), evolutionary sociology (Hopcroft 2005), evolutionary economics (Gintis 2009), evolutionary linguistics (Hurford 2007), evolutionary legal studies (Hoffman 2008), evolutionary literary studies (Gottschall & Wilson 2005), and other areas. In each of these disciplines, people are discovering that the Darwinian perspective adds value to what they do.

This new Darwinian revolution represents an educational challenge, since students of the various disciplines named above have generally had no training in, and almost no high school exposure to, evolutionary biology. Textbooks have begun to appear, for example of evolutionary psychology (e.g. Buss 2008). These are fine books, but their general strategy is to take the theory of evolution for granted (perhaps sketching it out in a short introductory chapter) and then spend the bulk of their pages reviewing its applications to their particular areas of concern. Whilst this is quite understandable, there is a potential pitfall: research shows that most people in even quite educated sectors of the population misunderstand basic aspects of the theory of evolution in the most fundamental of ways (Shtulman 2006). There is thus a danger that the evolutionary biology that appears in these new research areas will be under-specified, simplistic, or even underlain by misapprehension.

Misunderstanding of evolutionary explanations in the human sciences is rife, amongst their critics and their advocates in more or less equal measure. The frequency with which colleagues earnestly argue that 'X is good for the species because . . .' or that 'An evolutionary explanation of X is inappropriate because X is learned rather than genetically determined' continues to depress me. What is needed is conceptual precision about what evolution is and how it works. As Richard Alexander—characteristically ahead of the game—put it over 20 years ago, what is needed for Darwinian ideas to be fruitfully deployed is a 'broad understanding of basic biological principles', to obtain which, 'significant curricular revisions will be required, introducing biology where it is now often completely absent, in the training of human-oriented scientists, lawyers, philosophers, and others' (Alexander 1987: 5–6).

The aim of this book, then, is not to teach evolutionary psychology or behaviour genetics (there are existing texts for this), but rather to teach the fundamental principles of evolutionary biology that one needs in order to go on and apply evolutionary and genetic arguments to any particular subject matter. I have called the book *Evolution and Genetics for Psychology*, since it is psychology that I mainly teach, but it could equally well have been called *Evolution and Genetics for Anthropology, for Social Policy, for Doctors*, or whatever. I hope biology students might use it too, since its subject matter is the very same theory of evolution that is a central part of their curriculum. This book differs from standard evolution textbooks only in brevity, a special focus on behaviour, a neglect of macro-evolution, and a relative predominance of examples from animals and from humans in particular.

The book falls into two parts (echoing a similar recent approach by Wilson 2007). Part one (Chapters 1–5) lays out the fundamental toolkit one needs in order to think in evolutionary terms, including the necessary building blocks of genetics. Some of this is hard work, but it needs to be mastered in order to have fun with the toolkit later on. These five chapters all need to be tackled, and in order. Part two (Chapters 6–11) comprises applications of the toolkit to some of the key issues that preoccupy us in life: sex, life history, social relationships, learning, human evolutionary history, and our present situation. I could have chosen other applications, but these seemed to be the most compelling. The chapters in part two can be tackled in any order or combination, but only once part one has been digested.

I hope users of this book will find that it answers a need for general undergraduate education in evolutionary biology. The book is meant to be one long argument, rather than one long list. I make no attempt to provide a comprehensive literature review for the areas covered (and I hereby acknowledge those many authors whose work I read in preparation of this book but do not directly cite in the text). Rather, I have sought to discuss the key principles, each with one or two referenced examples from the recent literature, in the hope that readers might pursue areas of interest further for themselves. I apologize for all the simplifications, selectivity, omissions, and errors I have doubtless made. My aim has been clarity rather than exhaustiveness. I want to show how the various components of the theory—variation, heredity, selection, and adaptation—fit together, and how their joint entailments apply to our lives.

Evolutionary theory is the most important body of ideas in the life sciences, perhaps in any science. It is worth learning it for this reason alone. I hope it will be as exciting to readers of this book as it was for me. I did not know that I wanted to study evolutionary biology until I did so, and when I did, it is scarcely an exaggeration to say that it changed my life. I was a student of philosophy and psychology who took an optional course on animal behaviour with Marian Stamp Dawkins. The Darwinian imagination became a friend for life.

Once grasped, this way of thinking begins to infuse one's understanding of big issues and the tiny interactions, work and play, sickness and health, life and death. It does not reduce the complexity or beauty of human life. If anything, it enhances it. I believe that human beings are to a considerable extent the products of their social environment and cultural history, that political progress is desirable and possible, that art is desperately important, and that the universe is a sublime spiritual mystery. None of these is contradicted by my enthusiastic Darwinism. On the contrary, my Darwinism helps me make sense of why these things should all be true. That is its beauty and its power. There is grandeur in this view of life. Enjoy it.

D.N.
Newcastle, 2008

Short contents

Long contents

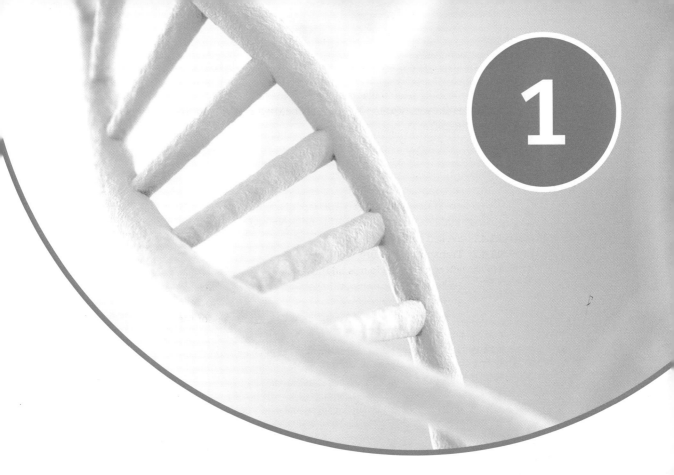

The significance of Darwinism

When John Simmons compiled a list of the most influential scientists of all time, he put Charles Darwin in fourth place (Simmons 1996). This is an astonishingly high ranking for a retiring Victorian gentleman naturalist who spent much of his life researching the habits of barnacles and earthworms (Figure 1.1). Darwin is well ahead of Copernicus, who established that the earth went around the sun, ahead of Galileo, father of the modern methods of physics and astronomy, and ahead of Lavoisier, father of the periodic table and modern chemistry. Darwin has a higher position than any other life scientist. The top three—Newton, Einstein, and Bohr—are all physicists who established the fundamental laws of the physical universe. Simmons ranked Darwin so highly because Darwin achieved for biology something comparable to what the top three achieved for physics: he set out, in his theory of evolution, the fundamental principles governing what happens. Interestingly, another 16 of Simmons' top 100 scientists are people

Figure 1.1 **Charles Darwin, 1809–92, father of the theory of evolution.** *Courtesy of The Library of Congress.*

who worked directly on developing the modern understanding of evolution. Thus, evolution gets 17 out of 100 of the top places. By comparison, the whole of psychology gets nine entries in the top 100, with anthropology contributing a couple more.

Of course, there are no completely objective criteria for compiling a list of the most import-ant scientists and if someone other than John Simmons had attempted it, they might have produced a different ranking. However, whoever did it would be certain to include Darwin in their overall top ten, probably as the most important life scientist, and would be likely to have included several of the other key figures in evolutionary thought as well. Almost all scientists agree that evolution by natural selection is one of the most important ideas in all of science. This is because it is the central explanatory theory of biology. It, and only it, explains why living things are as they are. As the great Brazilian evolutionist Theodosius Dobzhansky (number 67 on Simmons' list) put it, 'nothing in biology makes sense except in the light of evolution' (Dobzhansky 1973). Moreover, scientists accept that the truth of the theory of evolution is as well established as the truth of any major scientific paradigm.

How striking then, that the public at large is nowhere near as convinced. Careful survey evidence shows that only 15% of Americans and a little over 30% of British people think that the theory of evolution is 'definitely true' (Miller *et al.* 2006). Another 20% of Americans and 35% of British people think it is 'probably true', but the numbers believing it to be probably or definitely false are 30% in America and 15% in Britain. The rest are not sure.

There is one thing that unites those who reject the theory of evolution with quite a few of those who accept it: they are prone to misunderstand the very theory they are taking a view on. There are numerous cultural and educational reasons why misunderstanding of evolution persists, but there is also a deeper reason, which is that evolution is actually quite hard to understand. The theory is a strange combination of extreme simplicity and deceptively subtle consequences. It took all of human history until Darwin for anyone to grasp the concept of natural selection, and even once it had been grasped, biologists struggled to understand how it could be reconciled with the mechanisms of heredity, or what kinds of competition would occur. It was not for several generations after Darwin that these problems were solved. Given that some of the greatest minds of all times have had trouble getting these processes clear, the rest of us should not be too hard on ourselves for getting muddled about them.

When we try to understand evolution, we are pushing our minds somewhere that they may not be predisposed to go. For a start, the timescale is very different from the timescales that the human mind usually has to deal with. The cats you will see in your old age will look much the same to you as the cats you saw in your youth. Thus, common experience hardly lends support to the idea that cats are transforming through time. More particularly, the cats in your old age will seem exactly as different from dogs as the cats of your youth were from the dogs of their day. This just does not really square with what we know to be evolutionary fact, that is, that cats and dogs are diverging gradually from a common ancestor and that if you were transported back a quite short distance in geological time, cats and dogs, or rather their ancestors, would be indistinguishable from each other. The timescale of human experience is but an eye-blink compared with evolutionary time. Our senses seem to tell us that animals breed according to their kind, not that their kinds change. However, our senses also tell us that the sun moves around the earth and we know that this is an illusion, an artefact of perspective. Science moves us outside our normal point of view and thus its findings often seem unnatural to us.

We seem to think about each species of animal or plant as having a unique and distinct 'essence'. This might be useful, since we encounter different animals and plants in our environment and we need to learn generalizations about what to do with each class (run from it, eat it, avoid eating it, etc.) from a limited number of exemplars. Creating a mental record of 'essential qualities' for each species is an efficient way of dealing with this problem. However, this is a convenient device of the human mind, not the way nature actually is. Evolution requires us to accept that different types of animal have no fixed essence and are not different in kind from each other. They change over time and are connected by extinct intermediate forms. Evolution tells us, for example, that there was an individual living a few million years ago who has many great-great-great-great- . . . grandchildren alive in the world today and some of those descendants are human beings, whilst others are chimpanzees, and there was no point in the family tree where any catastrophic or abrupt change occurred. At no point did any two siblings need to be any more different from each other than siblings normally are. The divergence between a human and a chimpanzee is only quantitatively, not qualitatively, different from the difference between two humans—just a question of how far back into the family tree you have to go before you hit a shared grandparent. That is very counter-intuitive. However, evolution seeming alien to our usual way of thinking is not any kind of argument against its truth.

The purpose of this book is to explore the basic principles of evolution and evolutionary genetics, with the hope that their explanatory power, subtle consequences, and deep beauty will become clear. I hope to steer a path between too much detail on the one side and misunderstanding on the other. My hope is that if you are sceptical about evolution, your doubts will be dispelled. Many of your objections and also fears evaporate on a very clear examination of the case. If you are already comfortable with the theory, I hope you will come to understand more deeply the nature of the processes you have signed up to, the genetics that underlies them, and the ways they can be used to explain behaviour, especially the behaviour of humans. The rest of this chapter outlines in a nutshell why the theory of evolution is so important (section 1.1) and how it works (sections 1.2 and 1.3). The remainder of the chapter examines some of the most common objections to the theory which people come up with (section 1.4). I will argue that each of these is based on a misunderstanding.

In the rest of the book, Chapters 2–5 lay out in detail the key components of evolution—variation, heredity, competition, and selection—bringing in information about genetic mechanisms as it is required. The first five chapters taken together provide the 'conceptual toolkit' of how to think about problems in evolutionary terms. The remaining chapters apply the conceptual toolkit to some of the core issues of life, namely sex (Chapter 6), the lifespan (Chapter 7), social life (Chapter 8), and learning and culture (Chapter 9). Chapter 10 looks at the recent evolutionary history of our own species, and Chapter 11 concludes by considering how evolutionary thinking is best incorporated into the study of the human mind and behaviour—is it an alternative to the existing theories of psychology, an addition to them, or a way of linking them all together?

1.1 What problems does the theory of evolution solve?

Before we turn to the theory of evolution, it is important to be clear about why it is important. Charles Darwin laid out his theory in a book published in 1859 entitled *On the Origin of Species by Means of Natural Selection* (henceforth *The Origin*). This is one of the most influential books of all time. Why? In other words, what problem does it solve? It must be a pretty important one for the book to have become so renowned. In this section, I argue that Darwin's theory of evolution simultaneously and definitively solves two major problems in understanding living things, namely the problem of history and the problem of design. No other theory can solve either one, still less both. Solving them together makes Darwinian evolution the most important idea in the study of living things.

1.1.1 The problem of history

A brief look at the natural world reveals an obvious truth. Living organisms of many different types have commonalities and divergences. The commonalities are striking. For example, the limbs of different vertebrates contain a pattern of five bones that is recognizably similar, even though the limbs are serving functions as different as fins, legs, and wings in different species (Figure 1.2). Even where the adult creature does not have all five of the phalanges (that is, fingers), as in birds or the horse, five appear in the embryo, only for some of them not to develop fully. Why should these diverse limbs all have an underlying form that is so similar?

If we go down to a more microscopic scale and examine the biochemical makeup of cells, the commonalities are even more striking. Many of the proteins involved in keeping living cells going are very similar in organisms as distinct as humans and amoebae. Consider a small protein called cytochrome c, which is found across many plants and animals. Proteins are formed of chains of smaller constituents called amino acids and there are vast numbers of different forms that proteins can take, since there are 20 different amino acids that can occur in essentially any order in the chain. The main part of cytochrome c consists of a chain of 104

Figure 1.2 **The same basic pattern of bones can be identified in the limbs of many different vertebrates. This is a classic case of homology.** *From Wolpert et al. (2006), p. 506.*

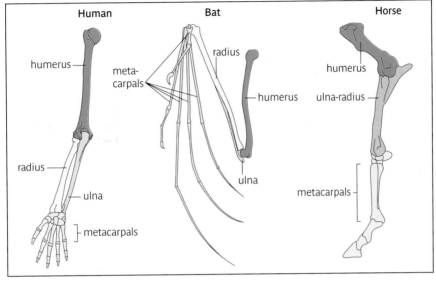

amino acids. If we, as it were, lay cytochrome c molecules from different species alongside each other, we find that the same amino acids appear at the same point in the chain more often than not. The number of 'matches' between amino acids in the main part of the cytochrome c molecule of some different vertebrate species is given in Table 1.1. As you can see, organisms as diverse as a fish and a monkey have exactly the same amino acid at the same position in the chain for most positions.

The same two examples discussed so far also serve as examples of how the natural world contains divergences. Although the bones are similar in all vertebrate limbs, they are not identical. The phalanges are enormously lengthened in bats, where they serve as stretchers for wings made of skin, and in the whale they make a flat flipper rather than a leg (Figure 1.2). Thus, from a common theme, many variations can be seen. We can make a similar point in respect of cytochrome c. Its exact structure varies somewhat, to a differing extent depending on which two species are being compared. There is a single difference in the chain between humans and rhesus monkeys, but over 20 discrepancies between the sequence in humans and that found in tuna. Again, we see a common theme and local variations.

Hierarchical organization in nature

Biologists realized a long time ago that the commonalities and divergences between different species had a very special property: hierarchical organization. By this, it is meant that two species that are similar in the details of system A tend also to be similar in the details of system B. For example, humans are very similar to rhesus monkeys in terms of cytochrome c (Table 1.1). When you study other proteins, such as α- and β-globins, or fibrinopeptides, the same pattern

Table 1.1 **The number of amino acids in common in the 104 amino acid chain of the cytochrome c molecule in various species of vertebrates. The number of matches is high in all cases, the human molecule is especially close to the rhesus monkey, and the whale is clearly closer to the other mammals than it is to the tuna.**

	Human	Rhesus monkey	Rabbit	California grey whale	Great grey kangaroo	Chicken	Tuna
Human		103	95	94	92	90	82
Rhesus monkey	103		96	95	91	91	82
Rabbit	95	96		102	98	96	86
California grey whale	94	95	102		99	94	86
Great grey kangaroo	92	91	98	99		92	87
Chicken	90	91	96	94	92		87
Tuna	82	82	86	86	87	87	

always emerges, with humans right next to monkeys and closer to all mammals than to any non-mammals (Penny et al. 1982). This could have been otherwise. Humans might have been most like monkeys in cytochrome c, most like kangaroos in α-globins, and most like tuna in β-globins. But this is not how it is. Humans look more like monkeys than they do rabbits in pretty much any aspect of their anatomy and physiology that you care to name, and look more like rabbits than they do kangaroos, and more like kangaroos than they do tuna.

As a consequence of this, we can group different species together on the basis of the amount of structure that they share. Humans belong in a group with monkeys, and humans and monkeys belong in a super-group with rabbits, and humans, monkeys, and rabbits belong in a super-super-group with kangaroos, and so on. At each level in the hierarchy of groupings, the amount of commonality becomes less and the amount of divergence greater (Figure 1.3). Why are we able to make hierarchical groups like this? More generally, where did all these different types of animal come from? This is what we will call the problem of history.

Solving the problem of history

One solution to the problem of history would be simply to say that the different animals have always been as they are. People have always been people, monkeys have always been monkeys, and kangaroos have always been kangaroos. However, this would not explain why structures within different animals are so strikingly similar to each other and, in particular, why structures in monkeys are consistently so much *more* similar to their counterparts in humans than structures in kangaroos are. In other words, it could not explain the hierarchical pattern of similarities and differences observed in nature. Moreover, that pattern of similarities and differences is sometimes quite surprising. For example, whales, although they live in the deep ocean, are much more similar to humans than they are to fish across numerous aspects of their anatomy and physiology. Their cytochrome c is more like that of humans than it is of tuna (Table 1.1), as is their habit of giving birth to live young and feeding them on milk. Why would this be?

Figure 1.3 **Species can be arranged hierarchically into groups, super-groups, super-super-groups, etc. on the basis of commonalities and divergences. In multiple aspects of their anatomy and physiology, humans share most with apes, a little less with other non-ape primates, a little less with other non-primate mammals, a little less with other non-mammal vertebrates, and so on. Not all the levels of grouping recognized by biologists are shown.** *Photos: woman, chimp – © Digital Vision; lemur – Vladimir Wrangel/Fotolia.com; rabbit – Photodisc; shark – Corbis.*

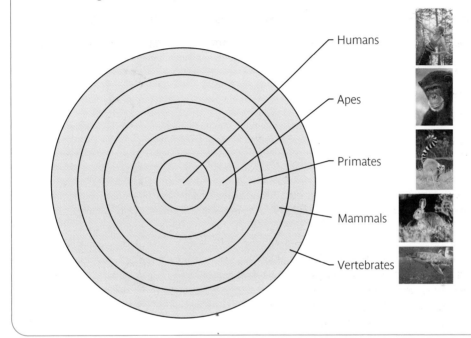

This brings us on to Darwin's solution to the problem of history. Darwin argued that the organisms we see today are derived from a creature ancestral to them all, by a process of gradual modification. The descendants of the ancestral form, each altering slightly over the generations, have split and split again, giving rise to many distinct branches on the tree of life. Going backwards from the present, some species' ancestors diverged relatively recently, whilst the ancestors of some others diverged further back in time. This has overwhelming merit as an account of the diversity of life. It explains patterns like the vertebrate limb, with its myriad variations on the same basic structure. The shared plan is there because all vertebrates descend from the same creature, which started out with a particular configuration of bones. The bones have been stretched, fused, reduced, or twisted within different lineages over the generations, but the common ancestry is still traceable.

Descent with modification automatically assures that organisms will be classifiable into hierarchical groups. Indeed, what these groupings represent is the closeness of the common ancestor of all the species involved—the time since the split, if you will. Thus, the diagram shown in Figure 1.3 can be replaced with a representation of what it more truly represents: a family tree or, to use the biological term, a phylogeny (Figure 1.4). As the phylogeny shows, humans and monkeys share a more recent common ancestor than humans and rabbits, who in turn

Figure 1.4 **The hierarchical relationships between species from Figure 1.3. are better captured by showing what they really represent—relationships on a family tree or phylogeny. Here, the distances between branching points are not to scale and many thousands of other branches are not shown.** *Photos: woman, chimp – © Digital Vision; lemur – Vladimir Wrangel/Fotolia.com; rabbit – Photodisc; shark – Corbis.*

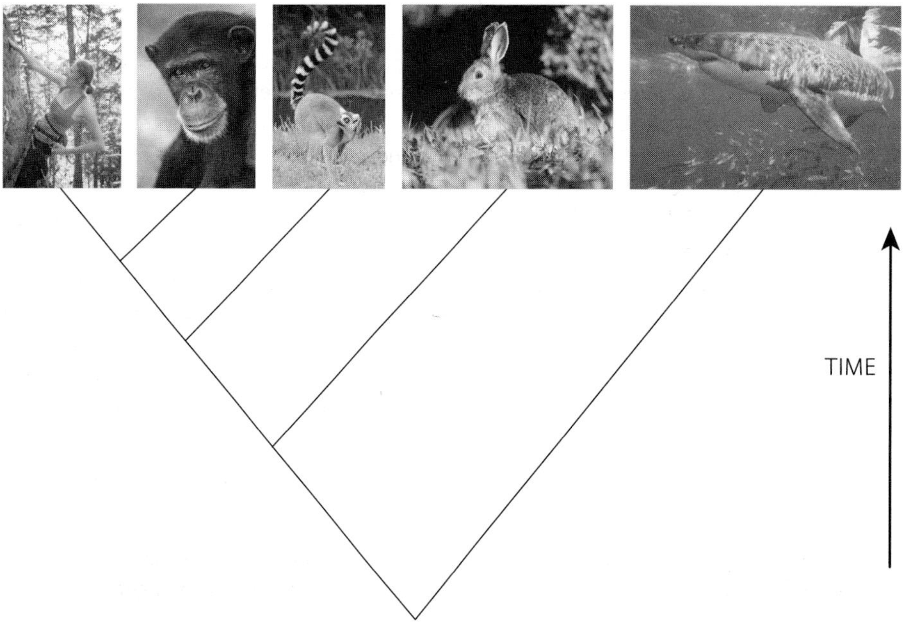

TIME

share a more recent common ancestor than humans and kangaroos, and so on. At a stroke, this explains the patterns of commonality in cytochrome c, globins, hands, blood, and many other characteristics. It also explains why whales, beneath the skin, are more like humans than they are tuna: because they have an ancestor that was a land mammal.

Several scientists before Darwin had proposed that the species we see today had not always been as they are, but had been produced by descent with modification from common ancestors. Darwin was not therefore unique in making this claim, although he did assemble a more thorough and persuasive case than anyone who went before him. More importantly, however, Darwin was the first to fully address another question. Why does modification occur? In other words, what mechanism is responsible for the development of whales from an ancestor that was a legged mammal? Darwin answered this question by proposing a novel mechanism—natural selection—and in so doing he solved another, even greater, problem in the understanding of life on earth: the problem of design.

1.1.2 The problem of design

A second striking aspect of the natural world is that the various creatures that we see seem quite well designed for the tasks that they have to perform. Bodies consist of a large number of distinct subsystems, each of which is efficient at solving a particular problem—the heart for pumping, lungs for transferring oxygen to blood, the gut for digestion, and the liver for detoxifying blood. All of these contribute to the continued functioning of the individual and, perhaps even more impressively, all the systems communicate and cooperate in an integrated way. Some of these systems are impressive feats of engineering. Let us take just one example—the echolocation of bats (Dawkins 1986; Jones & Holdereid 2007).

Design in nature: bat echolocation

Bats, of which there are a large number of different species, have to solve the problem of navigating and hunting insects at night (Figure 1.5). The system they use to do this is a breath-taking masterpiece of good design. They use what human engineers—who did not make use of the principle until well into the 20th century—would call sonar, but it is referred to by biologists as echolocation. The bats emit high-frequency sounds and then use the echoes returning to their ears to calculate the position of obstacles and prey.

The first piece of good design is that echolocation only appears where it is needed. Echolocation of some form has evolved several times in nature and only where the organism is active in the dark or in murky waters. Where light is available, creatures prefer the less energy-intensive solution of vision. Old World fruit bats can rely on good night vision for finding their less evasive dinner and they have no echolocation. Amongst bats that do echolocate, there are a number of different call types that can be used. Some are better suited than others to, for example, open versus cluttered spaces and bats use the form of signal most suited to their habitats.

Figure 1.5 **Many bats have sophisticated echolocation abilities that appear well designed for solving the problem of manoeuvring and hunting at night.**
© Dietmar Nill/Photolibrary Group.

Using echolocation causes many design problems. The pulse of sound needs to be short, with plenty of gaps for listening for echoes, and thus bat signals are a tiny fraction of a second long. To get a good fix on a rapidly moving target, you need a rapid succession of echoes and bats are able to produce up to 200 sound pulses per second. However, this is energetically very costly. Bats have hit on the sensible solution of a much lower 'cruising' rate of pulse emission—about 10 per second—which rises only when a potential target has been detected and needs to be pursued. For the echoes to be strong enough to detect at any kind of distance, the original signal has to be extremely loud. Bats have evolved two kinds of sound production, tongue clicking and calling using the larynx, both of which are capable of producing incredibly intense bursts of sound (in many species, these are outside the range of human hearing, so they have to be recorded with special equipment). However, this immediately causes other problems. The bursts of sound are so loud as to be much louder than the echoes (obviously), but also so loud as to deafen the bat when they are emitted. This is a real quandary. Make the ears any less sensitive and they would not be able to detect the echoes; keep them sensitive and they will be deafened.

Human radar and sonar engineers encountered exactly the same problem and solved it with a design that switches off the signal receiver as the pulse is emitted. It turns out that some bats had hit upon this exact principle millions of years earlier. A set of muscles damps the transmission of vibrations to the eardrum exactly as the outgoing pulse is produced, thus reducing the impact of the outgoing signal on the auditory mechanisms. These damping muscles can be contracted and relaxed up to 50 times per second, in exact synchrony with the production of sounds.

The marvels of bat echolocation do not stop there. Some bats use a fixed pitch of signal at the frequency at which they hear best (others have an even more complex system of frequency modulation over time). However, if the bat itself is moving, the apparent pitch of the returning signal will be distorted by the Doppler effect. (The Doppler effect is the reason a police siren appears to change pitch if you are in a car moving towards or away from it. The reason it occurs is that your own motion affects the rate of arrival of sound waves at your ear.) These bats therefore lower the pitch of the signal in proportion to the speed at which they are flying, so that echoes always return to them at the frequency they would if the bat was at rest, thus close to their auditory optimum.

Design oddities in nature

You cannot examine bat echolocation systems without being impressed at what appears to be good design. Several different systems, the larynx or tongue, the ears, the muscles connecting them, the brain, the mouth, and the wings, all have to interact to make the whole thing work and it does work astonishingly well. Again and again, the bat system turns out to embody solutions which human engineers have painstakingly worked out are really good ways of solving the problems of locating targets where there is no light. However, design in nature also has some puzzling limitations. Certain types of whale, it turns out, have a pelvis (Figure 1.6). The pelvis is a bony structure whose function in other animals is to attach the hind legs. As whales have no hind legs, this makes no design sense at all. Tuna and sharks make their livings in similar ways to whales and they do not need a pelvis. If nature can produce such brilliant designs, what is this useless structure doing there?

Solving the problem of design

We thus need to explain (a) how nature comes to be so full of generally good design solutions and (b) why those design solutions sometimes have curious non-functional features. The first point to make is that the designs we see in nature are far too complex and sophisticated to be

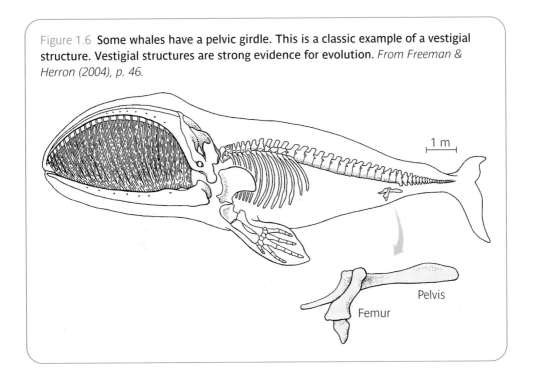

Figure 1.6 **Some whales have a pelvic girdle. This is a classic example of a vestigial structure. Vestigial structures are strong evidence for evolution.** *From Freeman & Herron (2004), p. 46.*

due to chance. You could shuffle all the proteins that make up a bat together at random billions and billions of times and never come with anything that functioned at all, let alone had the remarkable design features of an actual bat. So some process has structured the material that makes up a bat in a non-random way.

The traditional pre-Darwinian solution to the problem of design is that there must be a designing agent. In particular, the English theologian William Paley had argued in 1802 that, were one to come across a watch upon a heath, the intricacy and functional organization of its parts would immediately make one assume that, somewhere, there must be a watchmaker who had made it. So it was, argued Paley, with the design of living things. Their intricacy and functional organization is so great that we must accept that there is a designer (in Paley's case, God).

The problems with this hypothesis are several. First, it is what scientists call unparsimonious. Scientists always prefer the explanation that invokes fewest unknown forces and objects, a principle known as parsimony. The designer hypothesis invokes a being whom we have never been able to observe or measure, presumably outside the physical universe, who must have powers of some unknown kind to work on matter. Thus, it requires a lot of extra processes for which there is no independent evidence and which are currently mysterious. If we can find another candidate explanation that invokes only things that we already know to exist, that other explanation will be more parsimonious and should therefore be preferred.

The second problem with the designer hypothesis is that it is non-explanatory. Explanation in science is the business of showing how complex things that we do not understand arise from simpler things that we do understand. Logically, any agent with the capacity to design something as complex as an animal must be more complex than an animal. Thus, the question arises, where does the complexity of the designer come from? Do we need to then, following the very same argument that got us the designer, come up with a super-designer who created the designer, and a super-super-designer who created him or her? The problem with Paley's argument is that you end up having to postulate an infinite series of increasingly powerful designers. Since each of these is more complex than the previous one, and the properties of each are even less well understood than the last, then the promise of explanation is unfulfilled. In fact, you have gone from something we understand moderately well (the form of organisms), to something that we understand not at all (the properties of the infinite über-designer). This is not explanation in the scientific sense.

Finally, if these two problems with the designer hypothesis were not enough, there is the whale's pelvis. If some benign and omnipotent agent were brilliant enough to come up with the designs of all the creatures in nature from scratch, why would he or she put a pelvis in the whale?

Darwin's solution to the problem of design

Despite these problems, the idea of a designer persisted as the most widely accepted solution to the problem of design for a long time. This was because putting the design of living organisms down to chance was so obviously wrong and there just did not seem to be any alternative other than divine agency. What Darwin did in *The Origin* was to show that there is in fact a third possibility. He argued that organisms have the design-like features they do as a result of the cumulative effects of natural selection over the course of their evolution.

Natural selection is the process of non-random survival of useful innovations that, cumulatively, can lead to what seem to be well-designed structures without the involvement of a designer. We will be discussing how it works in detail later in the book, but the idea of natural selection is parsimonious, in that it does not invoke anything that we do not already know to occur, and explanatory, in that it shows how something complex (apparently well-designed organisms) arises from forces that are actually very simple. Crucially, it also predicts that designs will not always make current functional sense, as in the case of the whale's pelvis. This is because natural selection takes a long time and always works by modifying existing forms, not creating new ones from scratch. Thus, if a land mammal returned to living in the sea, it would not lose its hind legs instantaneously. Instead, they might gradually reduce in size over many generations, until they disappeared completely, but there would be many thousands of years when some parts of the now functionless leg structures—the pelvis, for example—would be visible. We call these 'vestigial structures' and they are an anatomical testament to the creature's evolutionary history.

The next section will examine what natural selection is in a nutshell, but the important point to note here is that Darwin's theory of evolution does not just solve the problems of history and design: it solves them in terms of each other. That is, the reason organisms have (mostly) good design is because of the historical modifications they have undergone, and the reason they have undergone historical modification is (mostly) because of differences in design functionality.

1.2 Evolution by natural selection in a nutshell

It is time to look briefly at what Darwin's theory of evolution by natural selection involves. It consists of four elements: variation, heredity, competition, and natural selection, and taken together these produce descent with modification and also build up what appears to be good design. The four elements work as follows:

1. *Variation*. Individual organisms may be similar but are not identical to each other. They have minor variations in their characteristics.

2. *Heredity*. Many of the characteristics that vary from individual to individual are passed on from parents to offspring.

3. *Competition*. Not all individuals leave the same number of offspring. Some die early in life, or do not manage to reproduce, or reproduce but have fewer offspring than others. As a consequence, not all individuals have the same representation in the next generation.

4. *Natural selection*. As long as individuals' success in survival and reproduction depends at least partly on the characteristics that they have and which they pass on to their offspring, then characteristics which confer an advantage will become more common and persist, whilst those conferring a disadvantage will disappear.

Darwin saw that the cumulative effect of these principles would be powerful. If in a certain generation there are some individuals who have a characteristic that is useful, then those individuals would do better in the competition to reproduce than the individuals who lacked the characteristic because they are more likely to stay alive, they are better at getting a mate, or they have more energy to produce and protect healthy offspring. Since their offspring would also have the characteristic (because of heredity) and would also then in turn be advantaged in their own competition for reproduction, it follows that the proportion of the population bearing the useful characteristic would increase from generation to generation, until the point where all individuals have it. Imagine this process repeated for thousands and thousands of generations. Any new variation in form that happened to arise and which conferred some advantage in survival or reproduction—was a better design, if you will—would be increasingly represented in the generations that followed. This solves the problem of design since, gradually, by the retention of advantageous characteristics, functional systems well designed for that particular environment could be built up in simple steps. The systems built up in this way are called adaptations.

Homologies and analogies

Natural selection solves the problem of history, since it suggests that lineages of organisms will gradually modify over the generations. In particular, cousins who go off and populate different habitats will experience different design problems, develop different adaptations, and thus become more dissimilar over time. However, distantly related species living in similar ways and thus experiencing the same design problems may come up with similar adaptations and thus become in some respects more similar over time. Darwin's theory thus predicts that we will be able to identify two types of similarity in nature. The first type are called homologies. These are similarities stemming from common origin. Thus, the phalanges of the bat, which it uses to stretch out its wings, may be said to be homologous to the fingers of the human, and the functionless pelvis of the whale may be said to be homologous to the altogether more useful version found in land mammals. Homologies, because they reflect history, are useful for constructing the phylogenies of different organisms.

Similarities of the second type are called analogies and the process that produces them is called convergent evolution. Whales and tuna have very similar, streamlined shapes. This cannot be explained by homology since, as we have seen, whales are more closely related to

Figure 1.7 **Similar streamlined shapes have been hit upon by sharks, fish, whales, and seals, none of which are particularly close phylogenetic relatives of each other. Human designers of submarines have also exploited the same principle.** *Left to right: © Klaas Lingbeek-van Kranen/istock.com; Corbis; Ingram; Corel; Megaport/Fotolia.com.*

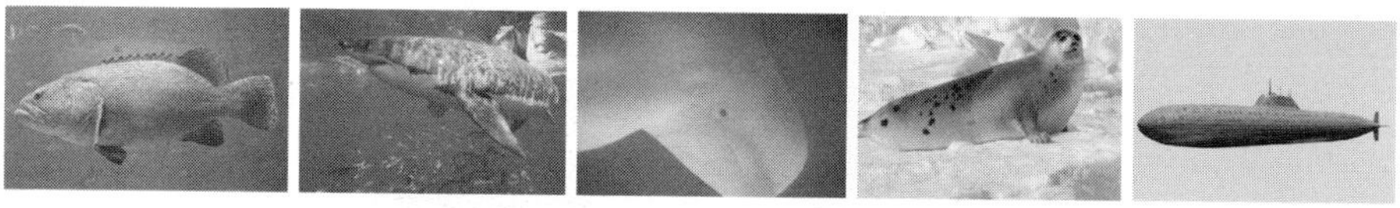

humans than they are to fish. This shape is efficient for moving through water, as fish and whales seem to have independently discovered. In fact, it is not just fish and whales. The streamlined hydrodynamic shape has arisen multiple times in organisms which are not closely phylogenetically related, but which have to move through water, and the designers of submarines have hit upon the very same principle (Figure 1.7). You can generally identify analogies because the analogical characteristic does not pattern with the other characteristics of the organism. We have already seen this with whales and tuna, and, to take another example, a bat has wings like a bird, but the rest of its anatomy and physiology has a closer resemblance to those of mammals. The natural world contains exactly the mix of homology and analogy in the form of organisms that Darwin's theory—and only Darwin's theory—predicts.

The conceptual power of Darwin's theory

The next four chapters will examine in detail how the process of natural selection actually works, but for now it is important to note that what Darwin is doing here is coming up with a new *kind* of process that is neither chance (as it is often wrongly characterized), nor has any goal, or intention, or conscious designer involved. A new kind of process means that we have a new kind of explanation available to us. We can explain a structure in a way which is neither 'X is as it is because it just worked out that way', nor 'X is as it is because of what its creator was trying to achieve', but 'X is as it is because ancestral versions which were slightly more X-like had an advantage relative to competitors that were slightly less X-like'. This opens the door to a whole new way of looking at things.

1.3 Incorporating genetics: the modern synthesis

At the time Darwin was writing, the mechanisms of heredity and variation were not really understood. Although common experience shows that individuals vary from each other in ways that are transmissible from parent to offspring, we did not really know why this was the case. More specifically, we did not know how, in sexual species, the characteristics of one parent combined with those of the other. Biologists assumed that the characteristics of the two were

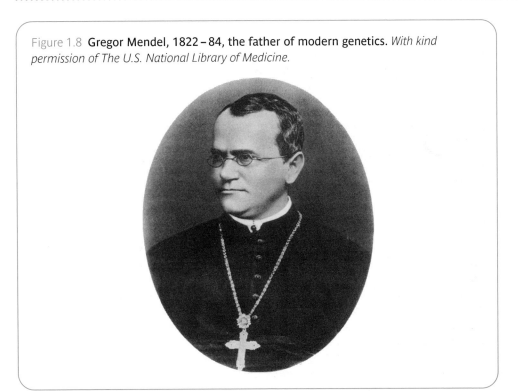

Figure 1.8 **Gregor Mendel, 1822 – 84, the father of modern genetics.** *With kind permission of The U.S. National Library of Medicine.*

simply blended, in much the way that cream and coffee are blended in a cup, and this looked like a problem for evolution by natural selection (see Chapter 3).

Almost simultaneously with Darwin, a brilliant Moravian monk called Gregor Mendel (Figure 1.8) was establishing that inheritance in sexual species does not work by blending in this way. Instead, each parent provides the offspring with a set of 'heredity particles', which we now call genes. The particles from each parent do not blend together, but are carried separately within the offspring and may in turn be passed on intact to subsequent generations. This is why a characteristic from a grandparent can turn up intact in some of the grandchildren, having not been at all visible in the intervening parental generation.

Mendel's ideas were not integrated with Darwin's until well after both figures had died. Indeed, for some decades, Darwinism, which emphasized gradual modification, and Mendelism, which emphasized that discrete particles are passed from generation to generation, were considered opposing schools of thought. Fortunately, in the 1930s, some of the great thinkers of modern evolution established, mainly through mathematics, that Darwinism and Mendelism were not just compatible, but that Mendelism provided the basis for natural selection to work in the way Darwin had envisaged it would. This step forward is known as the modern synthesis, after the title of a book published by Julian Huxley in 1942. (You may also see it referred to as neo-Darwinism.) The modern synthesis made possible the great growth of evolutionary knowledge that we benefit from today.

The modern synthesis clarified how natural selection actually worked. Natural selection amounts to changes in the relative frequencies of different forms of genes in the population over the generations. You therefore cannot understand natural selection without appreciating what it is doing at the genetic level. It is in the spirit of the modern synthesis that genetics and evolution are not presented separately in this book, but introduced in an intertwined way.

1.4 Common objections and misunderstandings

Despite the abundance of scientific evidence for the validity of Darwin's theory, resistance persists and misunderstanding abounds. This section discusses a few of the more common objections raised by doubters. I shall endeavour to show how they can be dislodged once the logic of evolution and the evidence for it are set out carefully.

1.4.1 Evolution is just a theory

From time to time, religious anti-evolutionists make efforts to have Darwinism taught 'alongside other theories' of the origin of life. Underlying this demand is the idea is that 'evolution is just a theory' and therefore might not be true. Partly, this rests on a confusion about the word 'theory'. In everyday usage, the word means an idea we have that we suspect may be true but is not yet supported by evidence, as in, 'My theory is that there will be a last-minute rush before the shop closes.' In everyday usage, there is an implied contrast between 'theory' and 'fact'. Evolution is *not* a theory in the everyday sense, since it is supported by a mass of factual evidence.

Scientists use 'theory' in a technical sense, to mean a body of principles that explain phenomena and can be used to make predictions about them. Evolution is a theory in this sense. Thus, in science, we have the theory of gravity, the theory of relativity, and so on. No one says, 'Gravity is just a theory; it ought to be taught alongside other alternatives.' The existence of gravity is supported by a mass of facts, but the laws of gravity can also be described as a theory in the scientific sense. So it is with evolution.

Related to the 'evolution is just a theory' objection is the assertion that we have never actually seen evolution happen. The evidence is only indirect, the doubters say, and the case for evolution relies on inference and conjecture. It is untrue that we have never seen evolution happen. Evolution can be directly observed. Most of the best examples come from organisms with fairly short lifespans, since these are easier to study in a reasonable time.

We will examine just one example here. This comes from the finches of the Galápagos Islands—Darwin's finches, as they are known, since Darwin spent time observing them during his voyage on *HMS Beagle*. There are several different species of finch across the different islands and they show a striking pattern of homology and local divergence. The biologists Peter and Rosemary Grant studied the finches, in particular the medium ground finch, *Geospiza fortis*, on the island of Daphne Major, for many years (Boag & Grant 1981; Grant 1986). This is an excellent population to study, not least since the island is small and few finches migrate on and off the island. Thus, the researchers are able to capture, measure, mark, and release essentially the whole population.

The Grants and their colleagues were able to show that the population of finches contained variation in beak size (element 1 of Darwin's theory). This variation is important, since birds with larger beaks can handle larger and harder seeds (seeds are what the finches feed on). Moreover, the Grants showed that beak size was heritable, or transmitted from parents to offspring (element 2). In 1977, there was a drought and, as a consequence, fewer seeds were available for the finches to eat and those that were available were larger and harder than usual. A large proportion of the finches died during the drought (element 3), but the larger an individual's beak, the more likely it was to survive. Because of this increased survival, when we compare the distribution of beak sizes of all living finches before and after the drought, we can see that the average has been moved to the right, or towards larger beaks. Since the large-beaked survivors tended to produce large-beaked offspring, the next generation of finches too had larger beaks (Figure 1.9). Over just a couple of years, the species had been shifted in its characteristics towards a form better adapted to the new environmental challenge (element 4). This is natural selection in action. We now have a substantial number of examples of natural selection going on in wild populations and biologists agree that selection in the wild is often strong and easily strong enough to account for all the changes we see over the history of life (Endler 1986).

People sometimes respond to examples like the finches by accepting that natural selection brings about change *within* a species, but disputing that it could account for evolution *between* species. That is, larger-beaked medium ground finches are still medium ground finches. The example has not shown how a new species could be produced by descent with modification. As it happens, there is another species of finch in the Galapágos, the large ground finch, which differs mainly from the medium ground finch in body and beak size. Peter Grant calculated that,

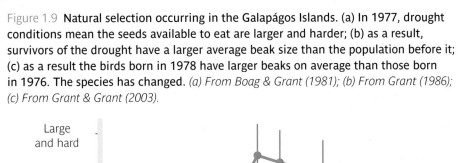

Figure 1.9 **Natural selection occurring in the Galapágos Islands. (a) In 1977, drought conditions mean the seeds available to eat are larger and harder; (b) as a result, survivors of the drought have a larger average beak size than the population before it; (c) as a result the birds born in 1978 have larger beaks on average than those born in 1976. The species has changed.** *(a) From Boag & Grant (1981); (b) From Grant (1986); (c) From Grant & Grant (2003).*

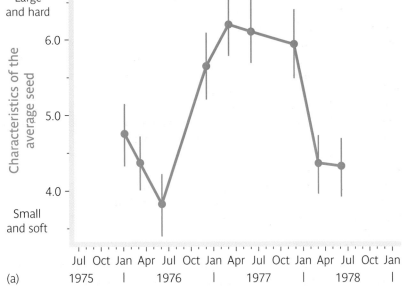

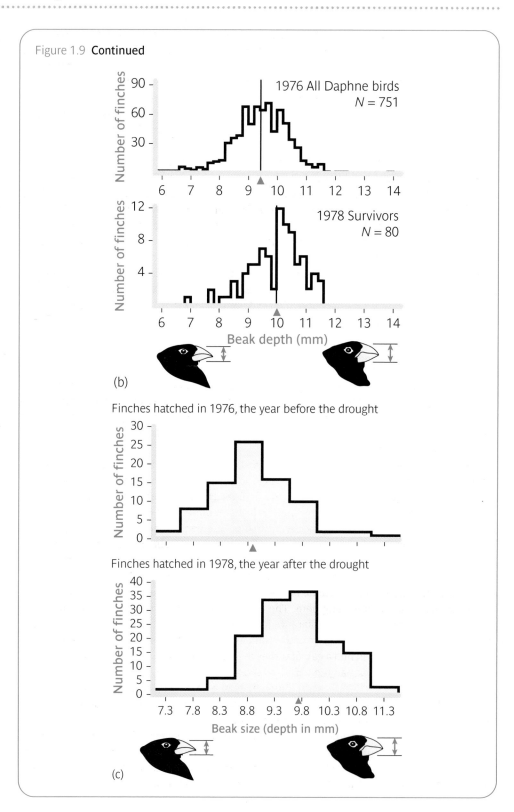

Figure 1.9 **Continued**

(b)

Finches hatched in 1976, the year before the drought

Finches hatched in 1978, the year after the drought

Beak size (depth in mm)

(c)

if the selective pressures of the 1977 drought continued, it would take no more than 46 years to produce the large ground finch from the medium ground finch. A different species of finch would have been produced in what is a mere eye-blink of time. (In fact, the drought conditions of 1977 did not persist and so the selective advantage of larger beaks was not maintained.)

Objectors might wish to say that a large ground finch may be a different species, but it is still a finch. They might continue to be sceptical that natural selection could produce an organism different *in kind*, like a mouse from an ancestor that is not a mammal. This objection implies that there are natural kinds in the history of life and it is difficult to cross from one kind to another. However, the theory of evolution says that there are no natural kinds. There is only one tree of life and all organisms differ only by degree from each other. The big difference between, say, a bird and a mammal, is just a long chain of little differences. To get to a large ground finch from a medium ground finch you go via a chain of intermediate forms, each one of which is just a little different from the last. To get to a mammal from some ancestor that is not a mammal is exactly the same, but the chain of steps is very much longer. If you were a very long-lived researcher who could observe the gradual evolution of mammals from a non-mammalian ancestor, you would observe a sequence of small, gradual changes in form. Sometimes they might go faster, and sometimes slower, but there would be no point at which there was a sudden leap or change of kind. It is only after the event, when the intermediates have died out, that you find yourself in a position to say, 'now we have a different kind of thing: a mammal'.

1.4.2 There are gaps in the record

A second area of objection to evolution by doubters concerns the fossil record. The doubters assert, correctly, that Darwinism requires that there have existed over time a continuous sequence of intermediates connecting any two living species. There must, for example, have been ancestral forms intermediate between humans and chimpanzees. Where is the fossil evidence that this is the case?

As it happens, there are fossils, designated with the generic name *Sahelanthropus*, dating from around the correct period of 6–7 million years ago, which are in many ways intermediate between humans and apes (Chapter 6). However, the doubter can then point out that *Sahelanthropus* is very different from modern humans and evolution requires that there was an intermediate form between *Sahelanthropus* and modern humans. We could then produce the remains of *Australopithecus*, living 2 or 3 million years ago and more human-like than *Sahelanthropus*. Aha, says the doubter, but where is the intermediate between *Australopithecus* and modern humans? We might produce a specimen from the genus *Homo* dating to around 1.5 million years. But again, the doubter asks for another intermediate between this latest candidate and modern humans.

You can see what is happening here. The doubter can keep on asking for more and more intermediates between the intermediate we just produced and the modern creature. Although the fossil record concerning human and chimpanzee ancestors is actually very abundant, eventually the doubters will find a gap that they feel pleased with. Indeed, logically, the doubter could only be satisfied if we had the fossilized remains of *every single* generation that had lived for the past 7 million years. Fossilization is incredibly rare. Most skeletons decay or are destroyed and probably only one in many millions survives to be discovered by palaeontologists. Thus, the determined doubter will always find a gap in the record.

The problem is that the doubter has misunderstood the evidence for evolution. An over-whelming case could be made for the truth of evolution even if there were not a single fossil in existence. The case would rest on the patterns of homology and adaptation we see in the organisms currently alive. The existence of fossils is merely a bonus.

Fossils do exist and, although the record is patchy, they often provide satisfying support for evolutionary relationships. The fossils in continents like Australia, which today have marsupials rather than placental mammals, are fossil marsupials, not fossil placental mammals. Extinct intermediate forms can often be found more or less when and where we predict they will be found. For example, there is a fossil form given the generic name *Ambulocetus*, dating from around 50 million years ago, that is rather whale-like in some ways but has hind legs, which are a little reduced compared with those of land mammals. Then there is *Basilosaurus*, dating from around 38 million years ago, which has a tiny pair of hind legs that could not possibly have supported the creature's body, but may perhaps have been used for grasping. Finally, in today's whales there are no hind legs at all, although there is, as we have seen, a vestigial pelvis. These three form a satisfying evolutionary progression (Figure 1.10), but even without the fossils, the evidence that whales had descended from a land mammal would still be overwhelming.

1.4.3 The theory of evolution says living things arose by chance

Richard Dawkins recounts numerous examples of doubters about evolution pointing to some of the complex structures like the eye and saying, 'Look at the intricacy of this. How could something so complex arise by chance?' (Dawkins 2006a). Well, they are quite right. It could not. But this is not an argument *against* evolution; it is an argument *for* it. The doubters seem to believe that evolutionists believe the form of organisms is due to chance, but nothing could be further from the truth. Evolution is a theory of the *non-random* persistence of particular characteristics, so the very complexity the doubters point to is amongst the best evidence for evolution.

1.4.4 It all happened so long ago, who knows, and who cares?

This objection tends to arise when researchers propose to take an evolutionary perspective on some aspect of behaviour, especially human behaviour. Critics respond by saying, 'Who cares what might have happened thousands of years ago? We weren't there and so we can never know. What I care about is explaining the behaviour in the present.'

This objection is misguided. Evolutionary explanations for behaviour are not explanations of what happened in the past. They are explanations of *why* things are as they are in the present. However, because Darwin solves the problem of design in terms of history, the explanation of the present must invoke selective pressures that have acted over evolutionary time.

To take an example, let us consider how an evolutionary perspective helps us to understand nausea and vomiting of pregnancy (also known as morning sickness and henceforth referred to as NVP). The majority of women, all over the world, experience NVP when they are pregnant. They often feel sick and develop strong aversions to some foods whilst craving others. Until recently, no one understood why this happened. Several evolutionary researchers have proposed that NVP evolved to protect the mother and embryo from dangerous substances contained in foodstuffs (see Fessler 2002; Sherman & Flaxman 2002).

What the evolutionary hypothesis entails is that, at some point in human history, women with a capacity for NVP (and their children) survived better than women with no such capacity and this is, of course, difficult to verify. However, it also makes predictions about the features and consequences of NVP *as it is experienced right here in the present* and these are amenable to empirical test.

Figure 1.10 **Fossils document the descent of whales from a legged ancestor.**
(a) From Freeman & Herron (2004), p. 46. (b) From Gingerich, Smith & Simons (1990), pp. 154–7. (c) From Thewissen, Hussain & Arif (1994), pp. 210–12.

(a) Contemporary whale (Bowhead, *Balaena mysticetus*)

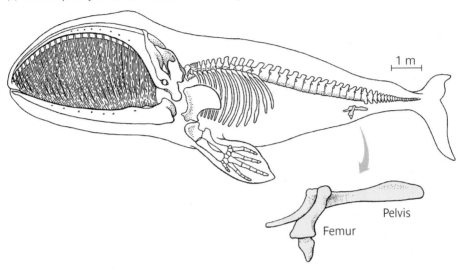

Pelvis

Femur

(b) *Basilosaurus isis* (38 million years ago)

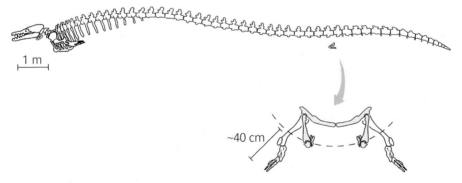

1 m

~40 cm

(c) *Ambulocetus natans* (50 million years ago)

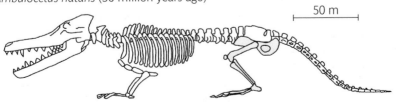

50 m

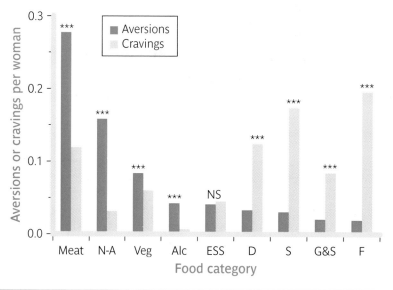

Figure 1.11 **Frequency of aversions and cravings to different food types during pregnancy, based on interviews with 5,432 and 6,239 women. The triple asterisk indicates that the difference in frequency between aversions and cravings for that food type is highly significant.** *Meat,* **includes fish and eggs;** *N-A,* **non-alcoholic beverages, including coffee and tea;** *Alc,* **alcohol;** *ESS,* **ethnic and spicy foods;** *D,* **dairy and ice cream;** *S,* **sweets and desserts,** *G&S* **grains and starches,** *F,* **fruits and fruit juices;** *NS,* **not significant.** *From Sherman & Flaxman (2002).*

If the evolved function of NVP is to protect women and embryos, then it should appear when they are most vulnerable. The embryo is most at risk from damage in the first trimester of pregnancy, when the major organ systems are differentiating, whilst the mother is also particularly vulnerable in this period, as her own immune system is down-regulated to stop her attacking the partly foreign tissue that is the embryo. And indeed, it is in the first trimester that NVP appears, largely disappearing after week 20 of pregnancy.

Furthermore, NVP should be particularly directed to foods likely to contain substances that are dangerous, particularly whilst the mother is immunosuppressed. The most obvious of these are meats, which can contain dangerous micro-organisms. NVP should thus switch food preferences away from meat and towards fruits and sugars as alternative, safer forms of energy. This is exactly the pattern that NVP shows—nausea is directed most strongly towards meat and animal products, whilst cravings are towards fruits and related sweet things, and to a lesser extent grains (Figure 1.11). Moreover, in cultures where the diet contains little meat and has as a staple a bland grain-like maize, women experience less NVP than in other cultures.

Another prediction of the evolutionary theory is that the presence of NVP should actually benefit the embryo, protecting it from the kinds of developmental abnormalities that could lead to it being miscarried. Here, too, the evidence is compelling. Pooling across nine studies involving 22,305 pregnancies, the presence of NVP reduces the miscarriage rate sharply (Figure 1.12 shows data from one such study; the more severe the NVP, the better).

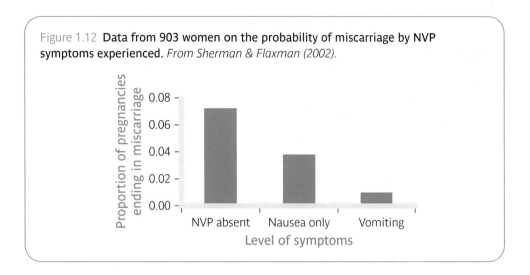

Figure 1.12 **Data from 903 women on the probability of miscarriage by NVP symptoms experienced.** *From Sherman & Flaxman (2002).*

Thus, the evolutionary theory explains why contemporary women experience NVP when they do, why it affects the foods that it does, and what the consequences of not having it will be. It reassures women that their experiences are neither abnormal nor bad for their babies, suggests reasons why physicians should not try to suppress NVP with drugs, and also reveals what diet will best minimize the symptoms, namely fruit and bland cereals. All of this understanding concerns the present, not the past, and none of it would have been derived without some idea, based on evolutionary thinking, of what NVP is for.

✔ Summary

1. The theory of evolution is firmly accepted by scientists as one of the most important bodies of scientific knowledge that we have.

2. Despite this, large numbers of people do not accept that evolution occurs.

3. Darwin's theory of evolution was set out in a book called *On the Origin of Species by Means of Natural Selection*, published in 1859. In it, he simultaneously solved two major problems in the understanding of living things: the problem of history and the problem of design.

4. The problem of history is the question of where all the different types of organisms come from. Darwin argued convincingly that the pattern of homology and divergence we see in the natural world is best explained by the idea that all the current forms have descended with gradual modification from a single ancestral form.

5. The problem of design is the question of why organisms have characteristics that are well designed for the requirements of living that they have. Darwin argued that the cumulative effect of non-random survival of beneficial characteristics would be to produce well-designed structures. He called this mechanism natural selection.

6. Natural selection solves both of the major problems in terms of each other. Organisms have the design they have because of their history, and the history they have primarily because of changes in their design, and natural selection is the causal mechanism.

7. The full power of Darwin's theory only became clear once it had been integrated with genetics in what is known as the modern synthesis.

8. Many of the objections to evolution evaporate once misunderstandings of the theory and the evidence for it are clarified.

❓ Questions to consider

Write a paragraph in answer to the following questions:

1. In *The Origin*, Darwin spent a lot of time considering issues such as the following: In several different parts of the world, insect species that live in deep caves and that lack eyesight are found. These blind forms are more similar to the sighted forms living immediately outside their caves than they are to the blind forms living in cave systems on other continents. What would a non-evolutionary approach to biology have predicted here? How might Darwin explain the pattern we actually do see and why is this such good evidence for Darwin's theory? Finally, what are the analogies and homologies in this case?

2. In the chapter, it was suggested that natural selection could explain how the whale, descended from a land mammal, has come to have a streamlined shape with no hind limbs, more like that of a fish. Explain exactly how this might come about through a long series of small changes, using the terms variation, heredity, competition, and selection in your answer.

3. Darwin argued that whenever there was variation, reproduction with heredity, and competition, a process of natural selection producing adaptation would ensue. This raises the possibility that there might be things in the world other than living organisms that evolve in a Darwinian manner. Can you think of any examples? What would be the similarities and differences between these systems and living organisms?

➡ Taking it further

One of the best modern writers on evolution is Richard Dawkins. Dawkins gives a readable introduction to the phylogenetic history of life (going backwards through time from modern humans) in *The Ancestor's Tale* (Dawkins 2004). On the modern doubters about evolution and the case against them, Kitcher (2007) is excellent. For a readable and short treatment of how natural selection works, and how it can change the way we think about many different aspects of our lives, try Wilson (2007). Dennett (1995) examines the revolutionary conceptual power of Darwinian theory at greater length.

Variation

As we saw in the previous chapter, all living organisms on earth are members of the same large family. They differ only in how far back you have to go to find the common ancestor—not far at all for any two humans, some way for a human and a monkey, and a very long way for a human and a bacterium. Like the members of any family, living creatures are all slightly different from one another. Most obviously, the creatures in one branch of the family, say chimpanzees, are rather different from the creatures in a parallel branch, say humans. However, within any one branch, the individuals also differ in all kinds of small ways. Human adults, for example, differ in height, even within the same population (Figure 2.1). This is the principle of variation and it is a crucial component of Darwin's theory. More specifically, it is crucial for understanding Darwin's theory to appreciate that there is variation both between species (humans are different from chimpanzees) and within species (humans are not identical to one another). This chapter examines where variation comes from, and why it is important.

The chapter will consider variation at two levels, the phenotype and the genotype. The phenotype means the observable characteristics of an individual, such as its size, shape, colour, internal structure, and physiological functioning. There is obviously variation in phenotypes. Just think about your friends—some have blue eyes and some brown, some have red hair and

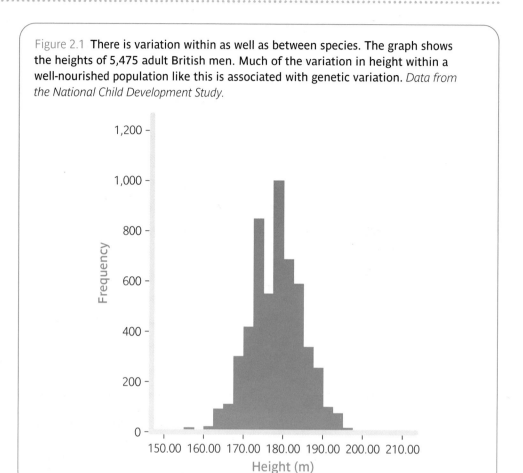

Figure 2.1 **There is variation within as well as between species. The graph shows the heights of 5,475 adult British men. Much of the variation in height within a well-nourished population like this is associated with genetic variation.** *Data from the National Child Development Study.*

some black, some have long legs and some short, some are allergic to peanuts and some are not, and so on. Why is there variation in phenotypes? An important answer, and the key one from the point of view of evolution, is because there is also variation in genes. Genes are not always identical even in members of the same species, and they differ much more markedly across species. We call the set of genetic variants that an individual bears its genotype. The genotype plays a crucial role in giving rise to the phenotype and, thus, to a significant extent, phenotypes end up different because the underlying genotypes are different. In this chapter, then, we will need to review what genes and genotypes are, how they give rise to phenotypes, and how they come to vary. Our focus will be on humans, but the principles are the same for other organisms.

2.1 The phenotype

Living things are, to a reasonable approximation, made of cells. The larger ones consist of many billions of cells, whilst the smallest ones are a single cell. In multicellular organisms

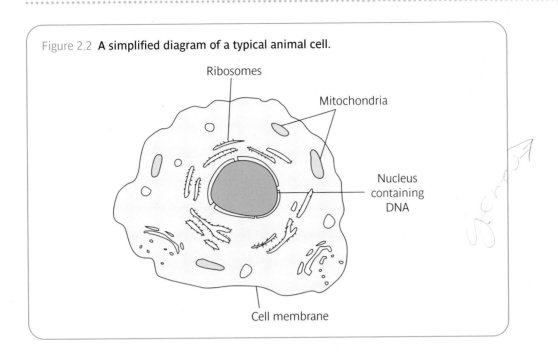

Figure 2.2 **A simplified diagram of a typical animal cell.**

Ribosomes

Mitochondria

Nucleus
containing
DNA

Cell membrane

like humans, the cells are of many different types, such as muscle cells, liver cells, skin cells, red blood cells, neurons, and so on. Although these cells are all specialized for different functions, they also share a great deal of common structure (Figure 2.2). Typical cells have an outer membrane enveloping the cell contents, chemical factories called ribosomes where substances needed in cell function are synthesized, and energy powerhouses called mitochondria where glucose and other fuels are broken down to release energy. (When we talk about cells and cellular structures in this chapter, we will have in mind the cells of eukaryotes, that is organisms such as animals and plants whose cells have nuclei and mitochondria. In prokaryotes, that is organisms like bacteria, many of the principles are the same but some of the details are different.)

If living things are made of cells, cells in turn are made of numerous different chemical compounds, including a large number of different proteins. Proteins make up about 12–18% of total body weight in humans, but they are particularly significant. They give cells their shape and structure, they form connecting tissues, they function as hormones and as antibodies, and, perhaps most importantly, they serve as enzymes, which control the many chemical reactions that are needed for a body to function and to create or acquire the other types of molecule that it requires (notably water, fats, and carbohydrates).

Proteins are large molecules made up of long chains of amino acids. There are 20 types of amino acid and the sequence of different amino acids along the chain determines what the properties of the protein will be. Thus, in simple terms, the phenotype of an organism is determined by the properties of the proteins in its cells, and the properties of proteins are determined by which amino acids are incorporated in what order as the protein chains are synthesized. This is where genes come in, since what genes do is to encode the amino acid recipes for particular proteins (Figure 2.3). We will learn how they do this in the next section.

Figure 2.3 **The logic of life. Genes make proteins, which make up bodies. More exactly, the sequence of bases along the DNA molecule determines the sequence of amino acids in proteins, and the structure of these proteins is a crucial determinant of what the body is like.** *Photo © Photodisc.*

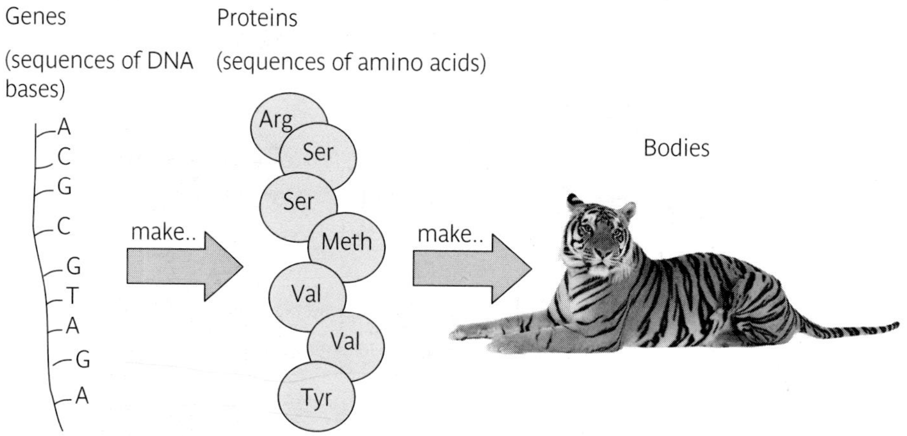

2.2 The genotype

This section reviews the basic concepts of what genes are and how they function. Progress in understanding the fundamentals of genetics has come in three main phases. The concept of the gene originated with Mendel, but the gene was initially an abstract entity that had to be inferred from the patterns of inheritance of characteristics. During this phase (classical genetics), progress was made in inferring what genes did, but what genes were actually made of was unknown. The principles of this era nonetheless remain valid, and some of them are reviewed in section 2.2.1.

The second phase began with the discovery that the DNA (deoxyribonucleic acid) molecule was the genetic material, and shortly thereafter British biologists James Watson and Francis Crick's resolution of the structure of DNA. Watson and Crick (Figure 2.4) published their work in 1953 and, along with Maurice Wilkins, received the Nobel prize in 1962. The DNA revolution paved the way for an understanding of how genes actually encode and transmit information. This body of knowledge is called molecular genetics, and we will look at some of its principles in section 2.2.2.

The third phase in genetics, genomics, is in full swing at the time of writing. For several decades after Watson and Crick, the characteristics of particular genes were worked out mainly indirectly from the proteins that they produced. From around 1980, techniques became available to 'read' large amounts of DNA sequence directly. This allowed biologists to describe the entire set of genetic material (the genome) of different organisms. The genome of some viruses was sequenced as early as 1977. The first genome of a free-living organism published was a bacterium, in 1995, followed closely by the first multicellular organism, a roundworm, in 1997. The first draft for the human genome followed in 2001, followed among others by the chimpanzee

Figure 2.4 **James Watson (left) and Francis Crick (right), the fathers of molecular genetics.** © *A. Barrington Brown/Science Photo Library.*

in 2005 and the rhesus monkey in 2007. There are currently many dozen organisms whose complete or nearly complete genome is established, and knowledge in this area proceeds at a staggering rate. The genomic revolution has established many unexpected things, not least that most of the human genome does not consist of genes, as we shall see in section 2.2.3.

2.2.1 Classical genetics and the central dogma

From his breeding experiments, which we will review further in Chapter 3, Mendel had worked out that certain general principles must be true. These principles were expanded over the few decades that followed his work. They form the principles of classical genetics and three of them in particular are relevant here:

1. There are particles of inheritance passed from parents to offspring, which determine particular phenotypic characteristics. These particles were first called genes by Wilhelm Johannsen.

2. Genes often come in alternate forms, called alleles. Thus, for example a gene might code for flower colour in plants, and there might be two alleles, one which produces red flowers, and one which produces white ones. By convention, we can distinguish between two alleles of the same gene by putting one in capital letters and one in lower case. Thus,

the flower colour gene might be said to have two alleles, *A* and *a*. If there are three or more alleles of a gene in the population, this notation does not work and the alleles are denoted a_1, a_2, a_3, and so on.

3. In many sexually reproducing species, individuals have two copies of each gene, with one copy coming from each parent. Organisms or cells with two copies of each gene are called diploid, as distinct from haploid organisms or cells that have only one.

Classical genetics established that genes have two functions. First, they influence the physical characteristics that the organism has (as we now know, mainly by causing particular proteins to be synthesized). In other words, the genotype influences the phenotype. Second, genes replicate themselves to produce new cells or new individuals with the same genotype. This is how information passes from parent to offspring in reproduction.

The central dogma

It is important to distinguish the two functions of genes, marked out as function a and function b in Figure 2.5. As the figure shows, the genotype gives rise to a phenotype with certain characteristics and also to the genotype of the next generation. However, the characteristics of the phenotype are not themselves transmitted to the next generation. Processes that the phenotype undergoes do not feed back into genes. For example, if all the proteins within a cell get altered by some chemical substance around in the environment, there is no mechanism for this to alter the genes that gave rise to these proteins. Genes affect the properties of proteins, but proteins do not normally affect the properties of genes. The flow of information is one way. This has been called the central dogma of genetics.

What are the implications of the central dogma? Imagine a person who develops huge muscles by weight training during his lifetime. Will his children be born with larger muscles than if he had not done the training? They will not. The training causes the man's phenotype, not his genotype, to change and what he gives his children is his genotype, not his phenotype.

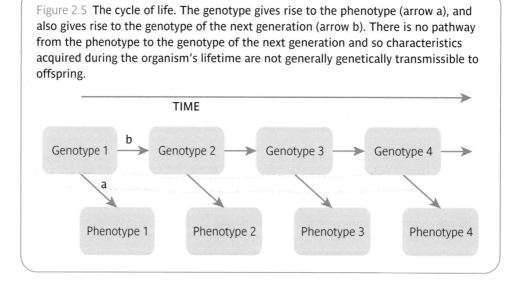

Figure 2.5 **The cycle of life. The genotype gives rise to the phenotype (arrow a), and also gives rise to the genotype of the next generation (arrow b). There is no pathway from the phenotype to the genotype of the next generation and so characteristics acquired during the organism's lifetime are not generally genetically transmissible to offspring.**

TIME

Genotype 1 → Genotype 2 → Genotype 3 → Genotype 4 →

Phenotype 1   Phenotype 2   Phenotype 3   Phenotype 4

Characteristics acquired during his lifetime will not be genetically transmissible. The central dogma also has a converse implication. Parents may carry genotypic characteristics that are not observable in their phenotypes, but which nonetheless show up in their offspring. This happens in the various hereditary conditions for which parents can be silent carriers (more on how this works is given in Chapter 3).

Somatic and germ lines

Related to the central dogma is the distinction between the somatic and germ lines of cells. In multicellular organisms, most cells are somatic, which means that they are only capable of function a (making more phenotype). Cells in the skin or the heart of a human being cannot produce another human being. At most, they can produce other skin or heart cells. They do this basically by splitting into genetically identical copies of themselves, a process called mitosis. Function b (making more genotypes) is handled exclusively by a special class of cells called germ cells or gametes, that is sperm in males and egg cells in females. Gametes are produced by a special type of cell division called meiosis. Note that the distinction between somatic and germ lines is not necessary for the central dogma to be true; in single-celled organisms, the same cell is the soma and the germ, and yet the central dogma remains true because of the way genetic material works.

2.2.2 Molecular genetics: genes are DNA

The nucleus of cells contains large amounts of a complex molecule called DNA (deoxyribonucleic acid). The DNA is wound around proteins called histones. At various points in the life cycle of cells, this DNA–protein configuration assembles itself into a number of linear chromosomes of different sizes. In diploid organisms, these chromosomes come in pairs, with the two pair members being physically similar to each other, but one coming from the father and one from the mother. In humans, there are 23 pairs, to give 46 chromosomes in all (Figure 2.6).

Any candidate for the molecular mechanism of genes would have to do two jobs. First, it would have to be able to produce proteins somehow (for function a, Figure 2.5). Second, it would have to be able to produce copies of itself (function b). Watson and Crick worked out that the properties of DNA would allow it to do both of these things. DNA is a long-chain molecule or polymer, consisting of two strands bound to each other and twisted around each other in a double helix (Figure 2.7). Each strand is made up of a backbone of sugars and phosphates. Along each backbone are strung sequences of four bases, which are compounds called nucleic acids. The four bases are adenine (henceforth A), thymine (T), cytosine (C), and guanine (G). Bases can occur in any order. Thus, even a very short section of DNA can have a very large number of different configurations or sequences. For example, even two bases have 16 possible sequences (AA, AC, AG, AT, CC, CA, CG, CT, GG . . .).

The chemical bonds *within* each strand are extremely strong (they are what chemists call covalent bonds). This means that if the two strands are caused to split apart from each other, each strand will maintain its integrity. The bonds *between* the two strands are weaker, hydrogen bonds. The structure of the bases is such that an A on one strand will only bond to a T on the other, whilst a C on one strand will only bond to a G on the other. This is the principle of base pairing. Because of base pairing, the two strands of a DNA molecule effectively mirror one another, with As on one strand always corresponding to Ts on the other, and Cs always corresponding to Gs. In light of their complementary properties, the two strands are often referred to as the sense and anti-sense strands.

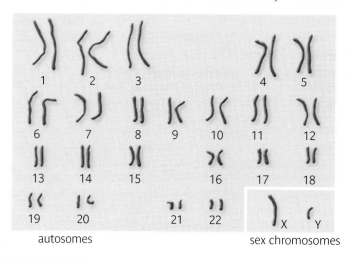

Figure 2.6 **An illustration of the complete set of human chromosomes. The first 22 pairs are numbered by size and are the same in both sexes. These are called the autosomes. The remaining pair are the sex chromosomes. Females have two copies of the larger sex chromosome called X, whilst males have one X and the smaller chromosome Y.** *With kind permission of The U.S. National Library of Medicine.*

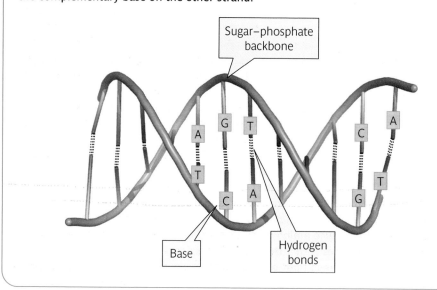

Figure 2.7 **Diagrammatic representation of the double-helix structure of DNA. The nucleic acid bases bind the two strands together by forming hydrogen bonds with the complementary base on the other strand.**

Watson and Crick, having established that this was the structure of DNA, realized that it could fulfil the two functions required of genes. To take function b first, DNA can make copies of itself. Suppose that a DNA molecule exists in a medium of free chemical constituents and is 'unzipped' by breaking the weak hydrogen bonds holding the two strands together. Free-floating As will bind on where there are Ts on the strand, free Cs will bind where there are Gs, and so on. These bases, once attached, form themselves a backbone and thus constitute a new complementary strand. Through this process you can come to have two copies of the DNA molecule where you had one. Given the right chemical environment then, DNA can self-replicate.

The other function of genes (function a, Figure 2.5) is to make phenotypic materials. DNA does this in an indirect way. Again the two strands are broken apart, but, under different chemical conditions from those of DNA replication, a single-strand molecule called messenger RNA (mRNA) forms along the anti-sense strand of the open DNA. RNA (ribonucleic acid) is chemically similar to a single DNA strand, except that the T base is replaced by another nucleic acid called uracil (U). U on the RNA binds to A on the DNA. Because of the base-pairing rules, the RNA molecule that forms is effectively a copy (or more strictly, a negative) of the sequence of bases on the DNA strand on which it has formed. The process of copying DNA sequences into sequences on an RNA molecule is called transcription.

The RNA, once formed, is separated from the DNA molecule, which zips itself back up, and is transported outside the nucleus into structures called ribosomes. These are chemical 'factories' where the raw materials for building proteins, amino acids, are available. Protein chains are then built up, and the order of amino acids added to the protein chain is determined by the order of bases on the mRNA. We do not need to go into details of how this is done, but it involves additional types of RNA. The process of turning base sequences in mRNA into amino acid sequences in proteins is called translation (see Figure 2.8).

The genetic code

You will have noted that there are four different bases, and 20 different types of amino acids. How, then, can mRNA code for proteins? There are not enough bases for each one to stand for an amino acid. There are not even enough for two bases to stand for an amino acid, since there are $4 \times 4 = 16$ duos of bases, four too few. The solution is therefore for a triplet of bases to stand for each amino acid. The triplets that do this are called codons. The mapping from particular codons in the mRNA to particular amino acids in the assembled protein is called the genetic code. The genetic code is almost identical across all living things. It was worked out in the decade or so following the resolution of the structure of DNA and is shown in Table 2.1. The code means that, for example, whenever the sequence of bases CGU is encountered on the mRNA, an arginine molecule is added to the protein chain. The codon AUG, as well as coding for methionine, initiates the process of translation. The codons UAG, UGA, and UAA indicate that the end of the protein has been reached and thus terminate translation.

There is considerable redundancy in the code and it has several significant properties. One of these is the fact that errors made in reading the third base of the codon will often make no difference to the amino acid produced (e.g. CC<anything> produces a proline). Changes in the sequence of bases that make no difference to the amino acid sequence are called synonymous substitutions. Another feature of the code is that codons differing by just the first base (e.g. CCU and GCU) tend to produce amino acids that are chemically similar to each other. The consequence of these features is that the code has considerable robustness to errors in transcription or translation. Errors in third positions often make no difference at all, and errors in first positions will tend to substitute an amino acid for one with reasonably similar behaviour.

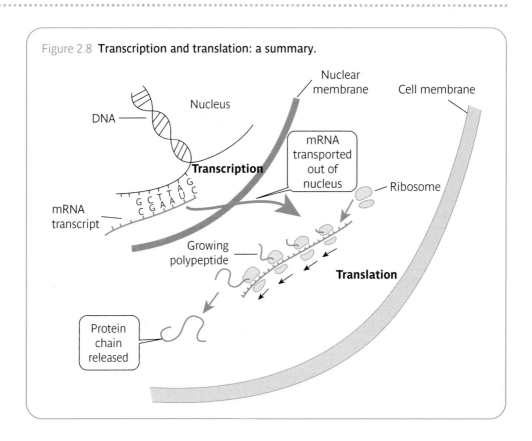

Figure 2.8 **Transcription and translation: a summary.**

The code is in the best one-millionth of all logically possible codings in terms of its robustness to errors in the sequence of bases (Freeland & Hurst 1998). Thus, it exhibits the property of good design that is so often a feature of living systems, from which we must conclude that it has itself undergone a history of natural selection (Sella & Ardell 2006).

2.2.3 The genome: most DNA is not genes

The human genome is estimated to contain around 3,200 million pairs of bases spread over 23 chromosomes (and since we are diploid organisms, each of our somatic cells carries about twice that amount over 46 chromosomes). You might be forgiven for supposing that all of this molecular space would be taken up with genes coding for particular proteins. In fact, this is far from the case. One of the big surprises of genomics has been that genes take up only a small fraction of the total genome.

There are around 25,000 genes in the human genome. A typical gene is about 27,000 base pairs long, although there are some much larger ones, such as the dystrophin gene with nearly 3 million base pairs, and some much smaller, such as SRY, the gene that turns a developing embryo into a male, which is barely 1,000 base pairs long. A quick calculation with these figures reveals that most of the genome (over 60%) does not consist of genes. Non-gene DNA is referred to as non-coding.

Table 2.1 **The genetic code translating triplets of bases into amino acids. By convention, the code is written for the mRNA sequence, so to get the DNA sequence that corresponds to each amino acid you have to further translate using the base-pairing rules.**

Amino acid	Codons
Alanine	GCA GCC GCG GCU
Cysteine	UGC UGU
Aspartic acid	GAC GAU
Glutamic acid	GAA GAG
Phenylalanine	UUC UUU
Glycine	GGA GGC GGG GGU
Histidine	CAC CAU
Isoleucine	AUA AUC AUU
Lysine	AAA AAG
Leucine	UUA UUG CUA CUC CUG CUU
Methionine/START	AUG
Asparagine	AAC AAU
Proline	CCA CCC CCG CCU
Glutamine	CAA CAG
Arginine	AGA AGG CGA CGC CGG CGU
Serine	AGC AGU UCA UCC UCG UCU
Threonine	ACC ACC ACG ACU
Valine	GUA GUC GUG GUU
Tryptophan	UGG
Tyrosine	UAC UAU
STOP	UAA UAG UGA

Even the 40% figure for genes is very misleading. Even within a gene, not all the bases code for proteins. Genes consist of an alternation of exons, stretches of codons that are translated into the protein, and many introns, which are non-coding sequences inserted within the gene. Introns are transcribed, but they are deleted from the mRNA and thus not translated. Human genes usually contain several introns, with the large dystrophin gene containing 79. There tends to be much more intron than exon overall (in dystrophin, 99% of the gene is intron). Once introns have been subtracted, the average length of coding sequence in a human gene is only about

1,340 base pairs. Thus, amazingly, we discover that only just over 1% of the genome actually codes for proteins.

Non-coding DNA

What, then, is the rest of the DNA? Some may be the remains of sequences that were once genes, but have fallen into disuse. Genes that have ceased to be translated are called pseudogenes. For example, humans have a large number of pseudogenes where monkeys have functioning genes coding for smell receptors (Rouquier *et al.* 1998). Monkeys have more sensitive senses of smell than humans, and humans have reduced their olfactory capacities during recent evolution. The pseudogenes are thus a kind of vestigial characteristic, like the whale's pelvis (Chapter 1).

However, most non-coding DNA does not consist of pseudogenes. A large portion of the genome, around 43% in humans, consists of multiple near-identical copies of particular sequences of bases. These are known as transposable elements, due to their ability to copy themselves into different parts of the genome as evolution proceeds. One transposable element, known as *Alu*, has nearly one million copies within the genome. Given that the *Alu* sequence is about 300 base pairs long, it accounts for over 10% of our DNA (much more than genes do!). *Alu* is unique to primates. It arose in the genome of an ancestral primate and has clearly been very successful at proliferating.

As well as large transposable elements like *Alu*, there are shorter sequences that are internally repetitive (like CCGCCGCCGCCG). These are called simple sequence repeats. The repeated motif varies from just a few base pairs up to a few dozen base pairs in length (the short ones are conventionally known as microsatellites and the longer ones as minisatellites). Simple sequence repeats do not jump around the genome in the way that transposable elements do, but they are very prone to mutations, which increase or decrease the number of repeats of the motif. There can be anything from a couple to many dozens of repeats of the motif.

Why is the bulk of the non-coding DNA there? It was thought initially to have no phenotypic function. We will return to the question of why functionless material would persist in the genome below. However, it is now clear that not all non-coding DNA is functionless. Some non-coding sequences are identical across such a wide range of living organisms as to suggest that they have been conserved by natural selection for some reason. Moreover, much more of the genome than just the coding genes is transcribed into RNA (Encode Project Consortium 2007). These RNAs may have regulatory functions, that is they may be involved in controlling the processes of transcription and translation. Overall, the proportion of functionless to functional non-coding DNA is still unknown.

Whatever its origins, there are suggestions that the presence of repetitive DNA enhances the genome's evolvability (in other words, its ability to generate novel phenotypes). Simple sequence repeats provide a source of small, slight changes in gene products from generation to generation, by the expansion or contraction of the number of repeats (Kashi & King 2006). Transposable elements promote the reshuffling and reduplication of genetic material during meiosis (we won't go into how they do this here), and they also move particular bits of genetic sequence around the genome where they are sometimes pressed into use (Nowak 1994).

Evolution of genome size

The fundamental questions about how this genome works are only now able to be addressed. One area of current interest is genome size. The amount of DNA in cells of different organisms varies enormously in ways that are not explicable by the complexity of the phenotype. For example, some worms, amphibians, and flowering plants have much more DNA than any mammals.

Table 2.2 **The size of the genome, the number of genes, and the number of chromosomes are all highly variable. Why, for example, do humans need ten times as much DNA to encode a similar number of genes as pufferfish do, and why do chickens need about twice as many chromosomes as humans?** *Data adapted from Coghlan et al. (2005). Genome size is in million base pairs, and numbers of genes are estimates.*

Organism	Genome size	Chromosomes	Genes
Roundworm	100	10 or 11	19,100
Fruitfly	180	8	13,600
Pufferfish	365	44	31,100
Chicken	1,100	78	20,000 – 23,000
Mouse	2,500	40	22,000
Human	3,200	56	25,000

Often such differences are due to polyploidy or the organism having more than two copies of the genome. However, even allowing for polyploidy, there are dramatic differences between species in terms of the overall configuration of the genome. The length of the DNA material, the number of chromosomes, and the estimated number of coding genes all vary significantly in ways that are not obviously related to each other or to phenotypic complexity (Table 2.2).

The number of genes tends to rise gradually with increasing phenotypic complexity, but its rise is very much slower than the rise in the amount of DNA. In prokaryotes, almost all of the genome is coding DNA. In eukaryotes, as phenotypic complexity increases, the proportion of the genome devoted to coding sequences tends to reduce and the proportion devoted to introns and to regions between genes tends to rise (Taft *et al.* 2007; Table 2.3). The proportion that consists of repeated sequences also tends to rise.

Table 2.3 **The proportion of the genome devoted to coding genes reduces as the complexity of the phenotype increases, whilst the proportion of the genome made up of introns and spaces between genes increases.** *Data from Taft et al. (2007). Figures are percentages of total genome, and due to simplification here do not quite add up to 100.*

Organism	Exons	Introns	Inter-gene
Amoeba	61.7	7.4	30.9
Roundworm	25.3	30.4	42.1
Fruitfly	16.6	29.1	49.4
Chicken	2.4	32.7	64.4
Mouse	1.1	29.3	68.6
Human	1.1	35.2	62.6

The interpretation of this pattern is still a matter of debate. More complex phenotypes with many types of tissue may require more regulatory sequences. This is because different genes will need to be transcribed and translated in different patterns in different tissues, and so the control mechanisms will need to be more complex. This could explain why the non-coding portion of the genome is expanded in complex eukaryotes.

However, most non-coding DNA may well be junk as far as the phenotype is concerned (i.e. it may not be doing anything at all in terms of building the phenotype). Any DNA sequence will persist to the extent that it can get itself replicated, and transposable elements in particular can be viewed as parasites that exploit the replicatory machinery of the organism to complete their life cycle, without doing anything to benefit it (Orgel & Crick 1980). In fact, one class of transposable element, human endogenous retroviruses, are literally parasites: they derive from a virus which, like human immunodeficiency virus (HIV), is adept at getting itself copied into the host's DNA.

Such parasitic sequences will impose some cost on the host, but the cost may be small and thus they can persist for many generations before they are weeded out by natural selection. Parasitic sequences may be more successful in persisting in more complex organisms than in simpler ones because the population sizes of say, mammals, are much smaller than those of, say, bacteria (Lynch & Connery 2003). When the population size is small, natural selection becomes less efficient at weeding out non-functional but non-lethal characteristics. Similarly, the genome is smaller in organisms with small cells and high metabolic or reproductive rates and, again, this could be because selection for genome compactness is stronger in such organisms (Jeffares *et al.* 2006). Note that thinking of transposable elements as parasites does not deny that they can acquire phenotypic functions, or that they increase the evolvability of the genome. It merely stresses that these are side-effects of their basically selfish proliferation (Burt & Trivers 2006).

Mitochondrial DNA

At the beginning of section 2.2.2, I stated that DNA was found in the nucleus of the eukaryotic cell. This is true, but it is not the case that the *entire* human genome is contained in the nucleus. Mitochondria have a small genome of their own: around 16,500 base pairs, organized in a loop. It is now widely accepted that this is because mitochondria were once separate organisms, namely bacteria. They were incorporated into early eukaryotic cells where they accepted the protection of the cellular environment in return for carrying out metabolic functions for the host cell. Their genome is recognizably similar to that of free-living bacteria, but much reduced. Over time, many of their genes have been lost, being no longer needed now the host cell performs so many of the mitochondria's vital functions, or else transferred into the nuclear genome. Mitochondrial DNA has special features that have led to its extensive use in the genetics of human populations. Its mutation rate is relatively fast and it is haploid, with our mitochondria being transmitted asexually down the maternal line. This occurs because only eggs and not sperm provide the mitochondria of the developing embryo.

2.3 Genetic variation

The theme of this chapter has been that phenotypic variation comes to a large extent from variation at the genetic level. However, where does genetic variation come from? In other words, why do individuals have different genotypes from each other? We will now look at two answers to this question, sexual reproduction (section 2.3.1) and mutation (section 2.3.2), and then go on to consider the extent of variation in the human genome (section 2.3.3).

2.3.1 Sexual reproduction shuffles the pack

In diploid organisms, there are usually two copies of each chromosome. One copy comes from each parent. The way this is achieved is by gametes, unlike all other body cells, having just one set of chromosomes. Then, when the gametes combine at fertilization, the correct total number of chromosomes is immediately present (Figure 2.9). This means that during the process of gamete formation, the gamete cell has to receive only half of the set of genetic material of the progenitor cell from which it is created. A human sperm, for example, receives 23 single chromosomes from a stem cell that has 46 paired chromosomes. This halving of the chromosome set is one of the properties of the specialized cell division process called meiosis that forms gametes.

Remember that the progenitor cell from which the sperm is formed has a copy of each chromosome from both of that individual's parents. Which one of the two copies goes forward to the sperm is basically random, with equal odds. Which of the two copies of chromosome 9 a sperm gets is independent of which copy of chromosome 4 it gets. Thus, if for each chromosome there is an independent 'choice' between the two available copies, it follows that a man can make 2^{23}, or over 8 million, genetically different types of sperm. Given that the same is true for women making eggs, then a given couple could produce 64 million million children who were all genetically different from one another.

Recombination

In fact, the calculation above understates the possible diversity. During meiosis, the paired chromosomes in the progenitor cell line up next to one another and may exchange DNA, such that a sequence that was originally on chromosome A ends up on chromosome B, and vice versa. This process is called **recombination**. Recombination events occur every time a gamete is formed and, since they occur somewhat randomly, the outcome of every episode of recombination is different. Thus, the number of possible types of sperm a man can produce is vastly higher than 8 million. The probability of recombination is very variable across different parts of chromosomes and between different chromosomes. It also varies between men and women. Nonetheless,

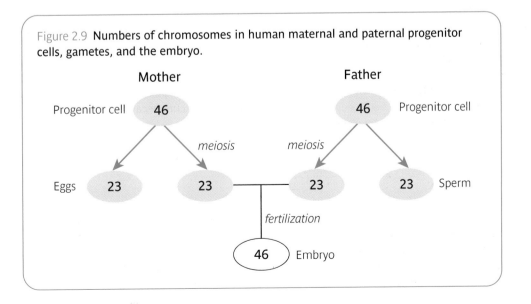

Figure 2.9 **Numbers of chromosomes in human maternal and paternal progenitor cells, gametes, and the embryo.**

some overall estimates have been made, according to which there are about 50 recombination events on average with every sperm formed and about 80 on average for every egg.

Independent segregation and linkage

When two genes reside on different chromosomes, they segregate independently. This means that which copy (grandpaternal or grandmaternal) a particular offspring receives from its father for the first gene is unrelated to which copy they receive of the second. (This follows directly from the fact that at meiosis, which copy of chromosome 4 goes forward to the sperm is independent of which copy of chromosome 9 goes forward.) However, where two genes reside on the same chromosome, there will be some degree of linkage (also called linkage disequilibrium) between them. When two genes are linked, the probability of having the grandpaternal copy of gene A is affected by whether you have the grandpaternal copy of gene B. You can probably see why being on the same chromosome leads to linkage; if my paternal copy of chromosome 7 is from my paternal grandfather, then all of the genes on that chromosome will have come from that source, and if my copy is from my paternal grandmother, then none of them will. They will travel—or fail to travel—together. The only thing that can disrupt this linkage is recombination. In general terms, the physically closer two genes are on a chromosome, the greater the degree of genetic linkage between them, since the probability of some recombination event breaking them apart is related to the distance between them. The principle of linkage is exploited in studies that hunt for the locations in the genome of alleles responsible for particular phenotypic differences (see section 2.4).

2.3.2 Mutation creates genetic variation

As we have just seen, sexual reproduction is a great shuffler. However, it cannot actually create diversity; it merely permutes what is there. The ultimate origin of variation is mutation, or the propensity of DNA sequences to alter during replication.

Single-base substitutions

The simplest type of mutation is the substitution of one base pair for another. This occurs occasionally due to what amounts to error when DNA is copied. C and T are chemically rather similar to each other, as are G and A. Changes between C and T or between G and A are called transitions, whereas those between dissimilar pairs of bases like C and G or C and A are called transversions. Transitions occur about twice as often as transversions. Deletions of a base also occur, but these are much less frequent than the substitution of another base. Rates of mutation are very much higher wherever a C is followed by a G in the DNA sequence. The occurrence of this sequence in the genome is less common than would be expected given the frequencies of C and G, but where it does occur, there is a 'hotspot' for single-base mutations. The average rate of single-base mutations has been estimated at about 1 in 4 million per site per generation. Although this is a very small probability, the large number of sites in the genome means that every individual will carry some new single-base changes.

Simple sequence repeat expansions and contractions

A slightly larger type of mutation is the expansion or contraction of a simple sequence repeat. In such a mutation, an extra copy of the repeat motif is added, or one lost. This occurs because of 'slippage' of one repeat as the enzyme responsible for DNA replication lines itself up to the DNA strand. Simple sequence repeats have mutation rates in excess of 1 in 1,000 per locus per generation (with some reporting rates as high as 1 in 7!). This is thousands of times higher than the rate of single-base substitutions and because it is so high, individuals are essentially unique in terms of their pattern of simple sequence repeats.

Transposable element insertions and segmental duplications

A larger mutation still is the copying of a transposable element such as *Alu* from one part of the genome to another. This is thought to occur about once in every 10 to 100 human births and, depending on where the element moves to, may or may not have a phenotypic effect. Even larger still are what are called segmental duplications. These are events where a chunk of sequence—often many thousands of bases, containing one or more genes—makes an extra copy of itself during replication, often but not necessarily adjacent to the original copy. There are also segmental deletions, and inversions of whole segments of sequence.

Segmental duplications are thought to be particularly significant in evolutionary change. Once a gene has evolved to serve some function, natural selection tends to preserve it exactly the way it is, since mutant forms are likely to perform the original function less well, even though they might have some other, novel, functional potential. Thus, evolution selection tends to 'get stuck' for any gene whose primary function is indispensable to the organism. Duplications give individuals extra copies of the genes involved. With one copy sufficient to perform the original function, the second copy can then acquire new mutations that may bring in new functions, without the organism ever being functionally impaired. Duplication is thus thought to be important in the creation of new genes and the elaboration of the phenotype. Many human genes belong to gene families, which are multiple genes descended from a common ancestor by duplication events. The members of the family often make similar products but have subtle functional differences, allowing the phenotypic systems they make to be highly complex. Examples of gene families are the *HOX* genes, 48 related sequences on four chromosomes that are involved in different ways in the development of the basic body plan, and the β-globin genes that contribute to red blood cells, with different members of the family being expressed at different stages of development.

Whole-genome duplication

The final type of mutation to be reviewed here is perhaps the most drastic of all: duplication of the entire genome. This is a very rare event, but it does happen, since it accounts for the polyploidy found in many parts of the living world (flowering plants and amphibians being examples). The whole-genome duplication acts like a huge segmental duplication, with the two copies going on to have divergent evolutionary histories. Whole-genome duplications also contribute to the origins of gene families, and it is thought that there were two rounds of whole-genome duplication early in the history of the vertebrate lineage (Panopoulou & Poustka 2005).

2.3.3 The extent of genetic diversity and its effects on the phenotype

As we saw in the previous section, there is a range of different types of mutation event, each with a characteristic range of probability of occurrence (Figure 2.10). Because these events are going on in every generation and then are reshuffled by sexual reproduction, there is considerable variation between the genomes of any two humans. We call a particular site in the genome a locus, and a locus at which two individuals of the same species have different sequences is called a polymorphism. Another way of saying this is that a polymorphism is a locus for which there is more than one allele in the population.

Figure 2.10 There is a range of different mutation events in the genome, with variable and very different probabilities of occurring. Probabilities are per locus per generation. The rate of single-base substitutions is generally higher in chromosomes coming from sperm than those coming from eggs, an effect that increases with increasing male age. This is because the progenitor cells of sperm, unlike eggs, divide continuously throughout a man's life, providing continual opportunities for mutation. Mutations other than single-base substitutions do not show this pattern.

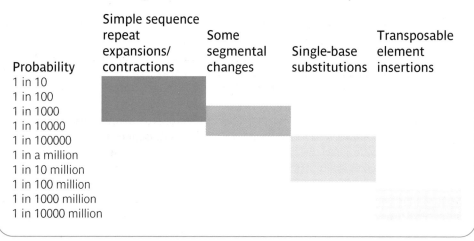

The extent of genetic polymorphism in humans is still being assessed, but the following are known:

1. Something like one in every 1,000 single bases in the genome varies from individual to individual. These are known as single-nucleotide polymorphisms (SNPs, pronounced 'snips').

2. Simple sequence repeats, with their high mutation rate, are extremely polymorphic, so much so that every individual is essentially unique. This is the basis of DNA fingerprinting.

3. Around 12% of the human genome consists of sequences of which individuals have varying numbers of copies due to recent segmental duplications (Redon *et al.* 2006). Some segmental duplications involve very long sequences, and there are hundreds of genes of which different individuals have varying numbers of copies. The phenotypic effects of this variable copy number could be quite substantial.

4. In a base-by-base comparison of the two copies of the genome within a single human, at least 0.5% of the genome was different between the two copies. Forty-four per cent of coding genes showed some difference between the maternal and paternal copies (Levy *et al.* 2007).

Most polymorphisms have no phenotypic effect

There is thus a reservoir of genetic variation from which phenotypic variation may result. However, most genetic variation has no effect on the phenotype. Why is this? Much of it will occur in non-coding DNA, a significant proportion of which has no apparent phenotypic effect (see section 2.2.3). For small polymorphisms, even where they occur in exons of genes, they often make no difference to the amino acid sequence because of the redundancy of the genetic code (see section 2.2.2). A SNP at the third position of a codon beginning GG will never have any effect, since GG<anything> produces a glycine. Even where a mutation causes an amino acid change, if the amino acid has fairly similar properties, the functioning of the protein may not be much affected.

Where mutations do have a phenotypic effect, they are usually deleterious

We have seen that only a small proportion of genetic diversity has any strong phenotypic effect. Of those mutations that do have a phenotypic effect, the overwhelming majority will be harmful (or deleterious, as geneticists usually describe it). This is why the majority of examples you read about in discussions of the effect of mutation on the phenotype are genetic diseases. Why are most mutations harmful? Recall that mutation, although not exactly random (it occurs predictably more often in some parts of the genome than others), is undirected. This means that the effect of mutation is totally unrelated to the physiological needs and functions of the phenotype. The phenotype, having been honed through billions of years of evolution, is already a fairly well-functioning system. Undirected modification of some part is thus much more likely to make it function less well than it is to make it function better. To take an example, imagine you reached inside a computer and, without any knowledge of how computers work, disconnected a wire from one component and reattached it to another. The overwhelming likelihood is that you would make the computer malfunction to a greater or lesser extent. It is just possible that you would actually improve its operation, but this is extremely unlikely. So it is with mutations.

There is direct evidence that mutations tend to be deleterious. Vassilieva *et al.* (2000) studied lineages of roundworms (*Caenorhabditis elegans*) that were removed from the effects of natural selection (by keeping individual worms on their own in nutrient-rich dishes and selecting one worm from the offspring to be the next generation in another such dish). They found that various measures of the worms' biological performance, such as longevity and reproductive rate, declined relative to comparison worms as the generations went along (Figure 2.11). They

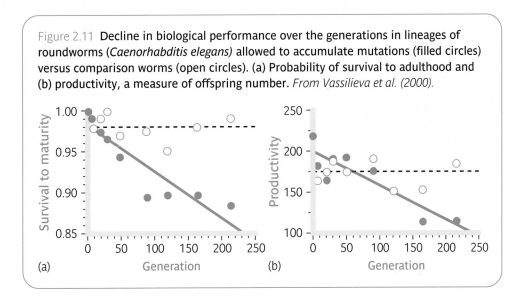

Figure 2.11 **Decline in biological performance over the generations in lineages of roundworms (*Caenorhabditis elegans*) allowed to accumulate mutations (filled circles) versus comparison worms (open circles). (a) Probability of survival to adulthood and (b) productivity, a measure of offspring number.** *From Vassilieva et al. (2000).*

were able to infer from this experiment that the net effect of mutation in this species is to reduce biological performance by around 0.1% per generation. Estimates for the extent of deleterious mutation have also been produced for humans, with an influential one being that each of us has 1.6–3 new deleterious mutations (Crow 2001). However, although most mutations are neutral, and most that are not neutral are deleterious, the occasional one will arise that actually improves biological performance in the animal's environment. These rare advantageous mutations are spread by natural selection, as we shall see in Chapter 4.

2.4 From genotype to phenotype

Now that we have reviewed some basic genetics, we are in a position to specify more completely how variation in the genotype gives rise to variation in the phenotype. The sequence of links is as shown in Figure 2.12. We start with a new, non-synonymous mutation in the coding sequence of a particular gene (1). This gives us a distinct allele of the gene. When transcribed and translated, this new allele will produce a different amino acid chain (2), which will mean a protein with different properties. (If the change is in a regulatory sequence, the difference may be in the amount rather than the sequence of the protein.) The protein in turn will do one of two things. Either it will be an integral part of some body system, in which case the new allele will make that body system different from usual, or it will serve as an enzyme to catalyse some other chemical reaction in the body, in which case that reaction will go a bit faster or slower or otherwise differently from usual. Either way, the constitution of the body ends up different (3) because of an initial difference in the sequence of DNA bases.

This helps us understand what is meant in reports which speak of scientists having found 'the gene for' some characteristic, such as red hair. What this means is that researchers have isolated a gene in which possessing one versus another allele is associated with having red hair versus not having it. Note that this does not mean only that gene is involved in making that part of the phenotype. Most genes have many functions and most functions in development involve many genes, so making hair of any colour doubtless involves dozens or hundreds

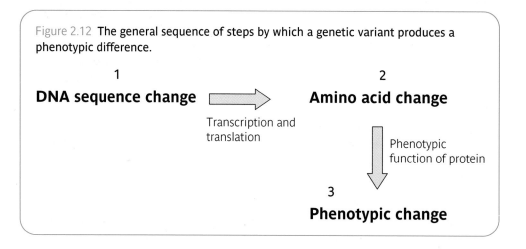

Figure 2.12 **The general sequence of steps by which a genetic variant produces a phenotypic difference.**

of genes. The 'genes for' statement means that, in the context of all the other genes and of a broadly normal environment, the *difference* between having red hair and not having it is related to which allele you have of that particular gene.

Single-gene and polygenic characteristics

Is it useful in this context to distinguish between single-gene and polygenic characteristics? For single-gene characteristics, the difference in phenotypes is determined by which allele the individual has at just one genetic locus. There are many human genetic diseases that are single gene in this sense. Cystic fibrosis, for example, is caused by having certain alleles of a particular gene on chromosome 7. Single-gene diseases are also called Mendelian diseases, since they follow Mendel's laws of heredity (see Chapter 3). Note that single-gene inheritance is not just associated with diseases, but also with aspects of healthy variation. For example, whether your blood type is Rhesus positive or Rhesus negative is determined by which alleles you have of a single gene on chromosome 1.

Gene hunting

You may be wondering how researchers find, in the vastness of the genome, which gene is responsible for cystic fibrosis or the Rhesus factor. There are two general approaches. Linkage studies examine large families where some members have and some lack the phenotypic characteristic of interest. They then establish the genotype of each individual for polymorphic genetic loci whose chromosomal location is known (the alleles at these loci are called markers). Next, they search the data for marker alleles that are shared by affected family members but not the unaffected ones. Thus, for example, if all the family members who were Rhesus negative shared most of the same marker alleles on chromosome 1, but not other chromosomes, then this would be strong evidence that the relevant gene was on chromosome 1 somewhere. More specific localization of the gene can be done, using the principle that the closer on a chromosome two genetic loci are, the less often they become separated by recombination. Linkage studies are useful for localizing the gene underlying single-gene traits using a fairly small number of genetic markers.

Association studies do not need to use members of the same family. Instead, they establish two samples of individuals from the population: those with the phenotypic characteristic of interest and those without it. The individuals are then genotyped to test for differences in allele frequencies between the two groups. Association studies can be very powerful, but require many hundreds of genetic markers per chromosome. This is because, since the individuals sampled are not closely related, recombination will have thoroughly shuffled their chromosomal contents, so significant allele frequency differences will only be found for the gene of interest itself or markers that are extremely close to it on a chromosome. For this reason, association studies can either focus in on a few genes which there is some prior reason to suspect might be involved (the 'candidate gene' approach) or, if they want to cover the whole genome in a prospective, non-hypothesis-driven way, use many thousands of marker loci (the 'genome-wide' or 'whole-genome' approach). Technological advances mean that whole-genome association studies are becoming increasingly feasible and economic. Given the large number of statistical comparisons performed in such a study, the criteria for what to count as a significant difference between the two groups have to be extremely stringent.

Identifying alleles involved in polygenic characteristics

Polygenic traits are those where variation in the phenotype is related to which allele is present across a number of genes. Classic examples of polygenic traits are things like height. Polygenic

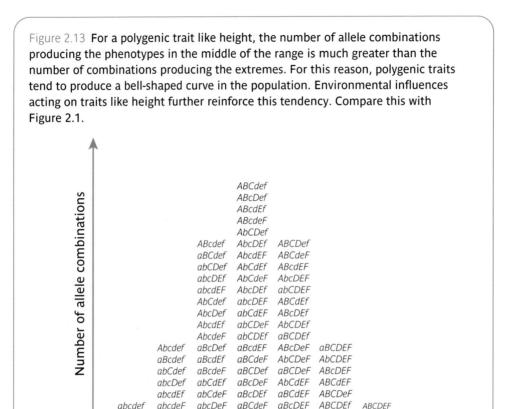

Figure 2.13 **For a polygenic trait like height, the number of allele combinations producing the phenotypes in the middle of the range is much greater than the number of combinations producing the extremes. For this reason, polygenic traits tend to produce a bell-shaped curve in the population. Environmental influences acting on traits like height further reinforce this tendency. Compare this with Figure 2.1.**

traits have a number of characteristics that it is worth noting and to illustrate these we will take a hypothetical example. Imagine in some species that body size is influenced by six genes (genes *a* to *f*; in reality the number would be much higher). Let us assume that each gene has two alleles and represent with a capital letter the allele of each gene that tends to increase size and the one that does not with a lower-case letter. There are 64 possible genotypes and body size will be related to how many capital-letter alleles the individual has. The largest individuals will be those of genotype *ABCDEF* and the smallest those of genotype *abcdef*. Individuals with three capital-letter alleles will be average in size, but note that there are many different genotypes leading to this outcome: *ABCdef*, *abcDEF*, *AbCdEf*, *aBcDeF*, and so on. In fact, because the number of allelic combinations giving the intermediate sizes is so much larger than those giving the extreme sizes, then we are more or less guaranteed a roughly bell-shaped distribution of size (like that shown in Figure 2.1), with most individuals in the middle. Figure 2.13 illustrates this principle.

From Figure 2.13 follow two conclusions. First, there are individuals of the same size who share no alleles at all (e.g. genotypes *abcDEF* and *ABCdef*), and second, although a capital-letter allele such as A is found on average in larger individuals than *a*, the difference is not great and there are many individuals with *a* who are larger than some individuals with A (compare, for example, genotypes *Abcdef* and *aBCDEF*). Because of these facts, the genes involved in polygenic traits are very much harder to detect using linkage and association than those responsible for single-gene traits. Essentially, much larger samples are needed to identify the genetic influence reliably. For this reason among others, the genes underlying the many diseases that are polygenic have been much harder to reliably localize than those underlying Mendelian diseases.

We conclude this chapter with some specific examples of how genotypic differences give rise to phenotypic ones, first in the domain of physiology, then in that of behaviour. All of the examples are ones for which each of the three steps shown in Figure 2.12 is understood to some extent.

2.4.1 Genes for physical characteristics

Huntington's disease is an incurable neurological condition which leads to a gradual loss of coordination and cognitive abilities, often beginning in the person's 40s. It is a single-gene disease, stemming from the possession of a disease-causing allele of a gene on chromosome 4, which codes for a protein called Huntingtin. Huntingtin is a sequence of around 3,000 amino acids, which is active in the central nervous system. Amongst its functions is involvement in making another protein that helps keep cells surviving and growing.

The Huntingtin gene contains a simple sequence repeat, with the motif CAG. Normal copies of the gene have between 11 and 34 copies of this repeat. Since CAG is the codon for glutamine, this means that after transcription and translation, the Huntingtin protein contains a chain of 11–34 glutamines in a row within it. The variation across the range of 11–34 repeats seems to be tolerated with no obvious phenotypic effect. However, since simple sequence repeats have such a high rate of mutation, every now and then an allele of the Huntingtin gene with 40 or more repeats of the motif appears. Such an allele will obviously make a protein containing 40 or more glutamines in its chain. Forms of the protein with this many glutamines start to have different reactive properties. They do not break down in quite the same way as the normal form, and they also have a knock-on effect on the other proteins involved in cell survival that are regulated by Huntingtin. As a consequence of these changes, although neurons form basically normally in a phenotype with the mutated form of Huntingtin, they do not survive as well. As the person gets older, the effect of this neuronal cell death becomes more and more marked, with coordination and cognitive abilities often declining precipitously by late middle age.

As mentioned above, it is more difficult to map out the pathway from mutation to phenotypic difference for polygenic traits. There are some examples where at least some of the genetic contributions are understood. A neat one comes from domestic dogs (*Canis lupus familiaris*) and concerns the polygenic trait of body size. Dogs show a vast range of sizes and since the smallest ones are a fraction of the size of the grey wolves with which they share an ancestor only a few thousand years ago, it is naturally of interest to uncover the genetic mechanisms behind their rapid evolutionary change. Sutter *et al.* (2007) first performed an association study on a large number of dogs of varying size which all belonged to the same breed, the Portuguese Water Dog. They found strong evidence for an association with size for a gene on chromosome 15, called *IGF1*. The researchers then showed that small dog breeds have a very high frequency of one allele (call it *a*) of this gene, whereas it is completely absent from giant dogs like the Great Dane (Figure 2.14). What does *IGF1* do? It codes for a protein, insulin-like growth factor, which is involved in turning on body growth. Mice with the gene deactivated altogether are born much smaller than normal. The naturally occurring *a* allele is not as drastic as this in its effects, but it does seem to do its job less efficiently than the A version.

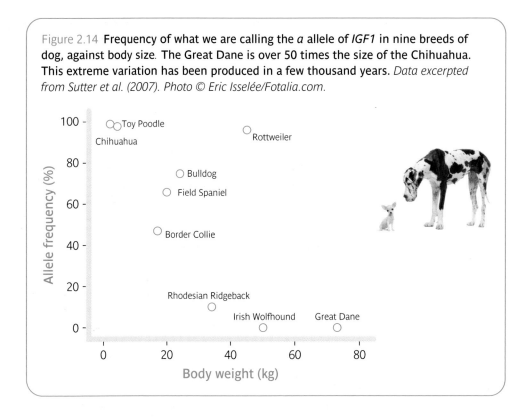

Figure 2.14 **Frequency of what we are calling the *a* allele of *IGF1* in nine breeds of dog, against body size**. **The Great Dane is over 50 times the size of the Chihuahua. This extreme variation has been produced in a few thousand years.** *Data excerpted from Sutter et al. (2007). Photo © Eric Isselée/Fotalia.com.*

Many other genes will contribute to size in dogs. Some large dogs have the *a* allele and some small ones have A, the size differences between them being accounted for by alleles at other loci. However, on average, *a* moves the phenotype towards small size, via the growth-regulating effects of insulin-like growth factor.

2.4.2 Genes for behaviour

The previous two examples have concerned what you might call 'structural' aspects of the phenotype—how big the body grows or how well neurons survive. We can equally well find examples where genotypic changes lead to changes in behaviour. You may find this counter-intuitive. After all, genes make proteins and the structure of the body is made of proteins in a fairly obvious sense. Behaviour, on the other hand, does not appear to be made of protein. How, then, can a genetic sequence code for it?

The answer emerges when we consider that all behaviour must have a physical mechanism underlying it, usually a neural or hormonal mechanism. If we respond to a particular stimulus by becoming aggressive, then there must be some physical circuit starting in the brain that coordinates this response. And that physical circuit is made of proteins. Thus, a genetic mutation could cause the mechanism to develop differently, which in turn causes a difference in the resulting behaviour. Once we adopt this perspective, then we see that behaviour is part of the phenotype—it is just one of the functional effects of particular patterns of protein, in fact—and that it is precisely as legitimate to look for 'genes for' behaviour, in the sense we used that phrase earlier, as it is to look for 'genes for' anything else.

A wonderfully neat example of genetic variation associated with behavioural variation comes from the study of prairie voles (*Microtus ochrogaster*). Prairie voles, unlike most rodents, form long-term pair bonds after mating. Whereas in other rodents, males are promiscuous and disperse after mating a female, in the prairie vole, the male stays with the female usually until one of the partners dies, and gets involved in caring for the pups. The immediate mechanism that causes a pair bond to form is that a hormone called arginine vasopressin is released in the male's brain on mating with a female. This is known because injecting the male with a chemical that blocks the action of arginine vasopressin causes pair bonding to fail. Arginine vasopressin is also produced in the brains of non-pair-bonding rodents. However, prairie voles have many more of the receptor molecules via which arginine vasopressin influences neurons over a number of different brain areas. Both arginine vasopressin and the receptor molecule (called V1a) are proteins.

The gene which produces V1a has been localized. Its coding sequence is almost identical across different types of rodent, but there is an associated regulatory sequence that is clearly different in the prairie vole. Young *et al.* (1999) created genetically engineered mice that had the prairie vole version of this sequence instead of the usual mouse one. These mice develop the same widespread pattern of brain V1a receptors that is seen in the prairie vole. More strikingly, their behaviour also changed, to closely resemble that of the pair-bonding voles (Figure 2.15). Thus, the additional V1a receptors created by the genetic variant made the animal respond to the arginine vasopressin hormone released by mating in such a way as to make it bond with the female. Thus, although many genes are involved in bonding behaviour, there is a legitimate sense in which V1a is a 'gene for' post-reproductive pair bonding in rodents.

Are there good examples of genetic effects on human behaviour? One such example comes from studies of the monoamine oxidase A (*MAOA*) gene. Monoamine oxidase is an enzyme that helps regulate (among other things) the functioning of a brain neurotransmitter called serotonin. Serotonin seems to be involved in the regulation of mood states. We know this because drugs that promote its activity either acutely (the drug of abuse Ecstasy or MDMA) or chronically (antidepressants like Prozac) increase feelings of well-being or relaxation.

Brunner *et al.* (1993) were studying a Dutch extended family in which many of the males had a severe conduct disorder, the symptoms of which included violence, rape, and arson. A linkage study implicated a gene on part of the X chromosome. *MAOA* was known to reside in this region and the role of MAOA in the regulation of serotonin and hence mood was already known. The researchers analysed the urine of the affected men and established that they were completely lacking in MAOA activity. To find out why, Brunner and colleagues sequenced the *MAOA* gene in some of the affected men. They found four single-base changes in the affected men. Three of these were synonymous and could therefore not be responsible. The fourth was a transition from a C to a T at the 936th base. This change was the first base of a codon, normally CAG, which now became TAG. CAG, when translated via mRNA, produes a glutamine, whereas TAG is the codon that stops transcription at this point. Thus, in these men, all of the rest of the *MAOA* gene was untranscribed, which is why they had no working MAOA at all.

This particular allele is extremely rare, and its effect is dramatic. However, in recent years, some evidence has emerged from association studies that less catastrophic allelic variation

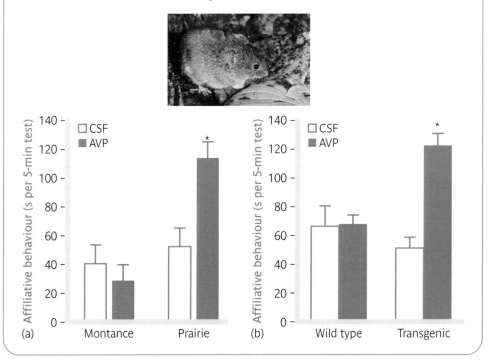

Figure 2.15 (a) When a male prairie vole (*Microtus ochrogaster*) is injected with arginine vasopressin (AVP), the amount of affiliative (bonding) behaviour is increased relative to after injection with a control substance (CSF). There is no such effect in the non-monogamous montane vole *(Microtus montanus)*. (b) Transgenic mice engineered to contain the prairie vole version of the *V1a* gene respond just like prairie voles. Normal mice, which are not monogamous, show no such response. Error bars represent 95% confidence intervals. *From Young et al. (1999). Photo © All Canada Photos/Alamy.*

(specifically, contraction of a simple sequence repeat in a regulatory sequence close to the gene) may be associated with attention-deficit hyperactivity disorder (ADHD) and with antisocial behaviour of various kinds (Craig 2007). In particular, there is some evidence that *MAOA* allele variation plays a part in determining whether individuals respond to aggressive maltreatment by becoming aggressive themselves.

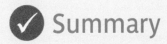 Summary

1.   Variation in the phenotypes of organisms arises from variation in their genetic material.

2.   The genotype gives rise to the phenotype, and also to the genotype of the next generation. There is no mechanism for the phenotype to permanently alter the genotype and thus characteristics acquired from the environment during life are not normally passed on genetically to offspring.

3.   Genes are sequences of DNA bases that code for proteins.

4.   Most of the genome in organisms like humans does not consist of coding genes, but of various kinds of non-coding elements.

5.   Sexual reproduction creates new genetic combinations, but the ultimate origin of genetic variation is mutation. Mutations of several different types occur during mitosis and meiosis. The net effect of mutation is that there is considerable genetic variation between individuals of the same species.

6.   Many mutations have no phenotypic effect. Amongst those that have an effect, most are deleterious.

7.   There is a legitimate sense in which there can be said to be 'genes for' a particular characteristic. What is meant is that variation in the characteristic is related to which alleles of those genes are present.

8.   Behaviour is part of the phenotype and thus the idea of 'genes for' behaviour is as legitimate as genes for any other characteristic.

? Questions to consider

1.   At the end of section 2.2.2, it was suggested that the genetic code of DNA triplets to amino acids had itself been subject to natural selection. How might such selection occur? Use the terms heredity, variation, and competition in your answer.

2.   If we humans are accumulating new genetic mutations at around three per individual and these are mostly deleterious, then why do we not experience gradual deterioration of biological performance over the generations like the roundworms in the experiment of Vassilieva *et al*. (2000)?

3.   Let us assume (arbitrarily) that pair bonding is a good strategy for rodents living on open plains, whilst promiscuity is a good strategy for those living in forests. An island with plains and forests on it is invaded by promiscuous rodents. After a few million years of evolution, we come back and find pair-bonding rodents living on the plains and promiscuous ones living in the forests. The rodents have adopted the best lifestyle for their habitat. Write down the steps involved in this process of adaptation, using the words mutation, phenotype, and competition in your answer.

4.   Does finding a 'gene for' aggressive behaviour, as in the *MAOA* example in section 2.4.2, mean that the environment plays no role in causing such behaviour?

→ Taking it further

More detailed information about genes and the genome can be obtained from a human genetics text-book, such as Strachan & Read (2003), which has a molecular emphasis, or Jobling *et al*. (2004), which has an emphasis on genetic diversity in the human population. Burt & Trivers (2006) is a compelling journey into the complexities of the genome, where different sequences cooperate and compete to get themselves replicated.

Genetics and genomics are progressing so fast that all textbooks are out of date as soon as they are published. Some of the papers in the special issue of *Nature* in which the first draft of the human genome was presented are accessible and well worth reading (volume 409, number 6822, 15 February 2001). Even these are ancient history. The journal *Trends in Genetics* publishes accessible and up-to-date reviews of new developments. The journals *Nature* and *Science* carry new breakthroughs in the study of genotype and phenotype almost weekly. The studies of voles, dogs, and MAOA used as examples in this chapter come from these sources.

Heredity

As we saw in Chapter 1, evolution can only occur if offspring tend to resemble their parents. This is the principle of heredity. Any variant that is not heritable will go no further than one generation and thus cannot feature in the evolutionary future of the lineage. Chapter 2 investigated how variant phenotypic forms arise, through mutation, but how are those variants passed on to future generations? In particular, when there are two parents with different phenotypes, what phenotype does the offspring have?

The answers to these questions were not at all clear at the time Darwin wrote *The Origin*. Biologists believed that the characteristics of the two parents were simply blended together and this looked like a problem for Darwin's account of evolution. Fortunately, Mendel's plant-breeding experiments established that blending at the genetic level does not occur. He instead established some basic laws of heredity for single-gene characteristics. In this chapter, we first look at the way heredity does *not* work, namely blending inheritance (section 3.1), then come on to how Mendel showed that it does work (section 3.2). We then consider how polygenic traits are transmitted (section 3.3), and finally look at the concept of heritability and its implications (section 3.4).

3.1 Inheritance does not work by blending

An attractive intuitive idea is that heredity in a sexually reproducing species works by blending the characteristics of the two parents to give some kind of average, just like you mix two paints to get the intermediate colour. Thus, if there was a population of red insects in which a green mutant cropped up and that green mutant bred with a red, the result would be a blend of red and green, presumably brown of some kind, and what that brown insect would in turn pass on to its offspring would be brownness.

If inheritance was like this, Darwin's theory could never work. To see why, let us continue with our insect example. Assume that the green mutant was better camouflaged than its red competitors, and therefore lived longer and left more offspring. Green is a better design than red in this scenario and natural selection requires that it therefore become more common. Now, the green mutant mates, obviously with a red individual because everyone else is red. If inheritance is by blending, the offspring have an intermediate red-green nature. Let us be generous and assume that these intermediates are also favoured by selection and again they survive and mate, mainly with reds since reds are still the overwhelming majority of the population. The offspring of this generation have a nature that is the blend of red-green and red, which is red-red-green. Again, assume they fare well. Their offspring in turn will be red-red-red-green, and the next generation red-red-red-red-green, and pretty quickly you have something that is pretty much indistinguishable from red.

You can see the problem. When rare advantageous variants arise, blending of characteristics averages them with the majority type and soon the advantageous innovation is diluted out. Everyone is the same as each other and very similar to the colour they started off. Such dilution of characteristics would mean that there would be essentially no change in phenotype even when selection for being green was very strong.

You may be puzzled at this point because certain characteristics do look as if they blend in sexual reproduction. For example, a couple's children will often have skin colours that are intermediate between the skin colours of the two parents. Appearances are deceptive here. Although the net *effect* of the two parents' alleles on the child's phenotype is a skin of an intermediate colour, the alleles themselves are unblended. They are carried on intact by the children. This means that variation is retained for natural selection to work on in future generations.

3.2 Mendelian genetics

The best way to appreciate why Mendel is so important is to review his experiments. Mendel worked mainly with ordinary pea plants in the garden of his monastery. He bred these in enormous numbers, taking careful control over which individuals pollinated each other and keeping meticulous track of the phenotypes in each generation. The result was a paper, one of the most important in the history of biology, entitled 'Experiments in Plant Hybridization' and published in 1866 in the *Proceedings of the Natural History Society of Brünn*, to almost no acclaim. Nobody, probably not even Mendel himself, understood the full significance of what he had done until decades later. This section reviews the principles that Mendel established.

3.2.1 Inheritance of single-gene characteristics

Mendel looked at various characteristics for which naturally occurring pea plants could be found in two forms. We will concentrate as an example on whether the pod is yellow or green. Mendel first established pure yellow-podded and pure green-podded lines of plants. He then crossbred yellow-podded individuals with green-podded individuals. The offspring of this cross, following the usual terminology of plant and animal breeding, are called the F_1 generation. He recorded the phenotypes carefully. He then bred the F_1s with themselves, producing the F_2s, whose phenotypes he again carefully recorded.

To see the significance of what Mendel found, let us think about the expectation we should have if inheritance is by blending (Figure 3.1a). The F_1s, with one yellow and one green parent, should be mid-way between yellow and green. The F_2s, with two parents both of whom are mid-way between yellow and green, should also be mid-way between yellow and green. In other words, the F_2s should be identical to the F_1s.

Compare this clear prediction with what Mendel actually observed (Figure 3.1b). The F_1s were all yellow. There was no trace of green or greenish-yellow at all. In the F_2s, three quarters of the plants were yellow and one quarter were green. The green of the green F_2s was completely green, not green diluted with yellow, and this despite the fact that none of the F_2s had a green parent.

Mendel needed to explain, then, how green could disappear from the phenotype but be carried intact somewhere, only to reappear in the next generation, and also why when it did reappear, it did so in a characteristic proportion of one plant in four. He deduced that you could account for the pattern using some simple assumptions, which we already met briefly in Chapter 2:

- Each individual must be carrying two particles of heredity (genes, as we now call them), one from each parent.
- These particles can be either 'green' or 'yellow' in form.

In more modern terminology, we can say that there are two alleles of the pod-colour gene, A and a, where A is associated with yellow pods, and a with green pods.

Homozygous and heterozygous genotypes
A plant could either have two copies of the same allele (i.e. genotype AA or aa), in which case we say it is **homozygous**, or one copy each of two different alleles (i.e. Aa), in which case it is **heterozygous**. When plants breed, it is a 50–50 chance which of their two copies of the

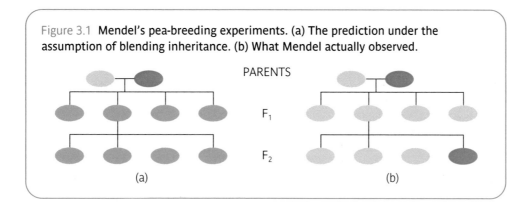

Figure 3.1 **Mendel's pea-breeding experiments. (a) The prediction under the assumption of blending inheritance. (b) What Mendel actually observed.**

PARENTS

F_1

F_2

(a)   (b)

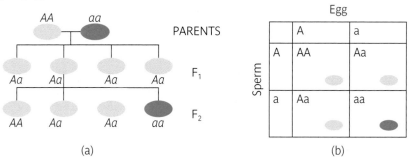

Figure 3.2 **(a) Phenotypes observed in Mendel's pea-breeding experiments, with the inferred genotypes indicated. (b) The Punnett square allows us to predict genotype frequencies in the offspring from those in the parents. Possible genotypes of sperm and egg are drawn in the rows and columns. Each square represents an equally likely offspring type.**

gene goes forward to any particular gamete. Thus, if the parent is a homozygote *Aa*, half its gametes will contain an *A* and half will contain an *a*. Homozygotes, obviously, produce 100% of gametes the same.

Dominant and recessive alleles

We can account for the pattern seen in the pea-breeding experiment with one further assumption. When the genotype is heterozygous, the phenotype is yellow. In other words, as long as there is at least one copy of A, yellow pods develop. We say in this case that yellow (or A) is dominant. For green to develop, by contrast, both copies of the gene have to be *a*. Thus, we say that green (or *a*) is recessive.

This allows us to interpret the pattern seen in the experiment (Figure 3.2). The parental generation, being from purebred lines, have genotypes *AA* (the yellow plant) and *aa* (the green plant). All of the F_1s get an *A* from their yellow parent and an *a* from their green parent, so they are heterozygous with genotype *Aa*. Since yellow is dominant, this means that they all have yellow pods. It also means that all of them produce half of their gametes with *A* and half with *a*.

Now we come to the F_2s. Half of the gametes that make them up are *A* and half *a*. Thus, the probability of them receiving two As is $0.5 \times 0.5 = 0.25$. The probability of two *as* is the same, 0.25. Thus, one quarter of the F_2s will be *AA* and one quarter *aa*. The remaining half will be heterozygous *Aa*. The odds of the different genotypes can be easily visualized using a Punnett square (Figure 3.2b). Because yellow is dominant, the quarter of the plants that are *AA* plus the half that are *Aa* will have yellow pods. Only the quarter that are homozygous *aa* will have green pods. This explains the characteristic 3 : 1 ratio of phenotypes in the F_2s.

Generality of Mendelian principles

These principles are not restricted to plants or to genes for pod colour. In fact, they describe how genetic material is passed on in all diploid, sexually reproducing organisms. Which allele is dominant and which is recessive depends on how the gene produces its phenotypic effect. Many genes code for enzymes and as long as *some* enzyme is made this will be sufficient to bring about the phenotypic effect. One copy will be sufficient to produce some enzyme. Thus, functional alleles of genes that make enzymes generally dominate over competing alleles where the capacity to make the enzyme has been lost. For example, human albinism is caused by mutations to the genes involved in making the pigment melanin that render them inoperative (these are called loss-of-function mutations). Almost all forms of albinism are recessive, as only if neither copy of the gene is functional will no melanin be made.

Genes can have more than two alleles; we consider only the two-allele case here, for convenience. It is not always the case that one allele is dominant and one recessive. Two alleles can also be co-dominant. This means that heterozygotes fully express the phenotype of both of their homozygous parents. You can see this in human blood groups. Your blood group depends on which alleles you have of blood protein genes. Some blood groups (such as AB) are found in people who are heterozygous with co-dominant alleles. Their blood contains two different versions of a blood protein, whereas the blood of homozygotes such as the A blood group contains only one. There can also be incomplete dominance, which means that the phenotype of the heterozygote is intermediate in form between those of the two homozygotes. For example, homozygous red carnations crossed with homozygous white carnations produce heterozygotes with pink flowers. Note that this is not the same as blending inheritance, since only the phenotypic effects have been blended, not the underlying alleles. In the next generation, such pink individuals could once again produce pure red or white offspring.

3.2.2 Independent segregation

The next important principle that Mendel was able to establish is the principle of independent segregation. This principle, which we met briefly in Chapter 2, means that phenotypic traits controlled by different genes can become separated from each other through the generations. Let us consider an example from Mendel's pea plants.

As we have seen previously, one characteristic, pod colour, is controlled by a gene with two alleles, A and a. Now, let us add another characteristic, pod texture. Texture can be smooth or wrinkled, and it is controlled by a second gene, with alleles B (dominant, producing smooth pods) and b (recessive, producing wrinkled pods). Suppose we establish two purebred lines, one of plants with yellow, smooth pods, and the other of plants with green, wrinkled pods. Now, we cross-fertilize the two lines with each other.

In the F_1s, we find only yellow, smooth pods, since both yellow and smoothness are dominant. If we then breed the F_1s to each other, we should find 3 : 1 yellow to green in the F_2s and also 3 : 1 smooth to wrinkled, for the reasons we have already reviewed. However, here is the critical question: will the 25% of F_2 plants that have green pods also be the 25% with wrinkled pods? Or will the colour of the pod be independent of its texture? We have, in effect, two predictions (Figure 3.3): colour and texture could travel together down the generations, in which case the quarter of plants with green pods would also have wrinkled pods, or they could travel independently. Mendel found that the latter was the case. In the F_2 generation of the experiment described above, four different phenotypes are observed: yellow smooth, yellow wrinkled, green smooth, and green wrinkled. These are found in the respective ratios of 9 : 3 : 3 : 1.

You can work out why these are the ratios using a Punnett square (Figure 3.4). It comes directly from the fact that which colour genotype you have is independent of which texture genotype you have. This principle is extremely important because it means that novel phenotypic combinations can arise through sexual reproduction (e.g. the combination of green and smooth was never seen in the first generation, but is present in the F_2s). It also means that

Figure 3.3 **The proportions of phenotypes in the first and second generation of offspring of purebred lines of plants with pods that are either yellow or green and either smooth or wrinkled, where yellow and smooth are dominant. (a) Under perfect linkage, green and wrinkled remain coupled through the generations. (b) Under independent segregation, novel combinations of phenotypes are generated.**

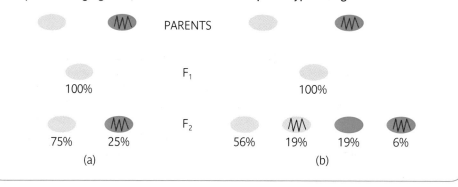

PARENTS

F$_1$

100%   100%

F$_2$

75%   25%   56%   19%   19%   6%

(a)   (b)

Figure 3.4 **Punnett square for two genes on different chromosomes, each with two alleles. We again list all the different possible genotypes of sperm and eggs in the rows and columns. Each square represents an equally likely offspring genotype. The frequency of a particular phenotype will follow from what proportion of the squares lead to that phenotype. (In this case, 9/16 yellow smooth, 3/16 yellow wrinkled, 3/16 green smooth, 1/16 green wrinkled).**

Egg

	AB	Ab	aB	ab
AB	AABB	AABb	AaBB	AaBb
Ab	AABb	AAbb	AaBb	Aabb
aB	AaBB	AaBb	aaBB	aaBb
ab	AaBb	Aabb	aaBb	aabb

Sperm

natural selection can change the frequencies of one phenotypic characteristic (say, colour) without also changing the frequency of another (say, texture).

Independent segregation occurs because, at meiosis, the selection between which of the two copies goes forward to the gamete is done independently for each chromosome (see Chapter 2).

The genes for pod colour and pod texture are on different chromosomes and therefore there is association between the alleles you inherit in the two cases. However, where two genes reside on the same chromosome, then, obviously, they do not segregate independently. Such genes exhibit linkage, which means that they tend to travel together down the generations until broken apart by recombination. As we saw in the last chapter, linkage allows us to identify which genes reside on the same chromosomes and even how far apart they are on those chromosomes.

3.2.3 Mendelian diseases and deleterious recessives

Mendel's principles apply to all single-gene characteristics, including, obviously, Mendelian diseases in humans. There are many known Mendelian diseases, including some we have met already such as cystic fibrosis and Huntington's disease. For most of them, the gene involved can be localized to a specific chromosome because of the clear patterns of linkage with other traits that they display and, in many cases, the molecular basis is known. The most serious Mendelian disorders, and the mutations underlying them, remain extremely rare, since most affected individuals die without reproducing. Indeed, the disease alleles can only attain anything higher than a vanishingly small frequency in two ways.

First, some Mendelian diseases, such as Huntington's disease, have their phenotypic effects after the individual has reproduced. This allows the disease allele to be passed on, even though it is dominant and ultimately lethal. Second, the disease alleles may be recessive. This means that, although being homozygous with the disease allele is lethal in early life, the disease allele continues to exist in the population by being carried in heterozygotes whose phenotype is normal. Only when two heterozygotes reproduce may two recessive alleles be passed on to an individual and the disease appear. For example, the most common type of haemophilia (inability of the blood to clot) is a Mendelian disease caused by recessive loss-of-function mutations to genes coding for blood-clotting factors.

Inbreeding

Lethal recessives, and deleterious recessives more generally, are the reason that organisms should avoid inbreeding. To see why, consider a lethal recessive whose frequency in the population is 1 in 1,000. Suppose a man who is heterozygous with this allele mates with a woman drawn randomly from the population. There is a 1 in 1,000 chance that she will also be a carrier, in which case one in four of the children would be affected. Thus, the risk per child of being affected is 1 in 4,000. Now suppose instead that the man marries his full sister. Since he has one copy of the disease allele, at least one of his parents must be heterozygous. This means that his sister's chance of also being heterozygous is at least one in two. Thus, children he produces with his sister will have a risk of the condition of one in eight or more, which is obviously very much higher than if he married a non-relative. The increased risk does not stop at full siblings. Instead, it declines with increasingly distant relatedness, basically reducing by a factor four with every generation since the common ancestor. This is why animals of all species tend to avoid mating with close kin: it dramatically increases the likelihood of suffering the ill-effects of deleterious recessive alleles.

3.2.4 Allele frequencies in Mendelian populations

Sexual reproduction can change the distribution of genotypes across the generations. For example, when homozygote AA men breed with homozygote aa women, the offspring have a genotype *Aa*, which is not present in either parent. When two *Aa* heterozygotes breed together, half of their children have the parental genotype and the other half have the novel genotypes *AA* or *aa*. As the generations go by, then, we might reasonably ask what will happen to the distribution of genotypes. Will the overall frequencies of A and a change? Will one displace the other? And what proportion of the population should we expect to end up heterozygous versus homozygous?

These questions are answered by a simple piece of genetic theory called the Hardy–Weinberg equilibrium (see Box 3.1). This theorem models the change in expected frequencies of alleles over the generations for a polymorphism in a diploid, sexually reproducing population, where all genotypes have the same probability of surviving and reproducing. It shows two things. First, the population frequencies of the different alleles do not change over the generations, even though sexual reproduction is constantly shuffling the combinations up. Second, as long as mating is largely random, the expected frequency of homozygotes and heterozygotes is predictable from the frequency of the allele in general. Basically, Hardy–Weinberg says what will happen in a sexually reproducing population where there is no natural selection. Alleles will become neither more common nor more rare over time, and the relative proportions of heterozygotes and homozygotes will be constant. Thus, if we observe changing frequencies of alleles, we can have a fair idea that there is some difference in the genotypes' ability to survive or reproduce—in other words, that natural selection is occurring.

Box 3.1 **Hardy–Weinberg equilibrium**

Let us think of a gene with two alleles *A* and *a* in a sexually reproducing population, where the proportion of *A* alleles in the population is designated *p* and the proportion that are *a* is designated *q*. There are only two alleles, so $p + q = 1$. It follows that:

$$q = 1 - p$$

and

$$p = 1 - q$$

Of all gametes produced in this population, then, *p* will carry *A* and *q* will carry *a*. What does this mean for the genotypes of the offspring formed? The probability of the different possible offspring genotypes is simply the product of the probability of an egg bearing one relevant allele and the probability of a sperm bearing the other relevant allele (Figure 3.5). For example, if 10% of all sperm carried *A* and 10% of all eggs carried *A*, then 1% (i.e. 10% of 10%) of all embryos would be *AA*. Thus, in general, the frequency of homozygotes *AA* will be p^2, of homozygotes *aa* will be q^2, and the frequency of heterozygotes will be $2pq$.

So far so good. Now allow those genotypes to mature and to reproduce themselves. Crucially, we assume that the probability of surviving and producing gametes to go forward to the next generation is the same for all three genotypes. What will be the frequency of *A* in the second-generation gamete pool?

All of the gametes of *AA* individuals have *A* and half of the gametes of *Aa* do. Thus, the proportion of second-generation gametes receiving *A* is:

$$p^2 + \frac{1}{2}(2pq)$$

Figure 3.5 Calculating the expected population frequencies (in blue) of different genotypes (in black) (left) for the general case when the frequency of the allele A in the gene pool is p and (right) for a specific example when the frequency of A is 0.1. Note that the frequency of the heterozygote is the frequency in the bottom left square plus the frequency in the top right square, i.e. $2pq$ or, in the example, 0.81.

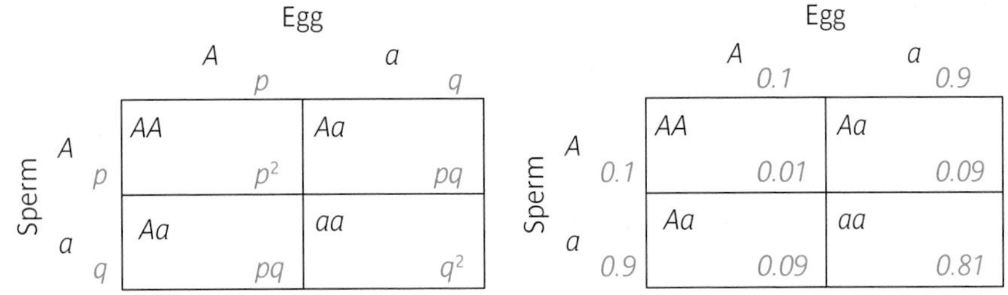

The two and the half cancel each other out, giving:

$$p^2 + pq$$

Substituting $(1 - p)$ for q, we get:

$$p^2 + p - p^2$$

which is just p. Thus, the proportion of gametes with A in the second generation is p, just as it was in the first generation.

Exactly the same reasoning applies to the proportion of second-generation gametes receiving a. All of the gametes of aa individuals have a and half of the gametes of Aa do. Thus, the proportion of second-generation gametes receiving a is:

$$q^2 + \tfrac{1}{2}(2pq)$$

The two and the half cancel each other out, giving:

$$q^2 + pq$$

Substituting $(1 - q)$ for p, we get:

$$q^2 + q - q^2$$

which is just q. Thus, the proportion of gametes receiving A and a remains the same in the second generation as it was in the first. By extension of reasoning, it will be the same in the third as it was in the second, the same in the fourth as it was in the third, and so on. Sexual reproduction leaves the underlying allele frequencies the same, and in each generation, genotypes AA, Aa, and aa will appear in the ratio $p^2 : 2pq : q^2$.

Genetic drift

The Hardy–Weinberg equilibrium depends on a further assumption. It treats the population as infinite in size, which clearly biological populations are not. The larger the population, the more similar its behaviour will be to an idealized infinite population. However, when populations are small, random events can lead them to behave atypically. It is like tossing a coin. Your expectation is that half of the time you will get heads and the more times you toss the coin, the closer to half the proportion of heads gets. However, if you toss just four times, you could well end up with 100% heads or indeed 0% heads. Similarly, the smaller the population, the more dramatically allele frequencies will fluctuate from generation to generation, even in the absence of natural selection. Such fluctuation is known as genetic drift.

We are not going to go into genetic drift in great detail here, but it is important for evolutionary biology in a number of ways. Because populations are in fact finite, some allele frequencies fluctuate up and down a little from generation to generation even if the alleles have no systematic effect on the phenotype either way (that is, even if they are neutral). In the course of such fluctuation, occasionally a new mutation will fluctuate all the way to the point where everybody in the population has it (an allele that everyone has is said to have reached fixation in the population). The probability of a new mutation drifting to fixation is $1/(2N)$, where N is the population size. For example, in a population of ten individuals, the probability of a new neutral mutation that arises drifting to fixation is $1/20$, whereas in a population of 100 individuals it is $1/200$. Thus, there will be a certain amount of change in genotype over time in a population, even in the absence of any natural selection.

Neutral theory and the molecular clock

Because of genetic drift, two populations that become isolated from each other will tend to diverge over time in terms of DNA sequences due to the occasional fixation of neutral or nearly neutral mutations in each population. Many mutations are neutral, since much of the genome is non-coding, and also because the redundancy of the genetic code means that many changes to coding sequences are synonymous anyway (see Chapter 2).

Rather surprisingly, the rate of divergence of two populations does not depend on their sizes. Although fixation of a new mutation is more likely in a small population, the number of new mutations arising is greater in large populations and these two opposite effects of population size neatly cancel each other out. Thus, to the extent that variation is neutral or near neutral, the amount of divergence between the DNA sequences of any two populations or species basically reflects the time since their common ancestor.

This is known as the neutral theory of molecular evolution and it is particularly associated with the Japanese geneticist Motóo Kimura (Kimura 1983). Because of neutral evolution, and the associated idea of the molecular clock, we can use the amount of molecular similarity of different species or populations to establish their family trees and times of divergence. The more difference there is between the DNA sequences of two animals, the greater the time since their common ancestor. We return to some applications of the molecular clock in Chapter 10.

There might appear to be some tension between the Darwinian view of evolution, in which changes spread because they are adaptive, and the neutral theory, which says that they just accumulate constantly over time by drift and have no effect on the phenotype. There is not really any contradiction. Neutral evolution accounts for much of the functionless molecular change that goes on in the history of life, but has little to say about change in design-like features of the phenotype. Darwinians do not need to claim that all or even most mutations affect the survival or reproduction of the phenotype. As long as *some* mutations have significant phenotypic effects, positive or negative, then Darwinian natural selection will occur. Neutral genetic changes, by contrast, provide us with a handy record of which populations diverged at which points during the long, branching history of life.

3.3 Quantitative genetics

Mendel's laws allow us to predict the phenotypes and genotypes of offspring from the phenotypes and genotypes of the parents for single-gene characteristics. When characteristics are affected by multiple genes, things become more complicated and we need to draw on a body of theory called quantitative genetics. This section reviews its basic principles.

3.3.1 Central ideas of quantitative genetics

Quantitative genetics is about predicting phenotype from genotype for traits which are not either/or (like being yellow or green), but continuously variable (like height or weight) and where large numbers of genetic loci are involved. In such cases, although all the alleles involved are being passed from generation to generation in a Mendelian manner, it is hard to isolate influences locus by locus using Mendelian cross-breeding experiments. Instead, quantitative genetics investigates the overall relationships between genetic and phenotypic similarity.

The central idea of quantitative genetics is that some of the variance in phenotypic characteristics is due to the influence of the environment (events unrelated to which genotype you have, which might include, for example, nutrition and disease) and some is due to the influence of genotype (in other words, which alleles you carry). Thus, the total phenotypic variation is the sum of the genetic variation and the environmental variation. How do we work out what the relative values of the genetic and environmental components are? The various techniques all rely on the basic insight that different categories of relatives share different proportions of their genetic material. Thus, if genotype contributes significantly to the phenotypic characteristic in which we are interested, then close relatives will be more similar to each other than are more distant relatives or randomly selected members of the population. In other words, the characteristics will tend to run in families.

Estimating heritability

People often resemble their relatives. We can quantify this resemblance by calculating the phenotypic correlation between a particular pair of relatives. For example, in the British population, a man's height is related to the height of his father. This relationship is shown in Figure 3.6. The fact that there is a non-random pattern of sons' and fathers' heights shows that there is a phenotypic correlation. How can we use this to calculate the genetic contribution to height in this population?

A commonly employed estimate of genetic contribution is called the heritability (sometimes represented by h^2). The heritability is the proportion of the observed phenotypic variation that can be accounted for by genetic variation. The maximum possible heritability is 1, which would mean all phenotypic variation was related to genotypic variation and a heritability of 0 means that none of it is. Since the total phenotypic variation is the sum of genetic and environmental influences, then, by exclusion, a heritability of 0 means that environmental effects account for all of the variation in the characteristic.

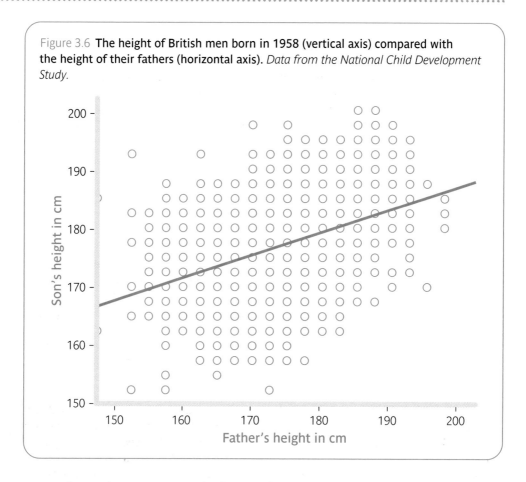

Figure 3.6 **The height of British men born in 1958 (vertical axis) compared with the height of their fathers (horizontal axis).** *Data from the National Child Development Study.*

Returning to Figure 3.6, a very simple way of estimating heritability would be to use the correlation between the father's and son's heights. We work out how much of the variation in son's height is accounted for by father's height and double that to estimate the heritability. Why do we double it? A man contains half his father's genetic material and so the correlation between them reflects the influence of half a genotype. To estimate the effect of a full set, you need to double it. Another way of saying this is that the **coefficient of relatedness** between a man and his father is one-half.

Coefficients of relatedness

Coefficients of relatedness for various types of relatives are shown in Figure 3.7. The coefficient of relatedness, also called *r*, is often wrongly described as 'the proportion of their genes' that two relatives share. This is nonsense since all members of the same species basically share all of their genes. Slightly more careful authors will sometimes describe the coefficient of relatedness as the probability that two relatives share the same allele at a particular locus. Unfortunately, this is wrong too. Consider an allele whose frequency in the population is 0.9. The probability that two randomly chosen, unrelated individuals from the population both have it is 0.9 × 0.9 = 0.81 and yet their coefficient of relatedness is 0. So what does the coefficient of relatedness actually mean?

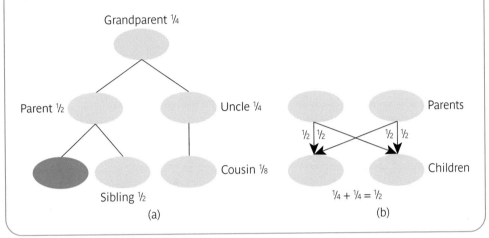

Figure 3.7 **(a)** Coefficients of relatedness (relative to the dark blue individual) for some family relationships. You will notice that the coefficient for a sibling is as high as that for a parent ($\frac{1}{2}$). This is because siblings are related in two ways. **(b)** Each sibling draws half of the father's genetic material, and since these two drawings are independent, the shared inheritance through the father is $\frac{1}{2} \times \frac{1}{2} = \frac{1}{4}$. The same is true through the mother, and so the total genetic material shared by descent is $\frac{1}{4} + \frac{1}{4} = \frac{1}{2}$.

A better way to think about the coefficient of relatedness is as the probability that any particular allele in individual A is **identical by descent** (i.e. derived from the same immediate source) as the allele in individual B. Let us say that I have a copy of an allele whose frequency in the population at large is 0.9. Half of the time (because $r = 0.5$) my brother will have got his version from the same parental chromosome that mine came from and so it will be the same as mine. The other half of the time, he will have got his copy from a different parental chromosome and therefore the probability that it will be the same as mine is equal to the background population frequency, in this case 0.9. So the probability he shares this allele with me is the average of 1 and 0.9, which is 0.95. You will notice that this figure is half-way between the background population frequency (0.9) and 1. So another way of describing r would as a measure of how much *more* similarity there is between me and my relative, above and beyond that which I would expect from a randomly selected member of the population.

Problems with simple heritability estimates

Returning to the height example, using the correlation shown in Figure 3.6 and the coefficient of relatedness of one-half yields a heritability estimate of 0.84, suggesting that 84% of the variation in height in this population is heritable. However, there is a problem with this figure as an estimate of genetic effects upon the phenotype. In calculating it, we assumed that the correlation between father and child was *entirely* due to the shared half of their genetic variation. However, this may not be the case. If a family in the parental generation was affluent, its members would be better nourished and healthier than average and thus become taller. The likelihood is that in the next generation, the family would still be relatively affluent, so the next generation would be taller than average for the same reasons. This would lead to a correlation between parent and offspring height, but it would actually be an effect of the environment, not of genotype. When simple estimates of heritability are made by attributing all of the parent–offspring (or sibling–sibling) correlation to shared genetic variation, they suffer from this inability to distinguish genetic effects and effects of persistent environmental influences.

This problem can be overcome by using family studies with more sophisticated designs. In the next two sections, we look at the ones most commonly used in studies of humans, namely twin studies and adoption studies. The branch of human science concerned with questions of heritability is known as behaviour genetics.

3.3.2 Twin studies

For humans, most estimates of heritability are produced using data from twins. This is because twins are a kind of convenient natural experiment, as we shall see. For species that do not produce twins, other types of family relationships (e.g. exploiting the difference in relatedness between half siblings and full siblings in the same family) have to be examined.

A small percentage of human births are twins, and twins are of two kinds. Monozygotic (MZ) twins grow from a single fertilized egg, and so are genetically identical to one another. Dizygotic (DZ) twins result from the implantation of two fertilized eggs in the same cycle. Since they come from different eggs and different sperm, albeit of the same parents, they are no more genetically similar to each other than full siblings are (their coefficient of relatedness is one-half). In both MZ and DZ cases, however, the twins grow up together and thus share a lot of environmental input. It is useful here to distinguish shared environment from non-shared environment influences. Shared environmental influences are things like parental social class, parental behaviour, diet available, the school attended, or the house grown up in. These influences will affect both twins, regardless of whether the twins are MZ or DZ. Non-shared environmental influences are things like childhood diseases or accidents that affected one twin and not the other. By definition, neither MZ nor DZ twins share any of the non-shared environmental influences.

A simple way of estimating heritability from twin data

The different types of influence on twin phenotype are summed up in Table 3.1. The only difference between the MZ case and the DZ case is the closer genetic affinity. Thus, if there is any difference in the phenotypic correlation between MZ twins and that between DZ twins, then this reflects the impact of the extra half in the coefficient of relatedness. An estimate of the total impact of genetic variation is therefore twice the difference between the phenotypic

Table 3.1 **DZ twins differ from MZ twins in having a coefficient of relatedness of 0.5 rather than 1. Twins of both types share all of their shared environmental influences and none of their non-shared ones. The only difference is in the extra half relatedness, and thus twice the difference between the phenotypic correlations of MZ and DZ twins can be used as an estimate of heritability. For example, if the height of MZ twins correlates at 0.9, and the heights of DZ twins correlate at 0.5, then the estimated heritability is $(0.9 - 0.5) \times 2 = 0.8$.**

	Genetic	Shared environment	Non-shared environment
MZ twins	1	1	0
DZ twins	½	1	0

correlation of MZ twins and the phenotypic correlation of DZ twins. This statistic is known as Falconer's estimate of heritability.

Twin data can also allow us to disentangle different types of environmental influence. MZ twins are genetically identical and their shared environment is also the same, so any phenotypic differences between them must be attributed to the non-shared environment. Thus, an estimate of the impact of the non-shared environment is one minus the MZ twin correlation. As for the shared environment, one estimate of its effect is the MZ phenotypic correlation minus the heritability. You can probably see why: the MZ phenotypic correlation is the sum of the resemblance due to shared genes and the resemblance due to shared environment, so if you subtract the heritability from the phenotypic correlation, what you are left with is the influence of the shared environment. Thus, we can use twin data to estimate all three of the major types of influence.

ACE models

In contemporary twin research, more sophisticated statistical model-fitting techniques than the ones described here are used, but their spirit is much the same. These models are called ACE models because they lead to the estimation of three parameters, conventionally labelled A (the heritability), C (the effect of the shared environment), and E (the effect of the non-shared environment). (Note that the usage of A here to represent the heritability is a completely unrelated usage to when we denote a dominant allele A in Mendelian genetics, as in section 3.2).

To take a recent example of the use of twin data to examine heritability, Geschwind *et al.* (2003) were interested in whether the size of various parts of the human brain was heritable. They examined the size of different brain structures in 139 pairs of adult twins (72 MZ and 67 DZ), using a non-invasive brain-scanning technique called magnetic resonance imaging. They found that the total volume of the cerebral hemispheres was correlated within MZ twins at 0.87, and within DZ twins at 0.56. This gives a Falconer heritability of 0.62 (0.87 minus 0.56 is 0.31; twice this is 0.62). A more sophisticated model-fitting analysis gave the following values: A = 0.64, C = 0.23, and E = 0.13. In other words, about 64% of the variation in cerebral volume is associated with genetic variation, about 23% with the shared environment, and about 13% with the non-shared environment. This seems plausible. Size (both overall, and of specific parts of the body) is controlled by many genes, and it is quite common to find polymorphisms in some of them, so A being substantial is no surprise. Environmental factors like the general amount of nutrition available to the developing brain both before and after birth are also going to be influential, so a greater than 0 value for C is expected. Finally, idiosyncratic events like an early-life disease or accident might well affect the growing brain, so we should not be surprised that E is greater than 0. However, the research is of interest because it does more than just identify that all three of these determinants exist: it allows us to quantify their relative strength in the population under study.

Results from twin studies

Twin studies have been used to give estimates of heritability for large numbers of characteristics in contemporary populations. The pattern emerging from these studies is that A is substantially greater than 0 for almost everything that has ever been studied, including physical characteristics, personality, attitudes, propensity to certain types of life events, and propensity to certain types of mental and physical illnesses (Bouchard & McGue 2003). Also, C tends to be much less important than E. That is, the important environmental events seem to be those that make us less like our siblings, not more like them (Plomin & Daniels 1987).

3.3.3 Adoption studies

A second natural experiment that can be useful in estimating heritability is provided by adoption. A consequence of adoption soon after birth is that one set of parents provides the genetic contribution and a different set the shared environment. Thus, a good test of the relative strengths of A and C will be the correlations between the adopted children and, first, their biological parents and, second, their social parents. You can also do a very similar analysis by comparing biological siblings who are adopted away into different families or unrelated children who are raised in the same household due to adoption.

A famous example of an adoption study was carried out by Heston (1966). Heston examined the incidence of schizophrenia in people who had been adopted as children and whose biological mothers had suffered schizophrenia. He matched them to control participants whose mothers had no known mental illness. Around 10% of the children of mothers with schizophrenia developed schizophrenia; none of the control group did. The rate of schizophrenia in the offspring of schizophrenia sufferers who are raised by their biological parents is also around 10%. Thus, Heston's data suggested that this mild tendency to run in families is due to genetic heritability rather than shared environment.

3.3.4 Problems with twin and adoption studies

Twin and adoption studies are not without their problems. For twin studies, a central assumption is that the environment is equally similar in the MZ and DZ cases. This might not be true, in which case heritability estimates could be inflated. Parents might treat their MZ twins more similarly than they do their DZ twins, and this could cause the MZ twins to become more similar for environmental rather than genetic reasons. Moreover, two-thirds of MZ twin pairs share a placenta, whereas no DZ twin pairs do. Thus, the prenatal environment of MZ twins could be more shared on average than that of DZ twins. This too could lead to overestimating A and underestimating C in standard twin designs.

These criticisms can be addressed to some extent. To take the intra-uterine environment first, twins that share a placenta have to compete directly for nourishment. This competition is usually resolved unequally and in fact MZ twins are more *dissimilar* in birth weight than DZ twins. Thus, aspects of the prenatal environment are actually less similar for them than for DZ twins. Post-birth, MZ twins are treated more similarly by parents than DZ twins are (Richardson & Norgate 2005). However, this could be largely because they are so much more alike! In other words, assuming parental treatment is at least partly a response to the characteristics of the child, which seems reasonable, then the more similar treatment received by MZ twins is in fact an effect of their more similar genotypes. Moreover, there are some studies of twins who are raised apart and in such studies the greater similarity of MZ twins than DZ twins is still observed, despite their not being raised in the same household as the other twin. For example, Bouchard et al. (1998) administered many different personality scales to 71 pairs of MZ and 53 pairs of DZ twins who had been raised apart, as well as 99 MZ pairs and 99 DZ pairs who had been raised together. They found that MZ correlations were higher than DZ regardless of whether the twins were raised apart or together (Table 3.2). Indeed, raising apart seemed to make little difference to the pattern of resemblance in twins. The estimates of A (about 0.5) and C (basically 0) they obtained from this data were much the same as those from studies using only twins raised together.

Table 3.2 **Correlations between twins in score on a personality scale measuring 'creative temperament', according to zygosity and whether they were raised together. MZ correlations are higher than DZ regardless of raising.** *From Bouchard et al. (1998).*

	Zygosity	
	MZ	DZ
Reared together	0.50	0.29
Reared apart	0.76	0.12

Finally, one clever study examined the correlation in language performance between pairs of twins, some of whose parents were mistaken in their beliefs about the zygosity of their children. MZ twins were more highly correlated even where the parents thought they were DZ and presumably treated them accordingly (Table 3.3; Munsinger & Douglass 1976).

Table 3.3 **Correlations in language performance between MZ and DZ twins correctly and incorrectly classified by their parents. These data suggest differential parental treatment is insufficient to account for MZ/DZ differences.** *Data from Munsinger & Douglass (1976).*

		True zygosity	
		MZ	DZ
Parents' belief about zygosity	MZ	0.86	0.47
	DZ	0.78	0.40

For adoption studies, one source of possible methodological problems is in the selective placement of children with adoptive families. Only a subset of families would be considered eligible to adopt and this restriction of range of social environments could affect generalizability of estimates of C. There could also be problems if adoption agencies tended to place children in families that seemed similar to the child or its biological parents. This would lead to inflating the estimate of C. However, since the major finding emerging from adoption studies has been how *unimportant* C, the shared environment, is, this problem seems moot.

The most serious problem for adoption studies is undoubtedly that, even if adoption occurs immediately at birth, the prenatal environment is provided by the biological mother. Thus, if a large part of the influence of C actually occurs *in utero*, C will be underestimated and A overestimated in adoption designs. This criticism can partly be met by comparing the correlation of the child with the biological father, who of course does not provide the prenatal environment, and that with the social father.

Because of issues like these, heritability estimates from twin and adoption studies must be treated as no more than that—estimates. A more accurate understanding will follow once we understand *which* genetic loci are involved. However, the results of behaviour genetic designs are generally quite consistent across many studies and it seems unlikely that the substantial heritabilities that they have identified are completely artefactual.

3.3.5 More complex models: epistasis and dominance

The ACE models considered in the previous section rely on some simplifying assumptions. Specifically, they assume that genetic effects accumulate in an additive manner, so that the effect on the phenotype of a coefficient of relatedness of 1 will be twice the effect of a coefficient of relatedness of 0.5. However, this need not be the case. The most obvious illustration of this comes from a recessive allele. If you have two copies, you develop the phenotype. If you have one copy, you do not develop half the phenotype, but none of the phenotype. Thus, the effect of a whole genetic load (having two recessive alleles) is more than twice the effect of half the genetic load (having just one recessive allele).

Such interactions may make the relationship between genetic and phenotypic similarity non-linear. Where such effects are due to the interaction between the pair of alleles at a locus, they are called dominance effects, and where they are due to interactions between alleles at different loci, they are called epistatic effects. There are model-fitting techniques that can estimate dominance and epistatic effects, but we do not consider them further here. It is worth noting that it is not possible to simultaneously estimate shared environment, non-shared environment, additive genetic effects, and epistasis and dominance effects using just MZ versus DZ twin data. Researchers are most interested in the additive genetic effects (A) and so the most commonly used study designs model just A, C, and E. Heritabilities based on additive effects only are also called narrow-sense heritabilities, whereas estimates that include dominance and epistatic effects are called broad-sense heritabilities. When I use 'heritability' unqualified in this book, it either does not matter or it is narrow-sense heritability I have in mind.

3.3.6 Heritability and its meaning

Heritability is a much misunderstood term. People often interpret it as meaning that the characteristic is 'genetically determined', but this is an error. (In fact, it is quite hard to say what, if anything, 'genetically determined' should be taken as meaning. If we had no genome, we would not be able to do anything at all, so at one level everything is genetically determined. However, on the other hand, if we had no food, we would not be able to do anything either, so everything is environmentally determined too!)

'Heritable' is not 'genetically determined'

To see why 'heritable' is a very different thing from 'genetically determined', consider two characteristics: having two arms and being able to read English. The first is surely a candidate for being genetically determined if anything is. It is deep in the body plan of all vertebrates that they have a limb on the left side matching the one on the right. All normally developing humans have two arms because the alleles that cause two to be made are at fixation and, indeed, have been so for hundreds of millions of years. Since the heritability is the proportion of phenotypic variation accounted for by genetic variation, and there is no genetic variation affecting arm number in the human population, it follows that the heritability of having two arms is zero. You do occasionally come across someone with fewer than two arms, but this is always because of some environmental event, such as an accident. They will not of course pass this on to their children. Thus, 100% of the variation in arm number is environmental, even though the characteristic is in some real sense genetically determined.

Now consider the example of reading English. Reading was only invented a few millennia ago and it was obviously a cultural invention, not a genetic adaptation. The alphabetic reading system we currently have is an even more recent cultural development restricted until very recently to only a small subset of humanity, who acquired it through a painstaking social learning process. Thus, we might want to say that being able to read alphabetic text is not directly genetically determined.

However, reading ability within the literate English-speaking population is highly heritable. Stromswold (2001) summarizes results of several twin studies yielding heritability estimates of around 0.42. How can a characteristic be heritable and yet not genetically determined? The important thing to remember is that heritability is about whether *variation* in a characteristic is associated with *variation* in genes. It could be that a cultural situation leads a group of people to all do something, but that once they do so, variation in their genetic background may lead to differences in how adeptly they perform it and this would mean a non-zero heritability.

Heritability is specific to a population and an environment

Heritability is specific to a particular population in a particular environment. It can go up and down as the environment changes. Furthermore, heritability within one population says nothing about what the causes of differences between populations are, nor has any implications for how the population will respond to environmental change. Let us examine these points through an example.

Imagine a population of plants growing in uniformly poor soil (Figure 3.8a). Since all of them are in equally poor soil, any differences in the heights they attain will plausibly reflect differences in their genotypes. Thus, heritability will be high. Now we come along and sprinkle fertilizer on some patches of the plot but not others (b). The plants on the fertilized spots grow enormously taller. Since heritability is the proportion of total variation attributable to genetic variation, and we have just introduced a load of non-genetic variation, heritability must go down. Thus, changing the environmental variation changes the heritability. (For this reason, geneticists sometimes prefer a statistic called the **coefficient of additive genetic variance** (CVA), to heritability. CVA is an estimate of the amount of genetic variation in a population relevant to a particular trait, independent of the amount of environmental variation.)

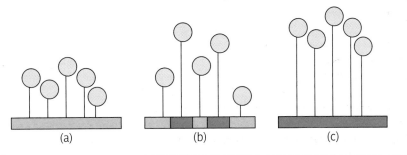

Figure 3.8 **In population (a), the soil is uniformly poor and the heritability of height is high. (b) The environment becomes more variable, and heritability goes down. (c) The environment is uniformly enriched, and heritability goes up again. Although the within-population heritability is high in both (a) and (c), the height differences between the two populations are entirely due to the environment.**

(a)   (b)   (c)

Now consider a third condition where we fertilize the whole plot (c). All plants grow much taller. Since they all experience equally rich environments, any variation in just how much taller they now grow will depend on their genotype. Thus, once again, within population (c), height is highly heritable. The heritability has gone up again. Moreover, although height is a heritable characteristic within population (a), and a heritable characteristic within population (c), the *difference* between populations (a) and (c) is entirely due to the difference in their environmental conditions. We can see that the heritability can change as the environment changes and also that the fact that a trait is highly heritable does not imply that a change in the environment will have no phenotypic consequences.

This is not just a hypothetical point. Within groups of relatively affluent children from developed nations, intelligence as measured by IQ tests is highly heritable. However, it does not follow that the differences in IQ scores between such children and children from less affluent populations is attributable to genetic differences. Turkheimer *et al.* (2003) studied IQ scores in 319 pairs of twins from different socioeconomic status (SES) backgrounds. They found that the estimate of A, the additive genetic effect, was large for the children of the highest SES, whereas it declined to zero for the most deprived children (Figure 3.9). C, the effect of shared environment, was substantial for the lowest SES children and declined to zero for the most affluent. This pattern is easy to interpret. In high SES families, the environment is uniformly pretty good. Thus, any remaining differences in IQ score between the children will be associated with genetic differences. In low-SES groups, by contrast, some but not all families are hit by problems of housing, disease, or deprivation. This introduces much more between-family variation, leading to a greater effect of C and a diminution of the importance of A. It follows that, although IQ is highly heritable within affluent classes, this does not hold for deprived communities, and it is environmental improvement that will have the biggest impact on IQ inequality. A more general conclusion is that the better we make the environment, the more heritable phenotypic characteristics will become.

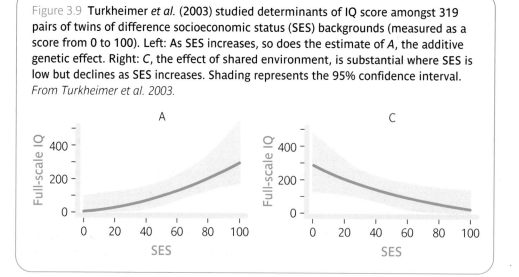

Figure 3.9 **Turkheimer *et al.* (2003) studied determinants of IQ score amongst 319 pairs of twins of difference socioeconomic status (SES) backgrounds (measured as a score from 0 to 100). Left: As SES increases, so does the estimate of *A*, the additive genetic effect. Right: *C*, the effect of shared environment, is substantial where SES is low but declines as SES increases. Shading represents the 95% confidence interval.** *From Turkheimer et al. 2003.*

3.4 Heritability and natural selection

Heritability is of interest in its own right. Behaviour geneticists study it because they want to understand what factors contribute to variation within the human population. Heritability is also very important to evolution. This is because the heritability of a trait determines how effectively natural selection can change it.

Imagine a population where the tallest individuals have a large survival advantage. They leave more offspring as a consequence. Imagine first that the heritability of height is high. This means that differences in height are mainly a function of differences in allelic makeup. The alleles in the next generation will thus be a very non-random selection of the alleles in the previous generation, since those with the 'tallest' genotype leave more copies of themselves than others do. The frequencies of alleles will thus be changed from generation to generation (with alleles that tend to make you taller increasing in frequency) and because of this the average height of the population will increase. This population is responding to natural selection by changing.

Now imagine instead that the heritability of height is 0. This means that there is no systematic relationship between which alleles you have and how tall you are. The taller individuals will still leave more copies of themselves, but the alleles represented in these copies will just be a random selection of all alleles. Thus, there will be no systematic change in allele frequencies across the generations and the average height in the population will remain the same (think back to the Hardy–Weinberg equilibrium). In such a population, although natural selection favours the tallest, no evolutionary change in the direction of greater height can occur.

Thus, a precondition of natural selection to be effective is that there is some heritable genetic variation for the characteristic present in the population. Heritability does not need to be 1 for natural selection to lead to evolutionary change. It just needs to be greater than 0, although the higher it is, the quicker natural selection can produce change. Natural selection also has effects on heritability, but we will see what those are in Chapter 5.

 Summary

1. Diploid individuals have two copies of each gene. When they reproduce, these copies are not blended with those of the other parent. Instead, one copy from one parent is selected randomly and paired with one copy from the other parent, potentially generating a combination not seen in either parent.

2. Individuals can be homozygous, which means they possess two copies of the same allele, or heterozygous, which means they possess one copy of two different alleles, for a particular gene. Alleles can be either dominant, which means their phenotype is expressed even when only one copy is present, or recessive, meaning two copies are required.

3. Genes on different chromosomes travel independently down the generations.

4. For single-gene characteristics, the frequencies of offspring genotypes and phenotypes produced by particular parental genotypes follow in law-like manner from the basic principles of heredity known as Mendel's laws.

5. For polygenic characteristics, offspring phenotype is predicted from parental phenotype using quantitative genetic models. These models are based on the greater genetic similarity between closer compared with more distant relatives and the idea that the net effect of genetic similarity is to increase phenotypic similarity.

6. Heritability is the proportion of phenotypic variation accounted for by genetic variation within a particular population at a particular time. Heritable is not the same as genetically determined.

7. For humans, heritability is mainly estimated using twin and adoption studies.

8. The heritability of a trait determines how strongly it can respond to natural selection.

? Questions to consider

1. The lack of tail in the Manx cat (Figure 3.10) is caused by a dominant allele of a single gene. Being homozygous for this allele is lethal, and embryos with this genotype are reabsorbed by the mother long before they are born. If two Manx cats are bred with each other, what will be the frequencies of tailed and tailless cats in the kittens born? Breeders are often concerned to make their prized strains 'breed true', that is to ensure that all the offspring are of the phenotype they want. What could Manx cat breeders do to stop getting kittens with tails in their litters?

Figure 3.10 **The Manx cat. Manxes are heterozygous for a dominant tailless mutation.**
© *blickwinkel/Alamy.*

2. The gene involved in the most common type of haemophilia resides on the X chromosome. This makes the pattern of inheritance a bit more complex than it is for most Mendelian diseases. Remember the disease allele is recessive. Suppose a woman is a carrier (i.e. a heterozygote with a normal phenotype), who marries a normal-type man. What will be the proportion of (a) her daughters and (b) her sons who will be affected by haemophilia? Now consider a man who is a haemophiliac, who marries a normal-type woman. What proportion of (a) his daughters and (b) his sons will inherit haemophilia?

3. Tay–Sachs disease is an incurable neurological condition caused by a recessive allele of a gene on chromosome 15. The risk of inheriting this disease is generally low, but for Ashkenazi Jews and Cajuns (people of French descent in Louisiana) it is at least ten times higher than for other groups. Why might there be such an increased risk?

4. In a number of twin studies, Falconer's estimate of heritability comes out greater than 1. What is a likely explanation for this pattern?

→ Taking it further

There is more information than given here about both Mendelian and quantitative genetics in evolution textbooks such as Ridley (1996: Chapters 5–9). Those wishing to explore human behaviour genetics further may wish to consult any of a number of good recent reviews. Bouchard & McGue (2003) provide an overview of behavioural genetic methods, and review the ubiquitous evidence of heritability in many psychological characteristics. Kendler & Baker (2007) review the fascinating evidence that many types of life events show heritable influence, which basically comes about because we choose our environments and interactions, and do so on the basis of our inherited temperament. Plomin & Kovas (2005) examine genetic correlation studies, which show that the same genetic differences are at work in several different heritable learning disabilities. Plomin & Daniels (1987) review the evidence that the non-shared environment tends to be much more important than the shared environment in determining human phenotype.

Competition

We have now examined two of the pre-conditions for natural selection: the tendency of individuals to be slightly different from one another (Chapter 2, Variation) and the propensity for variation to be passed from parents to offspring (Chapter 3, Heredity). These two principles will only generate natural selection if another condition is met: there must be competition to reproduce. This chapter introduces competition, examines the level at which it is best thought of as operating, and looks at what types of evolutionary outcomes it tends to favour.

4.1 Malthus: checks on reproduction and competition to reproduce

For the principles relevant to competition, we need to look beyond Darwin to an English thinker called Thomas Malthus (Figure 4.1). Malthus published a famous book called *An Essay on the*

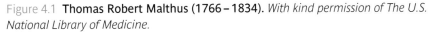

Figure 4.1 **Thomas Robert Malthus (1766 – 1834).** *With kind permission of The U.S. National Library of Medicine.*

Principle of Population in 1798. Darwin had read the essay and felt it to contain a fact of fundamental importance. However, Darwin's reliance on Malthus led to much misunderstanding of the implications of Darwin's thought. This is because Malthus' book contains many claims, only some of which are relevant to natural selection. In particular, we need to separate *general principles* that Malthus derived about the dynamics of populations from *specific claims* that Malthus made about the future of the human population in Europe. Only the former are necessary for understanding natural selection.

The general principles in Malthus' essay that we need be concerned with are that populations could potentially grow exponentially, but in practice cannot do so, and therefore must be limited by incomplete survival and/or reproduction. This means that there is competition between members of the same population to be in that fraction which manages to survive and reproduce. These are the principles that Darwin realized were key to his theory. Malthus' other claims, for example that the European population was expanding too fast for the available resources, that wars and famines were in prospect because of this, and that the lower classes should be discouraged from reproducing because of the poverty that their increase would generate, are not in any way part of the theory of evolution. Indeed, their factual aspects have not stood the test of time. The European population after Malthus continued to expand, but famine actually became less common, not more so, because agricultural productivity and economic growth went up even faster. Moreover, people in Europe eventually responded to the increased affluence by spontaneously decreasing their family sizes, leading to a stabilization of the population. However, none of this is in any way important for the general theory of natural selection.

4.1.1 Exponential population growth: an example

Let us then examine the key Malthusian principle that is correct—exponential potential population growth. To see what we mean by 'exponential growth' we will take as our example not humans but cats. We will assume the following, although the exact details are not critical: that a female cat can have two litters a year, of six kittens at a time; that the kittens are roughly 50% females; that the kittens mature sexually at 1 year of age; that cats can go on reproducing until they are 5 years old. We start from a single breeding pair.

You can probably see the trend that the population will follow. In year 1, the breeding pair produces 12 kittens. About six of these will be females. Thus, in year 2, there will be six new females plus the original one, giving seven. Seven females can produce 84 kittens, about 42 of which will be female. Thus, in year 3, there will be 42 plus seven, or 49, breeding females, who will produce 588 kittens between them. The number of kittens grows dramatically larger each year. From year 7, some of the older breeding females begin to die off, but this has a negligible effect. By then there are over 100,000 breeding females anyway!

If we graph the size of this hypothetical cat population, we see that it does not just increase, but increases at an ever-increasing rate (Figure 4.2a). Such a pattern is called exponential growth. Under exponential growth, the population would explode very quickly. Figure 4.2b illustrates this by calculating the number of years our cat population—starting from a single pair, remember—would take to reach various milestone sizes. Within a dozen years, the number of cats surpasses the current world human population, and within a century there are more cats than the estimated number of atoms in the observable universe. Of course, this could never actually occur, but it serves to illustrate what would happen if a population increased at anything like its theoretical maximum rate. The nature of exponential functions is such that it does not much matter which species we choose. Cod, which produce several million eggs a year, would reach the milestones a few years earlier and chimpanzees, with one offspring every few years, several decades later. However, this does not detract from the general pattern. Darwin himself calculated that even elephants, the slowest-breeding animals he could find, would ultimately follow a similar pattern, with one breeding pair having 19 million descendants in around 750 years.

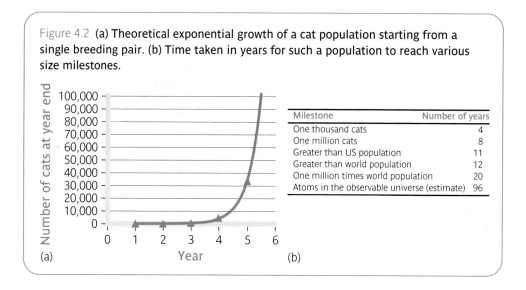

Figure 4.2 **(a) Theoretical exponential growth of a cat population starting from a single breeding pair. (b) Time taken in years for such a population to reach various size milestones.**

Milestone	Number of years
One thousand cats	4
One million cats	8
Greater than US population	11
Greater than world population	12
One million times world population	20
Atoms in the observable universe (estimate)	96

(a)   (b)

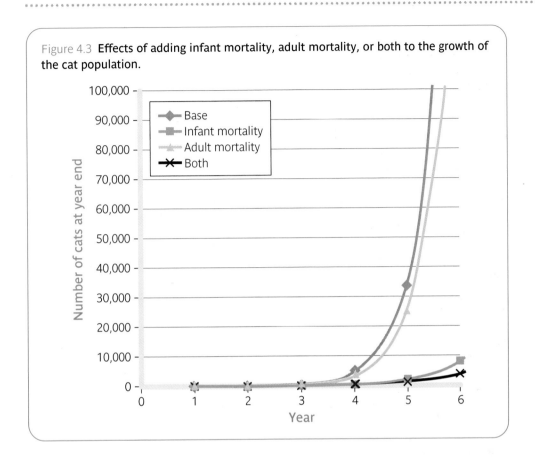

Figure 4.3 **Effects of adding infant mortality, adult mortality, or both to the growth of the cat population.**

Populations in nature do not, of course, follow such a pattern. If they did, the world would become excessively crowded with organisms in a few years. Populations are sometimes stable, sometimes contract, and sometimes grow, even quite fast, but they never attain anything remotely like the exponential pattern they could theoretically achieve. Why not? You will note that in the cat example, I assumed that all kittens conceived survived into adulthood and also that all adult females survived for the whole length of their breeding lifespan. What happens if we alter either of these assumptions?

Figure 4.3 shows how the population growth is changed by either giving each kitten just a 50% chance of surviving into adulthood (the infant mortality condition), or only a 50% chance of surviving from one breeding season to the next, once it is an adult (the adult mortality condition), or both. You will see that where there is both infant and adult mortality operating, the growth of the population is dramatically flattened off. The curve begins to look a little more realistic.

Mortality is not the only way the effect could be achieved. A similar flattening would follow if, for example, we made some cats unable to find a mate or to conceive. To flatten the curve just requires that for some reason or other not all individuals are reproducing at their potential rate. Malthus understood that such 'checks' on exponential growth must be operative in all biological populations.

4.1.2 There is a differential reproductive success

A biologically realistic model of the cat population requires, then, that not all individuals survive and reproduce as much as they might. Look at it from the point of view of all the kittens conceived. Not every one of them can make it into that much smaller fraction that will in turn have kittens. This means that the population in generation $n + 1$ is not a complete reflection of the population in generation n. Some cats have become ancestors of cats in the next generation and some have not. Another way of saying this is that there is competition amongst the cats in any particular generation to leave descendants in the next. Cats will thus differ in their reproductive success. Reproductive success is the number of viable descendants produced and so is obviously zero for any cat that dies at birth or in kittenhood.

4.1.3 Differences in reproductive success can lead to changes in the population

Natural selection requires that there is differential reproductive success, but differential reproductive success is not sufficient to produce any evolutionary change. To see why, let us introduce a further refinement to the model. Start with 100 cats, 90 of whom are black and 10 yellow. We will simplify things for now and assume that females always produce kittens of the same colour as they are. First consider a scenario where the mortality rate is 50% for the yellow cats and also 50% for the black cats. Although individuals vary in reproductive success, this has nothing to do with their colour and thus the proportions of the two colours in the population remain the same (Figure 4.4, black line). Where reproductive success is unrelated to a phenotypic characteristic, then that phenotypic characteristic does not change over the generations.

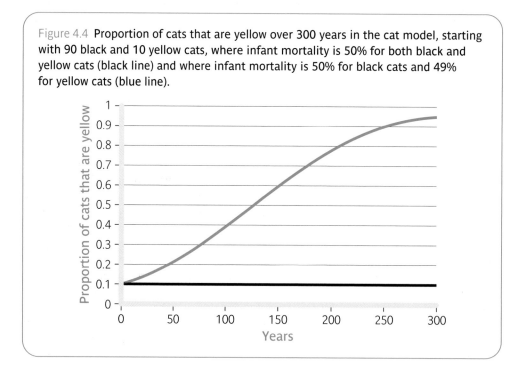

Figure 4.4 **Proportion of cats that are yellow over 300 years in the cat model, starting with 90 black and 10 yellow cats, where infant mortality is 50% for both black and yellow cats (black line) and where infant mortality is 50% for black cats and 49% for yellow cats (blue line).**

Now imagine the population moves into an open sandy environment. The yellow kittens are now slightly harder to see against the background and so slightly less likely to be killed by predators such as eagles. Eagle predation is only one source of infant mortality and the difference in camouflage is modest, so we will set the infant mortality rate for yellow kittens just slightly lower than for black ones—49% versus 50%. This means that yellow females in generation n, through greater survival to breeding, have a just slightly elevated chance of leaving offspring in the next generation, offspring who in turn will have a slightly elevated chance of leaving their own offspring, and so forth.

Observe what happens. The proportion of yellow cats increases generation on generation, until after about 300 years, almost all cats are yellow (Figure 4.4, blue line). Natural selection has driven a process of adaptation to the sandy environment. What we mean by this is that someone coming along at the end of the 300 years would be struck that these cats look well designed for the sandy environment in which they live. We know, of course, that there has been no designer, simply that the regime of differential mortality has caused a gradual change in the composition of the population, from mostly black cats to all yellow ones.

Note a few key features of the adaptive process. The sandy environment did not cause the yellow mutant to come about. Mutation is unrelated to the demands of the environment and in this example we just assumed that yellow was already around in the range of colour variation of the population. No individual cat changed in any way during the adaptive process. Black cats always had black offspring and yellow cats always had yellow offspring. However, whereas at the beginning we had a type of cat that was basically black, at the end we had a type of cat that was basically yellow. Note also that it took some time for adaptation to occur. Three hundred years is longer than the life of any one cat. On the other hand, it is a blink of an eye in biological timescales. Given the tiny survival advantage we gave the yellow cats (mortality of 49% rather than 50% in the first year of life, the same life history thereafter), it is extremely impressive how fast the adaptive process can occur. If we had made the competitive advantage of yellow cats bigger, then adaptation would have been even faster.

4.2 Natural selection at the genotypic level

In section 4.1, we examined the effect that natural selection was having at the level of the phenotype of the cats. This is a reasonable enough approximation to start off with. However, although success in competition is related to phenotypic characteristics, what natural selection is actually doing is changing the frequencies of the underlying alleles and so that is the process we should really model. This changes things somewhat, since the mapping between genotype and phenotype in a sexually reproducing population is not one-to-one (as we learned in Chapter 3). In this section, we examine how natural selection changes frequencies of alleles. We use the same cat example, but because we are now considering genotypes rather than phenotypes, we have to set up the model in a slightly different way. We track generations rather than years. Since there were two litters of kittens per year, we can think of each year in the previous model being approximately equivalent to two new generations in the current one.

4.2.1 Increase in frequency of an advantageous dominant allele

Let us assume that the colour of the cats is controlled by a single gene and that the allele A, which is dominant, gives the yellow colour, whereas a, recessive, gives black. The frequency of allele A in the population is equal to p and we will start where $p = 0.1$. This means that 10% of all copies of the gene in the population are A and thus that the other 90% are a. Where mortality is the same for yellow cats as for black cats, the allele frequencies do not change over the generations. This just follows from the Hardy–Weinberg principle (Chapter 3), given that genotypes AA and Aa (yellow cats) and aa (black cats) all have the same expected reproductive success.

Now let us return to the case where infant mortality is 49% for yellow cats and 50% for black cats. What we need to calculate is the fitness of each of the two alleles. The fitness of an allele is the number of copies in the next generation that a copy in this generation leaves. Where there are two alleles in competition with one another, their relative fitness determine how their proportions change over time.

The fitness of the dominant allele A is completely determined by the reproductive success of yellow cats, since all the cats in which A appears are yellow. The recessive allele a, by contrast, appears in yellow cats with genotype Aa and also in black cats with genotype aa. Thus, to calculate the fitness of a we need to do a weighted average of the reproductive success of the black cats it appears in and the reproductive success of the yellow cats it appears in.

Figure 4.5a graphs the way the allele frequencies change over the generations, given the slight survival advantage of yellow cats. It is the same general shape as in Figure 4.4, but you will note that the increase of allele A is a little slower than when we just considered phenotypes, particularly once A is common. This is because the allele a, although it is the allele 'for' black coats, spends quite a lot of its time in yellow cats (i.e. heterozygotes of genotype Aa), particularly when allele A is common. It is only exposed to its selective disadvantage, as it were, on the occasions when it is homozygous. The rest of the time it 'hides from selection' in the bodies of yellow cats. This means that it takes longer for natural selection to weed it out than it would otherwise.

Figure 4.5 **The change in frequencies of cat coat colour alleles over the generations, where yellow coats have a survival advantage. (a) The black line is the frequency of a dominant allele causing yellow coats and the blue line is the recessive allele associated with black. (b) The black line is the frequency of a recessive allele causing yellow coats and the blue line is the dominant allele associated with black.**

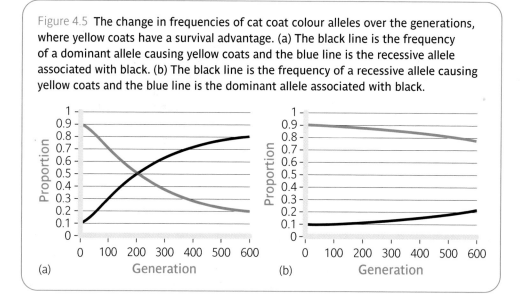

4.2.2 Increase in frequency of an advantageous recessive allele

Figure 4.5b considers the case where the advantageous yellow coat is produced by the recessive allele. The increase in frequency is much slower than the dominant case (although it would eventually become common if we ran the model for longer). This is also due to an imperfect association between alleles and phenotypes. Since the 'yellow' allele is recessive, it mainly appears in heterozygous black-coated cats, especially when it is rare. It can only show its advantage to selection, as it were, on those occasions where it happens to be homozygous and thus causes the cat's coat to be yellow.

4.2.3 Follow the alleles

The difference between Figure 4.5a and 4.5b illustrates a fundamental point that will recur in future sections. Natural selection changes allele frequencies, increasing the frequency of those alleles with high fitness and decreasing the frequencies of, and ultimately eliminating, alternative alleles with low fitness. The fitness of the allele is the weighted average of the relative reproductive success of all the different phenotypes it appears in. We will see in Chapter 5 that this means alleles with harmful effects in some individuals can persist, as long as they are causing reproductive success benefits in other individuals who also carry them. It is the *alleles* with the highest fitness that natural selection will preserve, not necessarily the *organisms* with the highest fitness.

  This means that when we are trying to predict evolutionary outcomes, we should always calculate the relative fitness of *alleles*, not organisms, species, or any other unit. In Figure 4.5, the advantage *to the individual* of having a yellow coat is exactly the same in (a) and (b). Just by thinking about individuals, we would have predicted the same result in the two cases. However, when we do our accounting at the level of the alleles involved, we make the correct prediction that yellow coats will increase much more slowly when the allele for them is recessive than when it is dominant.

4.2.4 Competition revisited

We have just seen that the allele should always be the unit of accounting in thinking about evolution. Adaptive evolutionary change—for example cats in a sandy environment becoming yellow—is a consequence of one allele defeating another in competition within the gene pool of a population. This is the part of Darwinian theory that is most often misunderstood. People tend to think that the competition that drives adaptation is competition between different species, or between different populations, or between different individuals. All of these types of competition do exist in nature, but the relevant question for the evolutionist should always be: why would the alleles for [the phenomenon of interest] have out-competed alternative alleles in the same gene pool?

4.2.5 Natural selection and polygenic characteristics

The cat example considered so far is a single-gene characteristic. When we consider alleles that contribute to a polygenic characteristic such as height, things become a little more complex. If many genes affect height, an allele X whose phenotypic effect is to increase height by some amount will appear in all kinds of phenotypes, including some phenotypes that are shorter than average. (If you need to review why this is, look back at Chapter 2, section 2.4.) However, the phenotypes in which X appears will on average be taller than those in which it does not and thus, if height is advantageous in a particular environment, X will gradually increase in frequency. Since, where tallness is advantageous, *all* alleles whose average effect is to increase height increase in frequency, the average height of the population increases over time. Thus, for polygenic characteristics too, differential reproductive success can lead to adaptive change in phenotype driven by changes in the underlying allele frequencies.

A general principle for thinking about change in polygenic traits as a result of natural selection is that the response to selection is the product of the selective pressure and the heritability. That is, evolutionary change gets faster with increasing selective advantage of whatever trait is being selected and also gets faster the more heritable the trait is (and, of course, the trait must show *some* level of heritability for there to be any evolutionary change at all).

4.3 Group selection

The power of natural selection is that, through a blind and goalless process of differential fitness of alleles, it can gradually create structures that appear well designed for their environment. This leads on to a difficult question, however: well designed for whom? For example, a tendency to attack neighbours might be a good design for an individual animal, but very bad for its neighbours. For whom should we expect adaptations to be optimized: the organism, the group, the population, or the species?

This issue is known in biology as the levels of selection debate and much has been written about the various approaches to it. For our current purposes, two points are important. First, it is generally best to do the accounting of advantages and disadvantages in terms of competing alleles, as we did in section 4.2, rather than any higher unit. The individual organism can often be used as an alternative, but this is best seen as a kind of approximation, for reasons we have partly touched on and to which we return below. Second, behaviours that promote the collective advantage of groups of organisms at the expense of individuals within the group will not usually evolve. This section shows why this is the case.

The idea that behaviours might exist because they benefit the group (e.g. the colony, the herd, the tribe, the population, or whatever) rather than the individual organism is known as group selection and it was widespread in biology until the late 1960s. The following example is from Wynne-Edwards (1962).

4.3.1 Wynne-Edwards and reproductive restraint in birds

In years when food is scarce, birds of various species lay a smaller number of eggs than they do in years when food is abundant. This is called reproductive restraint. Wynne-Edwards' interpretation of reproductive restraint was the following. If too many chicks hatch for the resources available, then there will be competition for food and all the chicks will be stressed. This can mean that the entire population dies out. Populations thus survive better if the birds within them restrain their reproduction, and reproductive restraint is thus an adaptation that has evolved because it increases the likelihood of the population surviving through bad years.

This argument is wrong, but to see precisely why it is wrong, we need to set up a model. Let us define a type of bird called an 'altruist' who, when times are hard for the population, limits

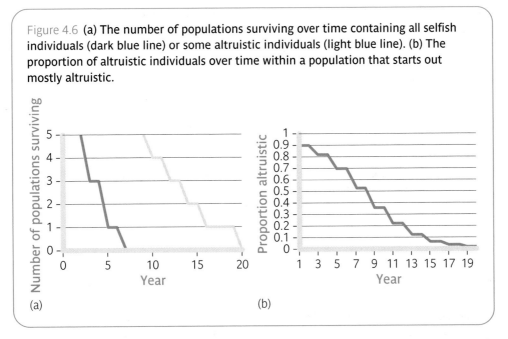

Figure 4.6 **(a) The number of populations surviving over time containing all selfish individuals (dark blue line) or some altruistic individuals (light blue line). (b) The proportion of altruistic individuals over time within a population that starts out mostly altruistic.**

the number of eggs it lays. Populations which contain altruists will suffer less stress and will thus be less likely to go extinct in bad years than populations made up of 'selfish' individuals, who always have as many chicks as possible. For Wynne-Edwards' argument to work, 'altruism' as defined has got to be able to out-compete 'selfishness' as defined.

Let us set up ten populations of ten breeding female birds each. (For simplicity, we do not worry about the males in this scenario and we assume females only live for one breeding season. These assumptions are not important for the conclusion.) Populations A, B, C, D, and E are all selfish. Populations F, G, H, I, and J all contain one selfish and nine altruistic females. In good years, all birds lay two eggs and all chicks survive. In bad years, selfish females lay two eggs, whilst altruistic females only lay one egg. In addition, we stipulate that, in a bad year, if less than half the population practises reproductive restraint, there is a 50% chance that the entire population dies out. We will alternate good and bad years.

Figure 4.6a shows that populations containing altruistic individuals (populations F–J) are indeed more likely to survive than populations A–E. This is as Wynne-Edwards argued. However, now consider what is happening *inside* any one of the populations F–J. The selfish individuals within these populations increase at a faster rate than their altruistic fellows. This is because they are laying two eggs every year, whereas the altruists are laying only one in the bad years. Thus, over time, populations F–J contain fewer and fewer altruists, until there are no altruists at all within them (Figure 4.6b). Thus, even though populations containing altruists do better than populations without them, altruists go extinct because they are disadvantaged as individuals relative to their competitors *within* their groups.

One might respond by saying that this outcome came about because, in my model, there were some selfish individuals in the largely altruistic populations from the start. A fairer test of group selection would be to start populations F–J with *only* altruists in them. Setting up the model this way might delay the inevitable, but it would not change the result. As long as even very occasionally a selfish individual migrates into a population composed of altruists, or alternatively a genetic mutation occurs which turns an altruistic individual selfish, then selfishness will always end up invading.

4.3.2 The evolutionarily stable strategy

Selfishness always out-competes altruistic behaviour in the above example because altruism is not an evolutionarily stable strategy (ESS). An ESS is a behavioural policy that, once common in a population, cannot be out-competed by any alternative behavioural policy. In our model, selfishness is an ESS, whereas altruism is not. Natural selection will always find the ESS in the end.

Follow the alleles, again

We could have come to the same conclusion about Wynne-Edwards' explanation for reproductive restraint in a simpler way by doing the fitness accounting at the allelic level rather than that of the bird. Assume simply that an allele causing selfishness is in competition with an allele causing altruism in the above scenario. We can calculate the fitness of these two alleles, in the three contexts in which they are each found: in good years, in populations that go extinct in bad years, and in populations that survive in bad years (Table 4.1). The fitness of the two alleles is the same in good years. They are also the same (i.e. zero) in bad years in populations that go extinct. However, in bad years in populations that survive, the selfish allele has twice the fitness of the altruistic one. Altruists are very slightly less likely to be found in populations that do go extinct (because being an altruist increases the number of altruists in your group by one), but this effect is not strong enough to offset the large fitness differential within the group. Thus, summed across the three types of situation, the average fitness of the selfish allele is higher than that of the altruistic allele. This means that it will eventually go to fixation and the altruistic allele will go extinct.

Implications

Let us summarize the implications of the results of this section. They mean that behaviours will not usually evolve if they benefit some larger group at a cost to the individuals performing them (although see section 4.5 for special cases). This is simply because the alleles underlying the behaviours will always have lower overall fitness than competitor alleles which do not take this cost. This is even true if the behaviours that evolve are damaging to the collective interest and drive the whole population extinct in the long term. Natural selection operates on immediate allelic advantage and cannot 'see through' this to the long-term viability or sustainability of a species or population.

Table 4.1 **Fitness of the 'altruistic' and 'selfish' alleles in the bird example. The fitness is identical in two out of three possible situations, and the selfish allele has higher fitness in the third, so the selfish allele has higher fitness overall and hence must spread.**

	In good years	In bad years when population goes extinct	In bad years where population survives
'Altruistic' allele	2	0	1
'Selfish' allele	2	0	2

This conclusion places important constraints on the types of evolutionary explanations we should consider. First, evolution does not produce outcomes that are 'good for the species'. There are many examples in nature of things that are clearly bad for the species yet still evolve. For example, amongst mammals, a major cause of death amongst young individuals is infanticide by members of the same species. This is clearly bad for the species and yet it has still evolved because alleles for committing infanticide can out-compete alleles for not committing infanticide. Wherever you encounter a 'good for the species to have members that do X'-type argument, you need to reframe it in 'good for individuals to do X because . . .' terms, or, even better, in 'the fitness of an allele causing X would be higher than competitor alleles because . . .' terms.

Second, we need to be cautious about arguments that invoke the good of a group or population rather than individuals. For example, if I wanted to explain the existence of celibate priestly castes in many human societies, I might be tempted to speculate that it is good for societies to have a group of individuals within them who have no partisan interests and are devoted to the broader social good. This speculation could well be true, but it is not sufficient to explain how celibacy evolved. Rephrase it in allelic terms. An allele that caused its bearers to become celibate would always be out-competed by competitor alleles that caused their bearers to reproduce, so the fact that celibacy would be good for the group is insufficient to make it evolutionarily stable.

4.4 Kin selection

The previous section has shown that adaptations that are better for groups than they are for individuals are likely to be out-competed and disappear. Thus, our general expectation should be that the organisms that we see today will be designed to behave in such a way as to promote their individual interests rather than anyone else's. A quick glance at animal behaviour provides plenty of evidence that this view is correct; individual animals feed themselves first, run away from predators, and may attack or even kill other group members when it seems to be in their interests to do so.

However, there is a striking group of exceptions to this picture, of which parenthood is the central one. Mothers gestate, suckle, and protect their offspring, and do not seem to receive anything in return. Intuitively, this is to do with the fact that a mother's offspring are genetically related to her. If you will, they represent the future of her genome and thus when she invests in them, she is investing in the fitness of her genotype (albeit, a different copy of her genotype than the one which happens to be in her body). This section formalizes this intuition and considers the extent to which it would be adaptive for individuals to invest in the copies of their genome that are inside bodies other than their own.

The part of evolutionary theory that deals with this issue is called the theory of kin selection and its development is largely credited to the English biologist William Hamilton (Figure 4.7). Kin selection theory does not just deal with parents and offspring. It can also be applied to sibling, nephew, and any other family relationships. The central component of the theory is one we have already met, the coefficient of relatedness (section 3.3.1). Recall that this coefficient represents the size of the expected increment of allelic similarity between two relatives above and beyond the similarity to be found between two randomly selected members of the population.

Figure 4.7 **William Hamilton (1936 – 2000), father of the theory of kin selection.**
© *James King-Holmes/Science Photo Library.*

4.4.1 Hamilton's rule and the concept of inclusive fitness

Let us consider a hypothetical example of a bird species, in which a mutant allele a_1 arises. The phenotypic effect of this mutant is to make younger sisters forego their reproduction and instead aid the reproduction of an older sister (e.g. by guarding the nest or bringing food for the sister's chicks). We assume that when this allele is found in an individual with no older sister, that individual reproduces as normal. The competitor of a_1 is a_2, which causes the younger sister to reproduce for herself. Under what conditions would a_1 out-compete a_2?

At first glance, the answer might seem to be 'never', because the effect of a_1 is to reduce its bearer's individual reproductive success. However, assume that, by helping, the younger sisters with a_1 increase the reproductive success of their older sisters. Because they are related, those older sisters are disproportionately likely *also* to be carrying a_1. More precisely, in around half of all cases (because the coefficient of relatedness is $^1/_2$), the sister will have an allele which is identical by descent and therefore bound also to be a_1. By helping a sister, bearers of a_1 are disproportionately helping other copies of a_1 to reproduce and therefore disproportionately not helping copies of a_2.

However, siblings are not genetically identical. The older sister could be carrying a_2 and thus aid given to a sister will sometimes benefit the competitors of one's own alleles. Thus, there seem to be both benefits and hazards to investing in kin rather than oneself. William Hamilton showed that a simple inequality described the circumstances under which such investment could be adaptive. This inequality is known as Hamilton's rule and it states that a kin-directed behaviour can be favoured by selection whenever:

$$c < rb$$

Box 4.1 Deriving Hamilton's rule

Where does Hamilton's rule come from? It is mathematically complex to derive it formally, but one simple way of showing where it comes from is to calculate the relative fitness of an allele for foregoing reproduction to help an older sibling. As before, allele a_1 causes younger sisters to forego reproduction, at cost c, to help their older sisters, who thereby get benefit b. It causes singletons and older sisters to reproduce normally. This allele is in competition with an alternative allele a_2, which causes younger sisters to reproduce for themselves. We can tabulate the expected fitness of alleles a_1 and a_2 in the three contexts in which they each occur: in the bodies of singletons, in the bodies of older siblings, and in the bodies of younger siblings (Table 4.2). In the table, w represents an average level of fitness for the population. The allele a_2 always has fitness w. The allele a_1 has fitness w when in an offspring with no siblings, $w + \frac{1}{2}b$ when it is in an older sister, since the younger sister may share the allele and thus give aid to the value of b, and $w - c$ when it is in a younger sister, since it causes the younger sister to forego reproduction to the value of c.

The sum of the top row of the table is $3w + \frac{1}{2}b - c$, and the sum of the second row is $3w$. The *difference* in fitness between a_1 and a_2 across all contexts is the difference between the top row and the second, namely:

$$3w + \frac{1}{2}b - c - 3w$$

This is equal to:

$$\frac{1}{2}b - c$$

When will a_1 have higher fitness than a_2? Exactly when the difference in fitness between them is greater than zero; that is, when:

$$\frac{1}{2}b - c > 0$$

or in other words, when:

$$c < \frac{1}{2}b$$

Thus, an allele for helping siblings has higher fitness than its competitors when $c < \frac{1}{2}b$, exactly as Hamilton's rule says. Similar reasoning could be followed for other coefficients of relatedness.

Table 4.2 **Fitness of an allele, a_1, whose effect is to make younger sisters forego reproduction to the value of c but provide aid to their older sisters to the value of b, compared with a competitor allele a_2 that has no such effect. w is the average level of fitness in the population. The situation has been simplified by assuming a_1 is dominant and rare. a_1 will out-compete a_2 exactly where $c < \frac{1}{2}b$.**

Allele	Fitness in offspring with no sibling	Fitness in older sister	Fitness in younger sister
a_1	w	$w + \frac{1}{2}b$	$w - c$
a_2	w	w	w

where c is the reduction in the actor's reproductive success, b is the increase in the recipient's reproductive success, and r is the coefficient of relatedness. For the sibling case described above, since $r = \frac{1}{2}$, the behaviour could spread whenever the average increase in the older sister's reproductive success caused by helping was more than twice the reproductive success foregone by the actor to help. For a nephew, where r is only $\frac{1}{4}$, the benefit to the recipient would have to be at least four extra chicks for every one forgone by the actor, in order for the behaviour to evolve.

Hamilton's rule means that if we want to calculate the reproductive success of an individual animal, we should not restrict ourselves to the animal's number of personal descendants, but also add in any extra reproduction by relatives that results from the individual's behaviour, adjusted by the coefficient of relatedness. Reproductive success so calculated is called inclusive fitness.

4.4.2 Applications of kin selection

Hamilton's rule has very wide applicability. Most obviously, it can be used to understand why and to what extent adults invest in their children and grandchildren (as we shall see in Chapter 8). There are also many phenomena in nature similar to the sibling example described above, where individuals under certain conditions invest in the offspring of their siblings or the later offspring of their parents, rather than reproducing for themselves. These behaviours are known as alloparenting and although there can be other benefits of alloparenting behaviour, kin selection appears to play a strong role (Griffin & West 2003).

Kin selection within the individual

Kin selection also has some less obvious applications. For example, most of the cells in your body have no chance of ever reproducing. Muscle cells, red blood cells, and cells in the immune system are all doomed to die out when you die and yet they go on working all your life to keep your body functional and intact. On the other hand, sperm and egg cells do nothing but sit around waiting to be used in reproduction. Thus, in one sense, cells in all other tissues forego their own reproduction to allow the gametes a chance to reproduce. How could such reproductive restraint be evolutionarily stable? The answer of course is kin selection. The cells in your body are genetically identical (coefficient of relatedness is 1) and thus investing in reproduction of gametes is, as far as a cell in your brain or blood is concerned, just as good as reproducing for itself.

It follows from this that genetic mutations arising in some cells of the body but not others are potentially disruptive to the body's functioning. This is because once such a mutation has occurred, there are allelic differences between cells (i.e. the coefficient of relatedness is no longer exactly 1) and there will be competition between different cell lineages. Selection within the organism then favours cells that proliferate differentially at the expense of other body tissues. For this reason, cells in which genetic mutations occur, for example during mitosis, are detected and destroyed by the immune system. However, sometimes such cells evade the immune system and then they become cancers—mutation-containing cell populations that favour their own expansion at the expense of the proper functioning and ultimate interests of the whole body. Cancers are related but non-identical organisms to their hosts, with their own fitness interests. Lest this seem a fanciful way of looking at it, there is at least one independent organism that started out as a group of cancerous cells within a larger one. Canine transmissible venereal sarcoma is a venereal disease of dogs. The cells of this parasite are descended from cancerous tissues of a particular dog, tissues that accumulated mutations allowing them to spread by mitosis, first through the body of the original host and later by genital contact with other hosts too.

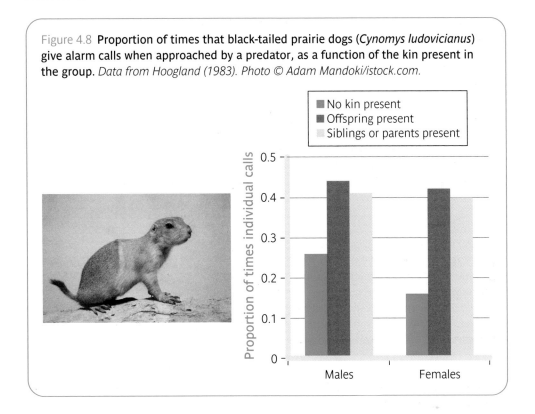

Figure 4.8 **Proportion of times that black-tailed prairie dogs (*Cynomys ludovicianus*) give alarm calls when approached by a predator, as a function of the kin present in the group.** *Data from Hoogland (1983). Photo © Adam Mandoki/istock.com.*

Alarm calling in prairie dogs

Kin selection is the key to understanding many behaviours that seemed to earlier genera-tions of biologists to exist for the good of the group. For example, in the black-tailed prairie dog, *Cynomys ludovicianus*, adults sometimes give alarm calls when they detect a predator. These calls alert other individuals to the presence of the predator, but will tend to draw attention to the individual giving the call. A classic group selection account would stress how groups containing callers might survive better than groups with no callers, but this account cannot be correct, as we saw in section 4.3. Instead, calling adults might disproportionately benefit their kin. If this was the case, then the behaviour could evolve by kin selection as long as Hamilton's rule is satisfied.

Hoogland (1983) approached prairie dog colonies with a model predator (a stuffed badger) and noted how often individuals gave alarm calls, having first established the kinship relation-ships between animals in the colony. The results show that calling is much more common when the caller has kin in the colony than when it does not (Figure 4.8). These kin can be offspring, but calling is also common when the kin in the colony are siblings or parents.

4.4.3 Conditions on Hamilton's rule

There are several constraints on the applicability of Hamilton's rule that should be mentioned. The first is that a behaviour will not evolve, even if $c < rb$, if c is too large in absolute terms. For example, an allele that made all its bearers sterile could not spread, even if they thereby helped their siblings have 100 extra offspring. This is because all the siblings they helped to reproduce would necessarily *not* possess the allele, otherwise they would be sterile too. Thus, Hamilton's rule is only applicable to behaviours with a weaker selective disadvantage, such as foregoing *some* reproduction under *some* circumstances. The second point is that the coefficient of relatedness alone is not sufficient to predict which behaviours will evolve. For example, parents nearing the end of their lifespan might well do more to protect their offspring than those offspring would do to protect the parents. The coefficient of relatedness is the same in the two cases, but the parents can have no more offspring, whereas the offspring have all of their reproduction ahead of them. We can incorporate such asymmetries into Hamilton's rule by assuming that *b* and *c* are not fixed for particular behaviours, but variable depending on the future prospects of the recipient and actor, respectively.

Finally, and unlike the case of cells within the same body, the coefficient of relatedness of one animal to another is always less than the coefficient of relatedness of that animal to itself. Thus, although the theory of kin selection predicts that there will be widespread investment in kin found in nature, it is perfectly compatible with there being behaviour directed *against* kin too. For example, in the very same prairie dogs that give calls to help kin, a major source of death in the young is being killed by the mother's sister when she has pups of her own (Hoogland 1985). Such females are reducing their inclusive fitness by killing their nieces and nephews. However, they are more closely related to their own offspring than they are to their sister's and so, providing that the benefit to their own pups is great enough, the behaviour can be adaptive. Behaviours that harm kin are particularly likely to evolve where there is local competition between relatives for finite resources. A clear example is found in certain bird species, such as kittiwakes, where nestlings may be killed by their siblings. Parental provision to the nest is limited and the benefit of receiving more of it is clearly greater under certain conditions than the inclusive fitness benefits through the lost siblings.

4.5 Advanced topics: evolutionary transitions, levels of selection, and intra-genomic conflict

We end this chapter by reviewing some more advanced topics in evolutionary theory, which build upon the principles set out so far. You can progress to Chapter 5 without covering these topics, but they are areas of great current research interest and since they go to the very heart of how evolutionary competition works, understanding them is useful for deepening your understanding of evolutionary theory more generally.

The argument so far in this chapter has been that behaviours that benefit groups at the expense of individuals will not be evolutionarily stable against competitor behaviours that benefit just individuals. This is usually true. However, there are examples in nature of collectives of unrelated entities who all work for the common good. In such groups, many of the adaptations that arise appear designed for the good of the collective, not the component elements.

The most compelling example of such a collective is the animal. Calling an animal a collective seems paradoxical, but the body of an animal such as yourself is created by around 25,000 genes working together and very few of these genes are closely related by kinship. Indeed, your mitochondria are descended from free-living bacteria that became incorporated within the cells of a very distant ancestor. They even have their own genome. Thus, your genome can be seen as a group of many unrelated elements, including some that originate from a different species, and yet they work together for the common good of making an integrated, functional body.

This leads us to re-examine the question of the level of selection. It seems plausible that genes, by working together on a common body, can do better at replicating than any of them could by trying to replicate alone and that this is the key to the evolution of complex organisms. However, it is not immediately obvious why this works for a group of genes uniting for the common good of the body and fails to work, for example, for a group of birds regulating their reproduction for the common good of the population. Would not the same issues of evolutionary stability that we saw in section 4.3 arise?

4.5.1 The Price equation

A useful tool for thinking about this problem (and all the others considered in this chapter) lies in a framework called the Price equation, developed by theorist George Price in the 1970s (for a non-mathematical introduction, see Okasha 2006). The Price equation is complex, but its message can be simplified for our purposes as follows. The evolutionary change we should expect in a characteristic depends on that characteristic's covariance with the fitness of the alleles coding for it. Covariance is basically like correlation; it is positive when increasing the characteristic increases fitness, negative when decreasing the characteristic decreases fitness, and zero when there is no association between the characteristic and fitness. Thus, characteristics that covary positively with fitness will increase over evolutionary time, those that covary negatively will decrease, and those with zero covariance with fitness will not change. Behaviours that benefit a collective are no exception to the general picture. They will evolve as long as the overall covariance between the fitness of the allele and the functioning of the collective is positive.

This condition explains why group selection did not work in the reproductive restraint example of section 4.3. There was a positive covariance between the amount of altruism in a group and the average fitness of that group's members, due to differential group survival. However, this was very weak, since selfish individuals that happened to find themselves in groups with altruists benefited as much as the altruists did. There was also a strong negative covariance between altruism and fitness *within* the group because altruistic individuals reproduced less than their selfish group-mates. This negative covariance more than offset the smaller positive one, leading to an overall negative relationship between altruism and fitness, and hence altruism's demise.

4.5.2 Suppression of within-group competition

Now consider genes cooperating to make a body. As our bodies are constituted, the only way for genes to replicate themselves is for the whole genome to get replicated (i.e. for the phenotype to survive and reproduce). If it does so, they all benefit to the same extent. If it fails to do so, they all die out. Thus, the covariance between phenotypic success and allelic fitness is positive and strong. This allows the complex genome to emerge as a functional collective; all of the individual genes have the same interest in making it work. In general, wherever there is a functional collective in nature, there are mechanisms in place that abolish any differences in reproductive success *within* the collective, so that any element's fitness is determined only by the functioning of the whole (Frank 2003).

In complex genomes, such mechanisms are in place. During meiosis, genes are bound together on chromosomes and either the whole chromosome goes forward to the gamete or none of it does. This abolishes any competition between neighbouring genes to, for example, replicate themselves faster than their neighbours. They are either all going forward together or all failing together. The way meiosis works makes it a lottery which of the two copies of a particular chromosome ends up in a particular gamete. It is not, for example, dependent on the chromosome's size. A fair lottery abolishes competition because no characteristic of the alleles on the chromosome makes any difference to the result. Because of mechanisms such as these, the fate of any one of your genes is highly correlated with the fate of all of the others and so the Price equation predicts that they will cooperate for the common good.

Earlier in the chapter I mentioned that, although the best way to keep track of fitness in evolutionary models is at the level of the allele, we can use the reproductive success of the individual as an approximation for many purposes. The Price equation allows us to understand why this works: because the covariance between the reproductive success of an organism and the fitness of all its alleles is substantial. We can also refine our conclusion concerning the level of selection. Rather than adaptations for the collective good never being evolutionarily stable, they will persist exactly where mechanisms have evolved that ensure high overall covariance between allelic fitness, the functioning of the collective, by suppressing competition between elements within the collective.

To visualize this more concretely, imagine two different reality television shows. In both shows, a group of people are taken to a remote location. They are set various challenges. In one show, the individual who completes more challenges than all the others gets £1 million. The Price equation predicts that people in this scenario will try to do each other down because there is a negative covariance between a participant's winnings and the success of the other group members in the challenges. In the second show, the participants each get £1 million if all of them complete all of the challenges. Here, you would expect the participants to help each other because there is a positive covariance between one individual's winnings and the performance of all the others. The former type of game is the one that free-living organisms are engaged in, which is why they look after themselves, not other group members. The latter type of game is the type that the genes in your genome are playing, which is why they work together to make a functioning body.

4.5.3 Evolutionary transitions

There are a number of points in the history of life where several previously free-living elements come together and start to operate as a collective. Each of these points heralds a major evolutionary transition, where a new form of biological organization arises (Figure 4.9; Maynard Smith & Szathmáry 1995). One transition is from independent molecular replicators to groups of genes. Another is the emergence of more complex genomes with chromosomal organization. Still another is the eukaryotic cell, which, you will recall, is formed by more than one species and then the multicellular organism from a single-celled ancestor. A further transition that a few species have undergone is the transition to eusociality. Eusociality is the situation where a whole colony of individuals work together to further the reproduction of one or just a few of their number. Eusociality has evolved at least 15 times in nature and is best known in, although not restricted to, the ants, bees, and wasps. Eusocial colonies have impressive functional organization, with different castes fulfilling different roles and a number of features existing that look well designed for the colony, not for the individuals within it. In fact, the eusocial colony of a

Figure 4.9 **The history of life shows a succession of transitions at which elements form collectives which then begin to evolve apparent adaptations of their own. The lower-level elements continue to exist independently alongside the new entities.** *Photo bottom left: © merrymoonmary/istock.com; photo bottom middle: © Graham Cripps/NHMPL.*

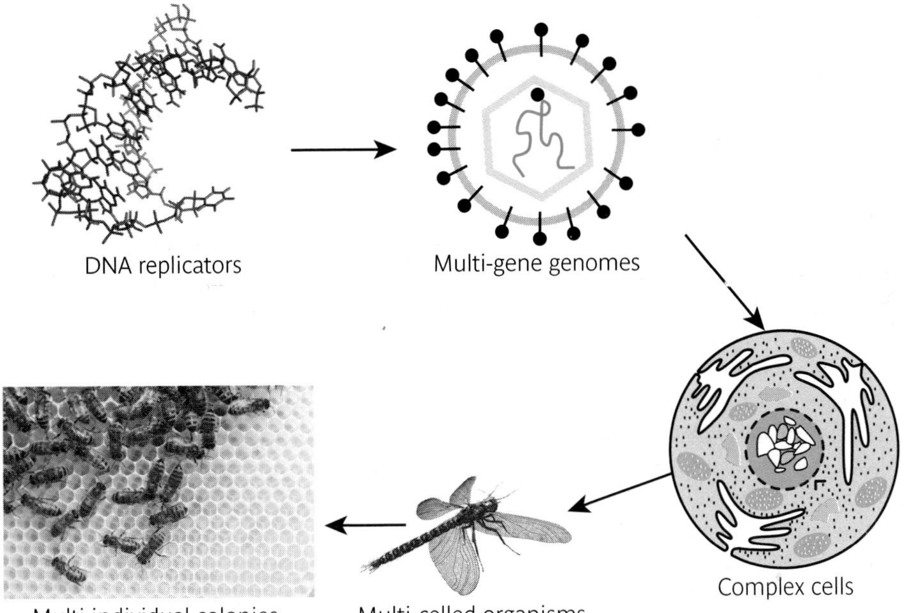

DNA replicators

Multi-gene genomes

Complex cells

Multi-individual colonies

Multi-celled organisms

bee or wasp can be viewed as a kind of organism rather than a group of individuals, just as you can be viewed as an organism rather than a coalition of genes.

The general pattern with the major transitions appears to be that the threshold from loose aggregation of independent entities to functional collective is very hard to cross because collective cooperation is usually disrupted by competition within the group. On those rare occasions that the threshold is crossed, there is a point of no return—it is very unusual for part of a multicelled organism to return to separate existence, as canine transmissible venereal sarcoma has done—and the resulting life forms can quickly radiate and prosper, forming a whole new branch of the tree of life.

4.5.4 Intra-genomic conflict

Even when the point of transition to collective functioning is crossed, there can still be some internal competition simmering away. Within the complex genome, for example, it has become clear that the mechanisms suppressing competition between genes are not always perfect and genes in the same individual can differ in fitness. Such competition and the effects it produces are known as intra-genomic conflict.

Transposable elements

Intra-genomic conflict arises whenever genes can favour their own interests above that of the whole. We have seen one example from transposable elements (Chapter 2). These are genetic sequences that have the ability to make extra copies of themselves at meiosis. Thus, over evolutionary time, they can proliferate within a complex genome—recall that humans have over 1 million copies of the *Alu* element. This serves *Alu's* own fitness interests, not the interests of the organism. Indeed, the rest of the genome probably incurs a small cost of the extra *Alu* material and is thus under selection to shut *Alu* activity down. The evolution of the genome will thus be a dynamic of cat and mouse between selfish elements like *Alu* and suppressive adaptations elsewhere in the genome.

Segregation distorters

Genetic variants that distort fair segregation provide another case. Under normal meiosis in a diploid organism, the chance of a particular allele going forward to the gamete is 50%. In mice, there is a genetic variant called the *t* haplotype. When one copy of the *t* allele is present in males, 90%, rather than 50%, of the viable sperm produced carry *t*. The *t* variant seems to achieve this by disabling most non-*t* sperm. It thus gives itself an advantage relative to all its competitors. The *t* variant is found in mice all over the world and has persisted for many thousands of generations. It has only remained at the low frequencies that it has—around 5% in many populations—because the homozygote is lethal in males and has negative effects in females too. It clearly persists because of its own ability to distort segregation rather than any beneficial effect at the level of the organism.

Cytoplasmic male sterility

Another example is the tug-of-war between the mitochondrial portion of the genome and the rest. Mitochondrial genomes are inherited down the female line only. This occurs because the larger, female gamete provides the cellular environment of the zygote, whereas the male gamete provides basically only nuclear DNA. This causes a conflict of interest between the mitochondrial genome and that of the cell nucleus. The former will only succeed in female offspring, whereas the latter succeed in offspring of either sex. Thus, any mutations within the mitochondrial genome that increase the proportion of females produced will have a selective advantage over their competitors.

Plants are usually hermaphrodite, producing both pollen and ovules. In many types of plant, there are mitochondrial genetic variants that shut down pollen production and make individuals only female. This phenotypic effect is called cytoplasmic male sterility (CMS). Genetic variants causing CMS can become extremely common and when they occur, there is selection on variants in the nuclear genome to counteract their effects and restore pollen production. Over evolutionary time, there is a dynamic arms race between mitochondrial CMS mutants and counter-mutants in the nuclear genome. Many plant species are found with a mixture of female-only individuals and hermaphrodite individuals as a result. Examples such as these remind us that organisms are not quite perfectly unified; their origins as groups of elements with distinct interests surface every now and then.

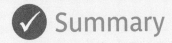 Summary

1. Within all biological populations, there are differentials in reproductive success and thus competition to reproduce.

2. Some phenotypes lead to greater success than others in reproductive competition in particular environments and it is these differences that drive the process of adaptation.

3. Although natural selection produces changes in phenotypes, it does this by changing the frequencies of the underlying genotypes. The best way to model evolutionary outcomes is to track the fitness of competing alleles, rather than competing individuals or competing populations.

4. Any behaviour or characteristic that, once common in the population, cannot be displaced by any competitor behaviour or characteristic is said to be an evolutionarily stable strategy (ESS).

5. In free-living organisms, innovations that enhance the interests of the group or species at the expense of the individual are unlikely to be evolutionarily stable. Adaptations in nature are generally well designed for the individuals' genes, rather than for groups or species.

6. Behaviours that benefit relatives at the expense of the actor can evolve, if Hamilton's rule is satisfied, because of allelic relatedness between kin.

7. Selection can favour adaptations for the good of higher-level collectives, as long as mechanisms have evolved that abolish differentials in reproductive success amongst their constituent elements, so that there is a positive overall covariance between the collective's functioning and the reproductive success of the elements within it. Complex organisms, which are coalitions of different genes, represent prime examples of such collectives.

8. The covariance between fitness of different genes within an organism is not always total, and this leads to intra-genomic conflict and patterns that benefit some genes at the expense of the whole organism.

? Questions to consider

1. In the cat model in section 4.1, adding mortality slows the exponential growth of the population (see Figure 4.3). You will note, however, that the population is still going to explode exponentially eventually. Adding a certain probability of mortality has merely retarded the increase by a few years. Such explosions do not tend to occur in real biological populations. Why do you think this is? How would you model the pattern more realistically?

2. Section 4.3 argued that Wynne-Edwards' group selection explanation for birds having smaller clutches in bad years must be wrong. However, birds do actually do this. How might it be explained without recourse to group selection ideas?

3. Identical (MZ) twins have identical genomes (coefficient of relatedness of 1). Thus, the theory of kin selection predicts that they will treat the offspring of the other twin exactly as they treat

their own. There is no reason for them to favour their own interests over those of the other twin. In fact, their behaviour is more like that of ordinary but close siblings. Why might the behaviour not follow the theory in this instance?

4. Shakespeare's plays frequently centre around brothers banishing or killing their brothers. For example, in *As You Like It* and *The Tempest*, one brother banishes the other and takes over the dukedom, and in *Hamlet* and *Richard III*, brothers kill brothers to become king. Are these aspects of Shakespeare's stories biologically implausible?

5. Given the section on intra-genomic conflict, how do the interests of the Y chromosome differ from those of the X? What kinds of alleles on the Y might be favoured, and what response from the X would be favoured?

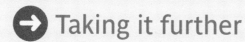 Taking it further

The models of evolution presented in this chapter are all simplified for expository purposes. For more mathematically rigorous approaches, the reader is referred to McElreath & Boyd (2007). The classic—and brilliantly written—statement of the need to consider the relative fitness of alternative alleles to understand evolution is Dawkins' *The Selfish Gene* (Dawkins 2006b). The debate about the correct level at which to conceive that selection is acting rumbles on—very different views are given, for example by West *et al*. (2006) and Wilson & Wilson (2007). The problem stems from the fact that frameworks using the language of group selection are formally equivalent to frameworks using the language of (inclusive) individual fitness. It thus becomes somewhat semantic which language is preferred.

The major transitions in evolution are reviewed by Maynard Smith & Szathmáry (1995), and the special role of repression of within-collective competition is modelled by Frank (2003). For a recent review on eusociality, see Wilson & Hölldobler (2005). Okasha (2006) shows how the Price equation and refinements of it can be used as an overall language for thinking about levels of selection and the evolutionary transitions. On intra-genomic conflict and selfish genetic elements, Burt & Trivers (2006) is the authoritative but vast source; shorter introductions are provided by Hurst *et al*. (1996) and Hatcher (2000).

Natural selection

All the components of natural selection—heredity, variation, and competition—have now been introduced and all the key concepts defined. This chapter concludes the review of fundamental evolutionary principles by looking in more detail at natural selection itself. Natural selection is often invoked as an explanation for why organisms are as they are nowadays, but if we are going to invoke it as an explanation, we need to be crystal clear about how it works, what results it tends to produce, and, especially, how we can identify that it has been operating. Thus, understanding the material in this chapter is a crucial prerequisite to offering an evolutionary explanation for any human characteristic or behaviour.

The chapter thus reviews the types of natural selection that exist (section 5.1) and the impact that selection has on the gene pool (section 5.2). We then consider what selection does to phenotypes—in other words, how and to what extent it leads to optimally designed individuals (sections 5.3 and 5.4). Section 5.5 examines the question of how we can determine scientifically why selection has produced the phenotypes that it has; in other words, how to study adaptations without being accused of telling 'Just-So' stories. Section 5.6 reviews the main conclusions of the first five chapters of the book by comparing some common misconceptions about how evolution works with how it actually works. Hopefully, having read the previous four chapters, you will quickly see why the common misconceptions are wrong.

5.1 Modes of selection

Natural selection occurs whenever the fitness of one allele at a locus is higher than any of its competitor alleles. However, selection takes a number of different forms, depending on the relationships of genotype, phenotype, and fitness.

5.1.1 Purifying selection

A fundamental, and often ignored, form is purifying selection. Purifying selection occurs when an allele that does something useful is fixed at a locus. Whenever mutations arise at that locus, they have lower fitness that the incumbent and thus are weeded out. If they are lethal, they will disappear in one generation. If their negative effect is more moderate, they will persist for some time before disappearing (exactly how long depends on chance and the magnitude of their selective disadvantage).

Purifying selection is ubiquitous. The lineages of the organisms we see today have managed to go many billions of years without dying out. Any genic form that is fixed in a contemporary genome has withstood millions of generations of competition and come out on top, so unless the environment has changed radically, it is likely to do what it does pretty well. Its effects will also be well integrated into the developmental programme of the organism, whereas those of any new mutant might be less so. Thus, most mutations that arise are less fit than the incumbent and most are removed by purifying selection. What we think of as Mendelian genetic diseases are largely deleterious mutations in the course of being eliminated.

At loci that have no phenotypic effect (e.g. because they are non-coding), there is no purifying selection. This means that new alleles can spread and occasionally displace the incumbent simply by chance. This is the basis of genetic drift, the neutral theory, and the molecular clock, all of which we have already met (section 3.2.4).

5.1.2 Stabilizing selection

For continuously varying characteristics, natural selection can take several forms. Stabilizing selection describes a situation where the current population average of the trait is also the optimum from a fitness point of view, with individuals higher or lower on the trait having reduced fitness.

Nettle (2002) looked for evidence for natural selection on height in a cohort of several thousand British women. The ideal measures to assess selection in progress would be the probability of survival to adulthood and reproductive success for those surviving. The former was not available, so instead he used proxies such as the probability of having a long-standing illness to assess the impact of phenotype on health. Health measures showed inverted U-shaped relationships with height, with costs at the extremes and the optimum somewhere in the middle of the distribution. Similarly, the highest likelihood of having children was found amongst women who were neither extremely tall nor extremely short. For British women, then, the data show that selection on height is basically stabilizing (Figure 5.1).

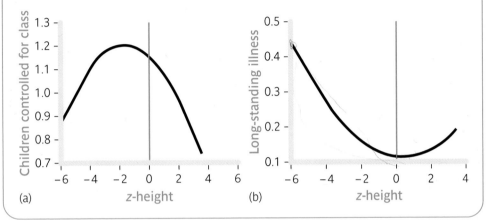

Figure 5.1 **Height against (a) number of children (with social class controlled for) and (b) probability of having a long-standing health problem, for around 4,000 British women. *z*-height represents actual height standardized so the mean is 0 and the standard deviation 1. The optimum height in terms of number of children is somewhat less than the mean, whilst the optimum for health is somewhat higher. Selection on height in this population is thus basically stabilizing overall.** *Data from Nettle (2002).*

Unless something else changes (e.g. nutrition), stabilizing selection will maintain the population mean for the characteristic exactly where it is. The reason this is so follows directly from the Price equation (section 4.5.1): there is no overall positive or negative covariance between the characteristic and fitness. Where a population has been in an environment for some time, it has probably reached the local optimum for traits such as size and shape, so selection on these kinds of things will often be stabilizing in form.

5.1.3 Directional selection

Purifying and stabilizing selection maintain the distributions of phenotypes as they are. The form of selection that leads to change is directional selection. Directional selection is operating wherever the optimum value of the characteristic from a fitness point of view differs from the average value of the characteristic in the current population. Phenotypes that are different from the average thus enjoy a fitness advantage and the composition of the population changes over time.

Several studies have suggested that taller-than-average men have increased reproductive success in contemporary human populations (Pawlowski *et al.* 2000; Mueller & Mazur 2001). In his study, Nettle observed no differences in numbers of children by male height. However, the men in his sample were not at the end of their reproductive careers, so it is not possible to ascertain their final reproductive success. However, taller-than-average men had more marital or cohabiting relationships than men of average height. This pattern is quite different from that for the women (Figure 5.2). This finding is consistent with there being directional selection on male height in this population, as has been found elsewhere. Note, however, that these data are not sufficient

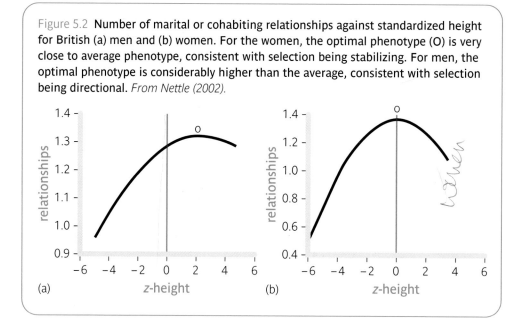

Figure 5.2 **Number of marital or cohabiting relationships against standardized height for British (a) men and (b) women. For the women, the optimal phenotype (O) is very close to average phenotype, consistent with selection being stabilizing. For men, the optimal phenotype is considerably higher than the average, consistent with selection being directional.** *From Nettle (2002).*

to demonstrate that directional selection is occurring. You would need to show that the extra relationships led to more children in the end and also that the increased reproductive success was not offset by a dramatic increase in mortality amongst taller men or their offspring.

The significance of directional selection is that, as long as the characteristic is heritable, it causes predictable change in the population, so that the population average moves towards the optimum for the characteristic. You can think of the optimal value of the characteristic for a given environment as being at the top of a hill. Directional selection causes the population to climb slowly, generation by generation, up the hill towards the summit.

Directional selection as an agent of change

Directional selection on polygenic characteristics can produce striking results and, importantly, can move the population average for the characteristic beyond the range observed in the starting population. This is possible because, as allele frequencies change across multiple loci, new genotypic combinations that were never observed in the starting population begin to be generated. In the long run, new mutations will also occur that introduce new variation for selection to work on. Thus, directional selection can eventually create phenotypes never seen in the ancestral population.

This principle can be illustrated by a long-term experiment conducted on *Zea mays*, the plant called corn in North America and maize in the UK, at an agricultural research station in Illinois (Figure 5.3). In 1896, researchers chose, from 163 corn ears, the 24 ears highest in oil content and the 12 ears lowest in oil content, and bred separately from these two groups. Every year they repeated the selection from within the corn resulting from each breeding. After around 75 generations, they had produced two radically different types of corn, one with about 1% oil content and one with about 20% oil content. These two values are both well outside the range of variation in oil level seen in the original stock (about 5%).

Figure 5.3 **Response to artificial selection for either high or low oil content in** *Zea mays*. **The selection moves the two lineages well beyond the range of variation seen in the starting population.** *Reproduced from Dudley (2007).*

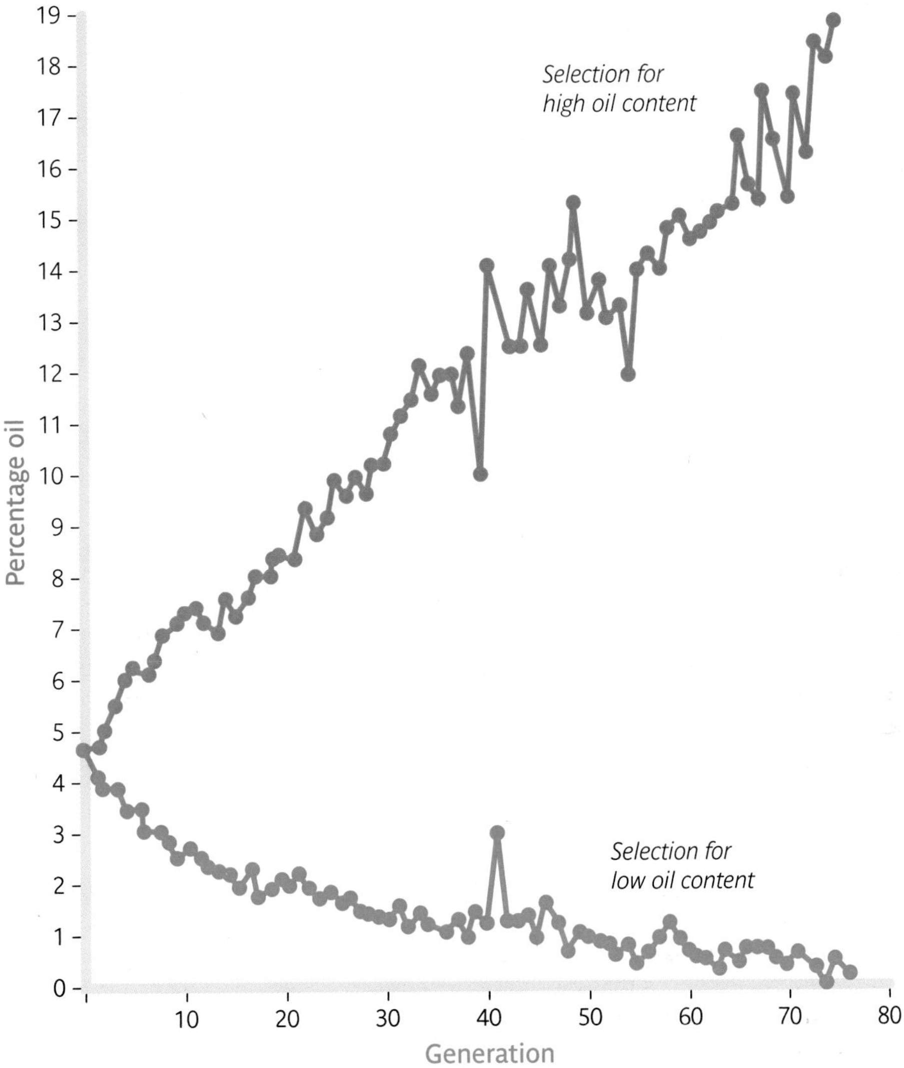

5.2 Selection and variation

Mutation constantly introduces new variation into a population, whilst selection is an opposing force, working to reduce variation. The variation-reducing effect of selection is most obvious for purifying selection, which weeds out new mutants and keeps the population homogenous, but directional selection also decreases variation, because it drives towards fixation those alleles associated with highest fitness, reducing and eliminating competitor alleles as it goes. Stabilizing selection, too, eventually eliminates genetic variation, for reasons that are slightly more complex.

Sustained selection can deplete population genetic variation, driving the heritability of the characteristic down towards zero. For example, in the Illinois maize/corn experiment, the heritabilities of oil content during the first ten generations were 0.32 in the high-oil line and 0.5 in the low-oil line. After the fiftieth generation, they had reduced to around 0.12 in the high line and 0.15 in the low.

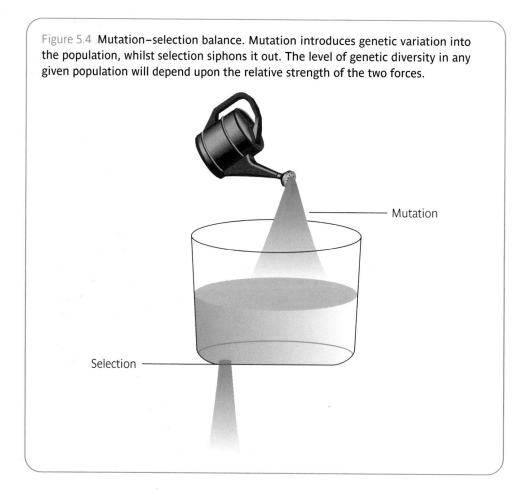

Figure 5.4 **Mutation–selection balance. Mutation introduces genetic variation into the population, whilst selection siphons it out. The level of genetic diversity in any given population will depend upon the relative strength of the two forces.**

Mutation

Selection

The amount of genetic variation in a characteristic in a population will thus be the resultant of the two forces, mutation and selection (the mutation–selection balance; Figure 5.4). Where selection is very strong and sustained, genetic variation can be used up and heritability is zero. This is the situation for number of arms or number of hearts in mammals. However, for many traits in many species, including traits directly related to fitness, substantial genetic variation is maintained. The ensuing sections describe five mechanisms that can lead to this happening: heterozygote advantage, negative frequency-dependent selection, force of mutation, inconsistent selection, and sexually antagonistic selection.

5.2.1 Heterozygote advantage

Heterozygote advantage is the situation where individuals with one copy of a particular allele have higher fitness than individuals with either no copies or two copies. The classic case is human sickle-cell disease. There is an allele of the human β-globin gene, which we can denote s, with the normal form called S. Homozygotes (ss) have red blood cells that deform in shape under conditions of low oxygen. This can lead to strokes and anaemia, and life expectancy is reduced. Heterozygotes (Ss) have some abnormalities in their red blood cells, but not sufficient to cause health problems. However, the red blood cell differences are sufficient to impair the survival of the malaria parasite, which spends part of its life cycle in these cells. Thus, heterozygotes enjoy increased resistance to malaria without any of the health problems that homozygotes suffer.

In an environment containing malaria, selection will increase the s allele when it is rare. At this point, due to its rarity, and assuming no inbreeding, s will mostly be found in heterozygotes and thus its fitness relative to S will be high. As it becomes more common, it will increasingly 'meet itself' in homozygotes. These have reduced fitness, so when s is very common, selection will reduce it. The predictable outcome of such a situation is that s stabilizes at an intermediate frequency determined by the relative magnitude of the health benefits of Ss and costs of ss. Where there is no malaria, there are no benefits and, indeed, s is found at appreciable frequency only where there is a history of malaria (Figure 5.5). Even where malaria is common, s has not gone to fixation because of the heterozygote advantage.

Figure 5.5 **The geographic distribution of human sickle-cell disease (a) closely maps to the historic distribution of malaria (b).**

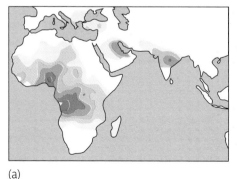
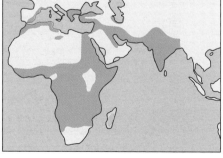

(a) (b)

5.2.2 Negative frequency-dependent selection

A second mechanism leading to the persistence of variation is negative frequency-dependent selection. This is the situation where a phenotype is associated with relatively high fitness when it is rare, but relatively low fitness when it is common. The result of such a selective regime is that the type stabilizes at an intermediate frequency.

Gross (1991) studied the bluegill sunfish, *Lepomis macrochirus*, in a lake in Canada. Males of this species come in two types. 'Parentals' delay maturity until they are 7 or 8 years old, then build nests, which they patrol, by creating depressions on the lake bed. Females may then visit their nests and deposit eggs for them to fertilize. 'Cuckolders' mature early and build no nests. They sneak into nests created by other males and deposit their sperm there.

It is easy to see intuitively that when all the other males in a colony are parentals, building nests, a rare cuckolder could do well, as he has plenty of target nests to sneak into. However, as cuckolders become more and more common, there are fewer and fewer nests available to sneak into, and more and more competition from other cuckolders. At the limit, where all males were cuckolders, their reproductive success would be zero, as there would be no nests at all to sneak into.

By carefully removing cuckolders from colonies, Gross manipulated their local density. He showed that the pairing success of cuckolders (which equates to success in fertilizing eggs) is highest when there are few of them in the colony and declines as they become common (Figure 5.6). Thus, it seems that what is maintaining the coexistence of the two types of male is negative frequency-dependent selection. Sneaking can neither die out (because it is so advantageous when rare) nor become universal (because it is disadvantageous when common), but instead the genes for male reproductive strategy remain polymorphic.

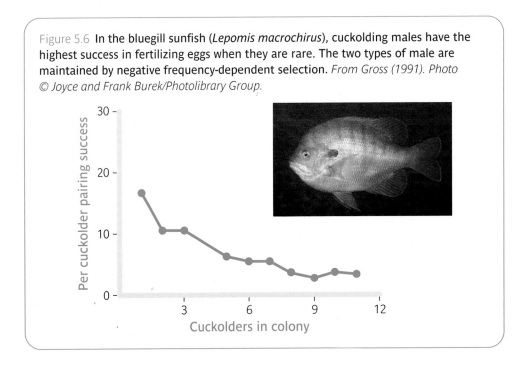

Figure 5.6 **In the bluegill sunfish (*Lepomis macrochirus*), cuckolding males have the highest success in fertilizing eggs when they are rare. The two types of male are maintained by negative frequency-dependent selection.** *From Gross (1991). Photo © Joyce and Frank Burek/Photolibrary Group.*

5.2.3 Force of mutation

The foregoing two mechanisms provide specific circumstances where genetic variation will be maintained. More generally, genetic variation will persist if the force of mutation is strengthened or that of selection weakened. For polygenic characteristics, the effective strength of mutation is proportional to the number of genes involved. This is because each of the genes involved has an independent chance of mutating at each generation and thus of introducing more variation into the characteristic. At the extreme, where a characteristic can be affected by mutations to more or less all genes, then considerable heritable variation can be sustained even in the face of very strong selection. For example, Houle (1992) showed that in the fruitfly, *Drosophila melanogaster*, many fitness-relevant traits such as number of offspring were substantially heritable. The most plausible general explanation is that mutations to almost any gene—genes creating the systems for growth, flight, mating, feeding, disease resistance, etc.—could affect number of offspring and thus there are many targets for mutational input.

5.2.4 Inconsistent selection

Another reason that variation may persist is if selection is inconsistent. In Chapter 1, section 1.4.1, we met the example of the medium ground finch in the Galápagos Islands. During a drought year, selection for larger beak size was very strong. However, the next year, the drought conditions were reversed and so was the gradient of selection. If the selective optimum moves around like this from time to time (or from place to place within the environment), then there is still an optimum phenotype (the one which on average across all times and places does best) and selection does still move towards it. However, the power of selection to counteract mutation and eliminate variation is effectively weakened. This is because it will decrease the frequency of particular alleles in some years or places and increase them in others.

Dingemanse *et al.* (2004) studied selection acting on a heritable behavioural characteristic called exploration score, which relates to how far the bird disperses from the nest it is born in, over 3 years in a population of the great tit, *Parus major*. In two of the years, food availability was poor and the females with the highest exploration scores were the most likely to survive. In the intervening year, large numbers of beech trees all seeded heavily and food was very abundant. Now, a different behaviour was optimal and it was the females with the lowest exploration scores who had the best chance of surviving (Figure 5.7). Female exploration score was systematically related to fitness, and thus under selection, in each year, but the direction of the evolutionary change varied from year to year. The pattern was different again for the males. It is thus not surprising that heritable variation in this behavioural characteristic is retained.

5.2.5 Sexually antagonistic selection

Both the great tit example and that of human height in sections 5.1.2 and 5.1.3 illustrate another reason that selection may not always eliminate variation. This is sexually antagonistic selection. The optimal phenotype may not be the same for males and for females. An allele that increases height may increase fitness when found in a male body, since there seems to be directional selection for height in males, but not when it is found in a female body, since selection on female height is stabilizing. Apart from the sex chromosomes, all genes have to spend an equal amount of time in male bodies and female bodies. The fate of alleles will be influenced by how they fare in both.

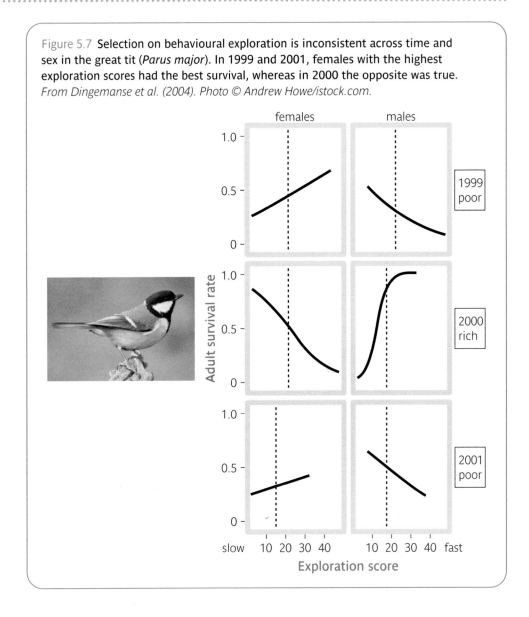

Figure 5.7 **Selection on behavioural exploration is inconsistent across time and sex in the great tit (*Parus major*). In 1999 and 2001, females with the highest exploration scores had the best survival, whereas in 2000 the opposite was true.** *From Dingemanse et al. (2004). Photo © Andrew Howe/istock.com.*

For example, in the red deer, *Cervus elephas*, such characteristics as female fecundity and male breeding success have considerable genetic variation, despite being under strong selection (Kruuk *et al.* 2000). Part of the explanation may be mutational input, since all the genes that make bones, muscles, antlers, the digestive system, the brain, and the immune system could indirectly contribute to a deer's breeding success. However, male deer with high fitness tend to sire daughters with relatively low fitness, and the genetic correlations of male and female fitness are negative (Figure 5.8; Foerster *et al.* 2007). This means that (at least some) alleles that improve the fitness of males harm the fitness of females, and vice versa. What this amounts to is another example of inconsistent selection. Instead of the selective coefficient varying

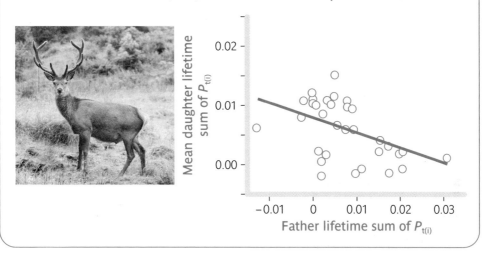

Figure 5.8 **Correlation between father and daughter fitness is negative in the red deer (*Cervus elephas*). This is because some alleles that improve male function have a negative effect on female function, and vice versa. The variable $P_{t(i)}$ is an estimate of fitness.** *From Foerster et al. (2007). Photo © Martin McCarthy/istock.com.*

from year to year, it varies according to whether the gene is in a male or female body. The net effect of sexually antagonistic selection, like other inconsistent selection, is to lead to the maintenance of more genetic variation than would otherwise be seen.

Because of the mechanisms reviewed in this section, it is not always the case that selection eliminates genetic variation. Where there is no genetic variation in a characteristic, unless there is another explanation such as very small population size, this may be an indication that selection has been at work. However, where there is genetic variation, it is not possible to say *a priori* whether there has been no selection on the characteristic, or whether one of the mechanisms described above has been maintaining variation in the face of selection.

5.3 Selection and adaptation

Section 5.2 considered what selection does to the gene pool. In this section, we look at what it does to the phenotype. The essence of the Darwinian world view is that natural selection produces good design (Chapter 1). But how exactly does it do this?

5.3.1 How selection produces design

We saw in the cat colour example (Chapter 4) that directional selection produces adaptation. However, in that example, there were just two colours of cat, yellow and black, and the adaptive state in the environment was to be yellow. Assuming a single mutation to a gene-making pigment can turn a cat from black to yellow, it is a relatively trivial achievement for selection to drive a yellow-causing mutation to fixation.

Many adaptations are not so simple, however. Consider the eye. In its current form, it is a good design, since it allows us to locate rewards, avoid dangers, and so forth. However, an eye could never come about by a single mutation arising in an eyeless population. It has far too many constituent parts: the lens, the liquid-filled eyeball, the retina, the retinal nerve, and so on. Each of these is built by different suites of genes and no one of them would seem to be of much use without the others. The question thus arises of how selection could produce an adaptive structure involving many interdependent parts underlain by many genes.

For such a feat to be possible, not only would the final structure have to be associated with higher fitness than the initial state, but also there would have to be a continuous sequence of intermediate phenotypes, each separated from the preceding one by a change small enough as to have been plausibly brought about by one genetic mutation. For the eye, this condition is quite plausible. In existing organisms, there are many types of eye. Some are just light-sensitive patches of skin. Some are open cups with a primitive retina inside them. Some are pinhole eyes, which are spherical in shape but have no lens, and then there are lensed eyes of several kinds. The eyes of other contemporary organisms are not the ancestors of our eyes, but they do show that it is possible to conceive a continuous sequence of kinds of eyes from very simple up to the lensed eyes that we have.

Natural selection must have driven eyeless ancestral organisms through a long sequence of intermediate phenotypes. Nilsson & Pelger (1994) use a mathematical model to explore how this might happen. They begin with a flat patch of skin with a photosensitive layer sandwiched between a dark underlay and a transparent protective layer. They calculate the optical resolution of this structure (basically, its ability to focus light). They then simulate all possible 1% changes in its shape and calculate the optical resolution of these. They adopt the descendant with the best resolution and then simulate all possible 1% changes in its shape, and so on. In just 192 changes of 1%, they produce a spherical lensless eye (Figure 5.9). They then allow all possible 1% changes in refraction of the tissue in the structure, again adopting that which gives the best resolution. In 300 more steps, they have produced a graded spherical lens, and, by allowing shape changes after this, in 300 more, they have produced an iris.

The final eye produced by their simulation is remarkably similar to that found in aquatic animals (their assumptions about resolution were based on vision underwater). The final lens has a focal length of 2.55 times its radius, which is the proportion actually found in many aquatic eyes. Their model shows that this proportion is optimal; no small changes from it in any direction can make resolution better. The optimal design has been reached by a sequence of tiny changes each of which was shown to be better than the last. Nilsson & Pelger calculate, using plausible assumptions about selection strength and heritability, that 350,000 generations would be sufficient for the complete evolution of the optimal lensed eye. This is a long period, but a tiny fraction of the time that aquatic animals have actually been evolving.

5.3.2 Adaptationist hypotheses

The example of the eye shows that the slow march of natural selection can push phenotypes up the gradient of design quality until the optimal design is reached. This conclusion allows us to take what we can call the adaptationist stance on the structures and behaviours that exist in nature. The adaptationist stance means reasoning in the following way: if some feature or behaviour is commonly found in a type of organism, then it is probably an efficient design solution to some problem that that organism has faced. If it were not, then the alleles building that feature would have been out-competed by alternatives that built a different feature.

Figure 5.9 **The model of Nilsson & Pelger (1994) begins with a flat photosensitive layer and simulates all possible 1% deformations of its shape, adopting the one which gives the best optical resolution and then simulating all 1% changes to this one, and so forth. A few hundred iterations are sufficient to produced an optimal lensed underwater eye.** *From Nilsson & Pelger (1994).*

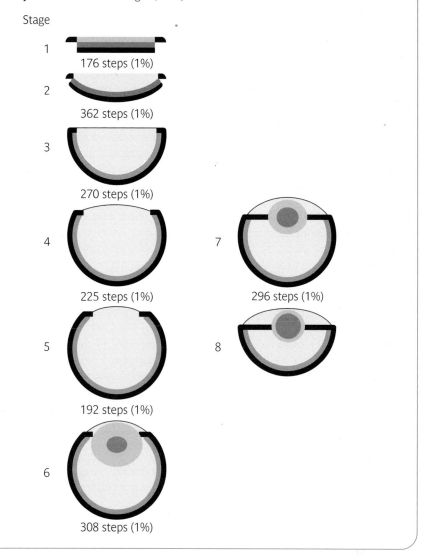

Stage

1
176 steps (1%)

2
362 steps (1%)

3
270 steps (1%)

4
225 steps (1%)

7
296 steps (1%)

5
192 steps (1%)

8

6
308 steps (1%)

Thus, the fact that we know all contemporary organisms are a result of a long history of selection allows us to assume that the phenotypes they have are adaptively constructed, and thus to generate hypotheses about the ways in which the particular characteristics they have increased ancestral fitness.

When we frame such adaptationist hypotheses, we can generally just look at the design of the phenotype and abstract away from the underlying genetics and development of the characteristic. For example, following Nilsson & Pelger (1994) we can conclude that the 2.55 ratio of lens focal length to diameter exists because it optimizes visual resolution underwater, without needing to know anything more about how the ratio is produced (how many genes are involved, which ones they are, whether they are recessive or dominant, etc.). We simply assume that whatever the genetic mechanisms, the alleles that have triumphed will be the ones that create the optimal ratio by some means or other. This strategy—of forming adaptationist hypotheses directly concerning the phenotypic design whilst remaining agnostic about the genetic mechanisms—is called the phenotypic gambit.

The phenotypic gambit seems reasonable, but its validity cannot be taken for granted. All natural selection can do is alter the frequencies of individual genes, but all biological structures are built by many genes working together and many genes have multiple effects in different body systems. However, in a series of mathematically difficult but fundamental papers, biologist Alan Grafen has shown that the gambit might be quite widely justified. Whatever the mapping between genes and phenotypes, if the organism is optimally designed for its environment, then there is no further selection and no potential selection. If, on the other hand, the organism is sub-optimally designed, then there is potential for selection and gene frequencies will change according to the covariance of their effects with fitness (Grafen 2002, 2006). Optimality of design refers in this context to design for that particular organism in the particular ecology it lives in.

5.3.3 Ultimate explanation and proximate mechanism

The conclusions of the previous section allow us to draw an important distinction between ultimate and proximate explanations of any particular phenotypic characteristic. The ultimate explanation of a characteristic is the explanation of how that particular design increased ancestral fitness. The proximate explanation is an account of the genetic or developmental mechanisms that led to the formation of that characteristic in individual organisms. Thus, the ultimate explanation for why the eyes of aquatic creatures have eyes with a length–diameter ratio of 2.55 is that this is the best design for seeing underwater, whilst the proximate explanation would be an account of the different genes, proteins, and growth processes involved in making such an eye.

Ultimate and proximate explanations are rather different from each other. Understanding the ultimate explanation for something tells you little about what the proximate mechanisms will be, and evidence of a proximate mechanism does not answer the question of ultimate adaptive value. In the end, the biologist needs both, but the pursuit of ultimate explanation can often be carried out in a way that is agnostic about what the exact proximate mechanisms are.

Ultimate and proximate explanations: an example

Let us return, for example, to alarm calling by blacked-tailed Prairie dogs (section 4.4.2). The ultimate explanation for the persistence of alarm calling is kin selection; those individuals who call when kin are in the colony have higher inclusive fitness than those who do not. Ultimate reasoning predicts that the animals will call wherever $c < rb$, since this will be the decision rule about when to call that will give the highest fitness. We also saw some data showing that the animals do modulate their calling according to the kin around, as the theory predicts.

However, note that this ultimate explanation tells us nothing about what the proximate mechanisms leading Prairie dogs to call are. The ultimate explanation does not claim that they actually know the rule $c < rb$, or calculate whether it is met before they call. The ultimate explanation only says that the animals will behave 'as if' following this rule, not what will be going on in their heads. In fact, there are all kinds of proximate mechanisms that could produce the required effect. For example, if close relatives (and only close relatives) clean each other's fur, then animals might use the rule 'call if you have been cleaned by another individual recently'. When kin were present, cleaning would be more frequent and thus this would lead to more calling. Alternatively, animals might be responsive to the smell of the urine produced by the colony. A mechanism that made animals call when they smelt odours similar to their own in urine could be adaptive. Another possibility would be learning from kin. Animals could have a default of not calling, but learn to call if they observed another individual that looked sufficiently similar to them doing so.

I am not claiming that any of these is the actual mechanism involved, but merely that the same ultimate function could be realized by many different proximate mechanisms. The candidate proximate mechanisms could use any sensory modality and might or might not involve learning. To find out which proximate mechanism is in fact operative, further careful experimentation would be needed. However, ignorance of the proximate mechanism does not invalidate the ultimate explanation.

Many means to the same end
Natural selection favours whatever proximate mechanism produces the optimal phenotype with the highest reliability and the smallest cost. Thus, which proximate mechanism evolves will depend on the prior biology of the organism, the type of variation that exists for selection to work with, and historical contingency. For example, deciduous trees have evolved mechanisms to sprout leaves whenever the available solar radiation is sufficient that the photosynthetic benefit of so doing exceeds the cost. However, the proximate mechanisms they have for doing this vary. As it happens, oak trees respond to cues of changes in temperature, whilst ash trees respond to varying day length. Historically, these two cues gave the same outcome, so no selective difference was made by the use of one or another. However, as the climate has warmed in the past few decades, spring temperature has started to be higher when days are still short. The result is that, whereas oak and ash trees used to come into leaf together, oak trees are now usually out at least a month earlier. Thus, it has taken changing conditions to tease apart the different proximate mechanisms that deliver the same adaptive function.

5.4 Constraints on optimality

This section considers whether the structures or behaviours that we observe in nature will always be optimal designs. We review four main reasons why they might not: time lags, the selective regime, genetic correlations, and the shape of the adaptive landscape.

5.4.1 Time lags

What we observe in a natural population is a snapshot of evolution at one moment in time. Adaptive change might be ongoing and might not yet have reached the equilibrium state. If the environment changes, the optimal phenotype changes, and it will take many generations for selection to respond. Thus, there is always a time lag between the environment changing and the organism's design responding.

For example, African-Americans have fairly high frequencies of the sickle-cell allele (section 5.2.1). Recall that this allele is maintained by its heterozygote anti-malaria benefits. However, North America has no malaria. Thus, this allele is purely deleterious in the African-American population. However, the few hundred years that this population has been in North America is insufficient for the allele to have disappeared. The prevalence of sickle cell makes no sense at all given the current environment; to understand its distribution, we have to consider the previous environment, in Africa.

5.4.2 The selective regime

A second group of reasons for possible sub-optimal design concerns the mode of selection. Let us imagine that selection is inconsistent, favouring green individuals in times of leafy vegetation and black individuals in rare years when the ground is scorched by fire. If we observe during one of the scorched years, we may see many green individuals who do not look well designed for their environment, when in fact selection has built up their number as a response to the now-absent vegetation cover.

However, if an oscillation between leafy and scorched conditions is a recurrent feature of an organism's environment, then we should ultimately expect selection to design a mechanism that can deal with both. This would usually be done via some kind of phenotypic plasticity, or ability of the phenotype to alter depending on the context. For example, in many grasshopper species, placing the nymph or even the adult on a dark background induces a change from green to a dark or black colour (Burtt 1951). Mechanisms producing learning or other types of plasticity evolve as adaptations to maintain phenotypic optimality in the face of fluctuating environments. However, such mechanisms do not appear instantaneously and so we may observe apparent sub-optimality for a considerable time before they do so.

A specific case of inconsistent selection considered in section 5.2.5 was sexually antagonistic selection. Such a selective regime can maintain characteristics in the females that are only useful in the males and the other way around. In the very long term, we should expect selection to produce mechanisms that express the characteristic only in the sex for which it is adaptive and not the other. However, it is not always straightforward to do this. For example, as argued in sections 5.1.2 and 5.1.3, selection on height may be directional in British men and stabilizing in British women. However, the inheritance of height goes equally through both sexes, so that tall men have tall sons as well as tall daughters. This does not maximize their reproductive success. One might expect selection to have favoured mechanisms for men to switch off their alleles for extra height in their daughters and just switch them on in their sons. However, this causes further difficulties, since a major component of adult height is size at birth and small women will have birth complications giving birth to very large babies. Thus, it may be that there is no achievable alternative, given the overall pattern of human development, but for male and female size to be yoked closely together.

5.4.3 Genetic correlations

The next sources of possible sub-optimal phenotypic design come from the genetic architecture of the traits under investigation. Changes to most genes will have not just one, but many different phenotypic consequences. This is known as pleiotropy. For example, an allele of a gene that makes pigment might cause changes in fur colour and changes in eye colour too.

Because of pleiotropy, and also because of linkage between different genes on the same chromosome, heritable characteristics do not always change independently. A genetic correlation between two traits is a situation in which selecting for one trait changes the population average of the other as well. Genetic correlations are abundant and have two important effects on the design of organisms. The first is that changes which are not in themselves fitness enhancing, but happen to be genetically correlated with other changes that are, can evolve by hitch-hiking along with the useful traits. The second is that phenotypes come to reflect trade-offs between different genetically correlated traits. There now follows an example of each of these.

Hitch-hiking traits in fox domestication

A classic example of hitch-hiking traits comes from a long-term experiment on artificial selection with foxes (Trut 1999). Researchers began with captive but undomesticated foxes from a fur farm in Estonia. They carefully assessed tameness through a series of measures of behaviour towards humans in standardized situations and bred from the tamest individuals. This procedure was repeated and 30–35 generations later the descendant foxes were unafraid, responsive, and affectionate towards humans, much as domestic dogs are. They sought out human contact and when some of them escaped from the experimental farm, they came back.

The interesting thing about this experiment is that, although selection was strictly on the basis of tameness, many other features changed too. Many of the descendant foxes had floppy ears, like dogs, and shortened tails that often curled over. Their heads became shorter and broader, and many of them became piebald, with large sections of their bodies lacking coloration. Piebald colouring is also seen in domestic cats, horses, cows, sheep, and goats, but not in the related wild animals. These changes occurred because all of these traits have significant genetic correlations. The correlations exist because tameness, in wild mammals, is a characteristic of juveniles. Selecting for tameness selects for alleles of genes controlling the progression of development that retain juvenile characteristics further into adulthood. The floppy ears, curly tail, and shortened muzzle are all juvenile traits that get dragged along as tameness is extended. Piebald colouring emerges for a similar reason. Pigmented cells migrate to the coat during embryonic development. Selection to alter the timing of maturation delays this migration and some of the cells die before they reach their target, leading to the white patches.

Trade-offs in guppies

The second consequence of genetic correlations arises when two traits have opposite effects on fitness but are genetically correlated with each other. For example, male ornamentation in the guppy, *Poecilia reticulata*, is negatively genetically correlated with male survival (Brooks 2000). Male ornamentation is used by females in choosing mates. This means that selecting for more attractive males produces males that tend to live less long, whereas selecting for long lifespan produces males that are more drab. As long as such a genetic correlation exists, then selection cannot maximize both lifespan and attractiveness. Instead, the organismic design we observe will reflect some kind of trade-off between the two quantities, and possibly be optimal for neither one considered in isolation. There is a whole branch of biology, life-history theory, which deals with the expected outcome of such trade-offs. The optimal compromise between the two conflicting selection pressures will depend on the details of the local selective regime.

In the very long run, where there is a genetic correlation between a beneficial characteristic and a deleterious one, then we should often expect natural selection to operate in such a way as to remove the correlation. For example, the presence in a population of the sickle-cell allele, with its malaria-resistance benefit and its homozygous costs, is not an evolutionarily stable strategy. This is because such a population would always be invaded by any mutant form that managed only to have the malaria-resistance benefit and not the health costs when homozygous. It might, however, be a long time before a mutant arose that happened to have this effect, during which time the sickle-cell allele would persist at substantial frequency.

Selection does alter the genetic correlations between traits. However, not all trade-offs can be eliminated. You cannot become smaller without changing your surface area to volume ratio, whatever mutations arise. Thus, at least some trade-offs and design compromises are simply given by the physical and embryological constraints on organisms.

5.4.4 Shape of the adaptive landscape

One condition for selection to be able to reach an optimal design is that there must be a continuous series of intermediate phenotypes between the starting point and the optimum (section 5.3.1). A second condition is that, at every point in the series, the successor eye type must have caused higher fitness than its predecessor. If this is not the case, then the next form may not evolve. Another way of putting this is in terms of the hill that selection is causing the population to climb (Figure 5.10). Not only must the optimal adaptation be higher in fitness than the initial state, but the slope must also generally be going upwards along the way. If there is a big valley in between the current state and the optimum, selection may not be able to cross it and the design may get stuck at the local peak.

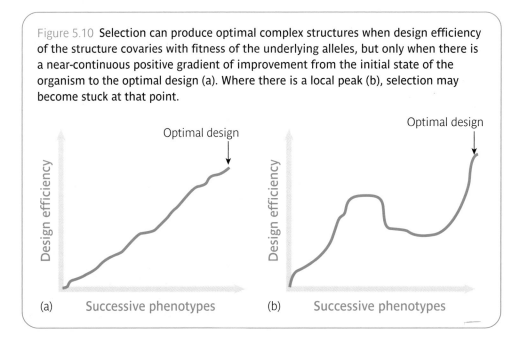

Figure 5.10 **Selection can produce optimal complex structures when design efficiency of the structure covaries with fitness of the underlying alleles, but only when there is a near-continuous positive gradient of improvement from the initial state of the organism to the optimal design (a). Where there is a local peak (b), selection may become stuck at that point.**

The vertebrate eye provides an illustration of this point. The photosensitive layer of the retina lies behind the blood vessels and neurons that service the retina. This means that the blood vessels and neurons have to disappear through a hole in the retina, causing a blind spot for which the visual system has to compensate (and which you can localize using some well-known tricks with one eye closed). A retina where the structure of photoreceptors and neurons was reversed would have no blind spot and be a better design in absolute terms—indeed, the eyes of squids and octopuses are structured this way. However, any single-gene mutation in a vertebrate eye which moved towards that design, for example by inverting the photoreceptor cells, would probably reduce fitness because the neuronal, circulatory, and brain changes that need to go with such an innovation are not in place. Thus, moving *part* of the way towards a better-designed retina will always *reduce* fitness and so the mammalian eye is stuck where it is, at a local peak in the adaptive landscape, which is not the highest possible point. Presumably, it was chance or some aspect of the prior structure of the ancestral organism that determined that this was the path taken, but once taken, it is irreversible.

5.5 How to test adaptationist hypotheses

In section 5.3, I argued that we could take an adaptationist stance towards the structures and behaviours that we find in nature, that is assume they are there because they promoted ancestral reproductive success in some way. However, in section 5.4, we saw reasons why patterns that are not in fact adaptive for the organism could be found—by-products of pleiotropy and genetic correlation, local peaks in the adaptive landscape, and time lags, to which we should also add genetic drift (section 3.2.4) and intra-genomic conflict (section 4.5.4). How do we square these two arguments? In other words, when is it justified to assume that a pattern is in fact an adaptation? This is an important question because there are many characteristics of con-temporary humans for which we might wish to seek an ultimate evolutionary explanation and we need to know if such an explanation is a valid objective to strive for.

In a well-known paper, Gould & Lewontin (1979) criticized the adaptationist stance. They pointed out all the non-adaptive reasons that characteristics could evolve and thus argued that making an assumption of adaptiveness was unjustified. They compared adaptationist hypotheses to the 'Just-So' stories written by Rudyard Kipling, in which a characteristic such as the elephant's trunk is given an ingenious rationale in some past situation, a rationale that can never be verified one way or another.

Gould and Lewontin's critique is misplaced because it misunderstands how adaptationist thinking works. It is always useful to assume adaptiveness, but not because everything actually is adaptive. Rather, assuming adaptive function is a useful way of *generating hypotheses* that will allow us to find out more about the biology of the characteristic in question. If we just generated an adaptive hypothesis and then stopped there, we would indeed be fair targets for criticism. However, that is not what adaptationists do. Their adaptive hypothesis is the beginning of their investigation, not the end. The bulk of their effort is (or should be) devoted to testing predictions from the hypothesis. If these predictions are met, this is good indication that the trait has been shaped by selection in the hypothesized way. If the predictions are not met, then the hypothesis is wrong and competing hypotheses need to be investigated.

Assuming adaptiveness as a starting point and then testing it works well because, fortunately, adaptationist hypotheses are generally easy to test. By contrast, non-adaptive hypotheses, such as genetic drift, by-products, local peaks in the adaptive landscape, and so on, are extremely difficult to test. If anything, it is these that tend to be 'Just-So' stories, apart from rare cases where the underlying proximate mechanisms are exceptionally well understood. Paradoxically, it is usually by testing an adaptive hypothesis and finding its predictions not to be met that we can detect that some other process is at work.

How, then, do we go about testing an adaptive hypothesis? There are three main ways, reviewed in each of the next three subsections. The example we will consider concerns the melanin pigmented layer of the human skin, and more particularly the hypothesis that the adaptive function of this layer is to regulate the amount of ultraviolet light entering the body, so as to optimize the conflicting demands of tissue protection and vitamin D synthesis (Jablonski & Chaplin 2000).

5.5.1 Reverse engineering and optimality models

If an engineer was given a piece of unfamiliar machinery, his first response would be simply to examine carefully how it worked and try to come up with ideas about what function would have led it to be designed in this way. This process is called reverse engineering. The melanin layer of the skin is on the outside of the body and it absorbs ultraviolet light so that less reaches the tissues underneath than is falling on its surface. We know from basic physiology that ultraviolet light is needed for the synthesis of vitamin D, an important substance whose deficiency is associated with rickets and osteoporosis. We also know that excessive ultraviolet light destroys sweat glands, burns skin, causes cancerous mutations, and destroys folate, which is a crucial substance in cellular functioning. Lack of folate can cause problems including abnormalities of developing embryos. The idea that skin might serve to regulate the amount of ultraviolet absorbed thus seems a good candidate for an adaptive function. However, there are other possibilities. It could, for example, be for camouflage against predators. Thus, reverse engineering might identify several hypotheses rather than just one.

Reverse engineering is often accompanied by a kind of virtual forward engineering called **optimality modelling**. Here, the investigator takes one of the candidate hypotheses identified by reverse engineering and derives an expectation of the following form: *if* the characteristic had been honed by natural selection for the function under test, what design features would we expect it to have? These expectations are often made using mathematical models, but we can use a kind of informal optimality thinking for present purposes.

Both sexes need vitamin D, but women tend to need more than men because of the demands of lactation and pregnancy. Thus, for women, the optimal amount of ultraviolet entering the body will be higher than the optimum for men. Also, the amount of ultraviolet hitting the skin will depend on the local weather conditions. These simple facts, using optimality assumptions, lead us to the following predictions. *If* skin coloration has evolved to regulate ultraviolet absorption, then (a) women should have lighter skins than men and (b) there should be some mechanism for making skin darker when the amount of sunlight hitting it increases and lighter when it decreases.

For (a), Jablonski & Chaplin (2000) show that, in every one of 86 human populations that they studied, women have paler skins than men. As for (b), there is such a mechanism, namely tanning, that precisely fulfils this function. Note that a different adaptive hypothesis, such as camouflage, makes testably different predictions. Women are no less vulnerable to predation than men (conceivably more so) and therefore, *if* skin pigmentation had been selected for camouflage, women should be at least as dark as men. Moreover, predators do not care whether it is sunny or not. People would need the same amount of camouflage whatever the weather. Thus, the tanning mechanism makes no sense if we assume skin pigmentation is optimized for camouflage.

5.5.2 Experimental manipulation and experiments of nature

A second methodology for testing adaptive hypotheses uses comparisons between individuals with more and less of the characteristic. Sometimes this variation is experimentally produced. In a famous experiment, Tinbergen *et al.* (1962) tested the hypothesis that black-headed gulls remove empty eggshells from the vicinity of their nests because shells attract predators. This was not the only possible hypothesis; used shells might harbour disease, for example. To test their hypothesis, the researchers made artificial nests with or without eggshells in the vicinity and simply measured the rate of visitation by predators. They showed that eggshells do indeed attract predators, a crucial prediction of their adaptive hypothesis.

With human skin, it is impossible to produce pigmentless individuals artificially. However, here we can rely on natural experiments furnished by existing variation within the human population. Our adaptive hypothesis predicts that people with unusually pale skin will have impaired survival and reproduction due to problems caused by excessive ultraviolet in their body tissues, whereas people with unusually dark skin will have problems caused by insufficient vitamin D.

One natural experiment is albinism. Albinos have a very pale skin and much lighter eyes and hair than is normal for their population. In Africa, one in every few thousand people is affected, a condition due to mutations in a number of genes controlling pigmentation. In one study of 1,000 Nigerian albinos over the age of 20 years, not a single individual was free of malignant or potentially malignant skin cancer (Okoro 1975). Many of these albinos die very young. Albinism is an extreme departure from normal pigmentation, but the effects show up in smaller differences too. Within the British 'white' population, for example, there is variation in skin colour, and those with relatively fair skin, blue or green eyes, and red or blond hair have at least a 60% increased risk of skin cancer (Lear *et al.* 1997).

Another damaging effect of excessive ultraviolet penetration is destruction of folate. Studies of neural tube defects (a type of developmental abnormality of the embryo that is linked to folate deficiency) have shown that in both the southern USA and South Africa, the defects are more common in the white population than in people of African descent (Grace 1981; Stevenson *et al.* 2000; Figure 5.11).

It also turns out that African-American women have dramatically higher levels of vitamin D insufficiency than white Americans (Nesby-O'Dell *et al.* 2002; Bodnar *et al.* 2007; Figure 5.11), leading to increased rates of serious diseases such as rickets in their children. Thus, the pattern of health problems caused by relatively lighter or darker skin colour is exactly as our adaptive hypothesis predicts.

5.5.3 Comparative evidence

The third class of evidence used to test adaptive hypotheses is known in biology as the comparative method. If eggshell removal evolved because of predation, then it should only be found in populations that have evolved in an environment containing predators. Satisfyingly, this

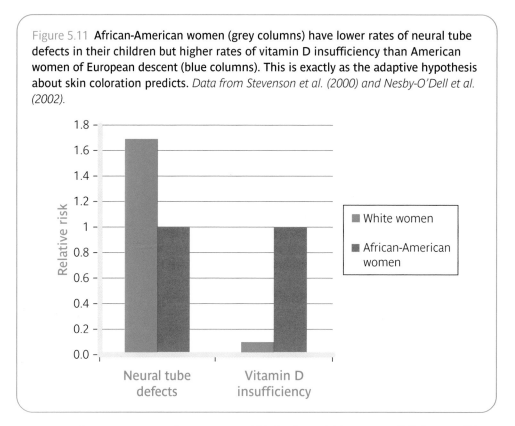

Figure 5.11 **African-American women (grey columns) have lower rates of neural tube defects in their children but higher rates of vitamin D insufficiency than American women of European descent (blue columns). This is exactly as the adaptive hypothesis about skin coloration predicts.** *Data from Stevenson et al. (2000) and Nesby-O'Dell et al. (2002).*

turns out to be true. For example, gannets and kittiwakes, which nest on cliffs inaccessible to predators, do not remove their eggshells.

The comparative method generally involves testing across different species that have experienced the selection pressure to differing extents. This is not really applicable to the human skin colour example, since we do not have furless apes of other species available for comparison. However, given that humans live all over the world in very different climates, we can test across different populations within the same species. The adaptive hypothesis predicts that the skin colour of the population should be related to the average amount of sunshine that population is exposed to. Jablonski & Chaplin (2000) show that this prediction is strikingly confirmed (Figure 5.12). Ultraviolet radiation predicts the distribution of human skin coloration. This would not be the case if it had been shaped by predation or was a non-adaptive by-product of something else.

Note that the comparative pattern supports the hypothesis, but in itself says nothing about the proximate mechanism involved. Without further investigation, we could not know whether the inter-population differences were due to genetic differences or to tanning or some other kind of early-life plasticity. As it happens, genetic differences are implicated, as shown by the fact that parents of African descent have deeply pigmented babies whether they have them in Abuja or Oslo. The fact that genetic changes are involved means that considerable time will be required for local adaptation to occur. Jablonski and Chaplin observe that populations tend to have skin coloration closest to the predicted value for their habitat when they have been living in the current location for more than 10,000 years.

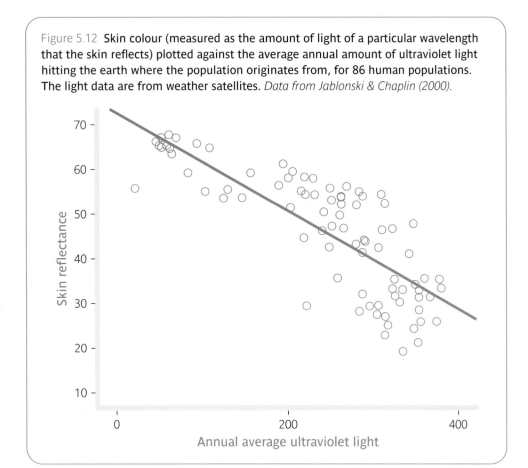

Figure 5.12 **Skin colour (measured as the amount of light of a particular wavelength that the skin reflects) plotted against the average annual amount of ultraviolet light hitting the earth where the population originates from, for 86 human populations. The light data are from weather satellites.** *Data from Jablonski & Chaplin (2000).*

Thus, the adaptive hypothesis that skin pigmentation optimizes the amount of ultraviolet light absorbed is far from a 'Just-So' story. It is subject to critical tests on the basis of global distribution, tanning, sex differences, health problems of albinos and blonds, respective health problems of African-Americans and white Americans, and so on. A vast amount of seemingly disparate data is brought together satisfyingly as a result of thinking about what skin coloration is *for*.

5.6 Getting natural selection clear

We have reviewed all the fundamental mechanisms of genetics and evolution. The rest of the book is concerned with the application of the toolkit provided so far to a set of specific topics concerning the behaviour of humans and other animals. Thus, this is a good point at which to make sure we are clear about how natural selection does and does not work.

In Chapter 1, I argued that many of the people who do not accept the theory of evolution misunderstand it. Unfortunately, the same is true of those who *do* accept it. Misunderstandings are very common, even in published literature. Shtulman (2006) has recently shown, using carefully constructed examples to probe people's conceptions of how evolutionary change occurs, that many educated students who say they believe in evolution actually believe—quite systematically—in a mechanism that is utterly different from the real one.

This section outlines a few of the more common misconceptions, some but not all revealed by Shtulman's experiments. The material in Chapters 1 to 5 should be sufficient for you to see why they are wrong, and the relevant sections are highlighted so you can take this chance to review any principles you are not sure of.

5.6.1 The good of the species

Many people believe that characteristics evolve because they are 'good for the species'. I recently reviewed a book by a very eminent psychologist who argued that the liability to manic-depressive illness, which is heritable, persisted because it was good for the human species to contain a few such individuals, who can be sources of new ideas and so forth.

From section 4.2, you should be clear why arguments on the basis of the good of the species are logically wrong. Good of the species arguments are extreme forms of group selection arguments, since a species is just a very large group of similar individuals. Competition is between alleles at the same locus and consequently an individual that did something that was better for the species than it was for its own reproductive success would be benefiting its competitors at its own expense and thus would become extinct.

People also tend to believe that what changes during the course of evolution is a kind of 'essence', which is shared by all members of the species. To illustrate using the yellow versus black cat colour example (section 4.1), people tend to believe that if yellow cats are doing better in the current environment, then all cats will start to have yellower offspring. This is not true. Only yellow cats have yellow offspring. It is just that, because of differential survival, there are more and more of them around, until all cats are yellow. The non-yellow cats go on having non-yellow offspring until the end. You can characterize this process exactly without making any reference to the species or its essence.

What is a species anyway?

In fact, the species as a unit plays a remarkably small role in natural selection. Sometimes it is possible to clearly and unambiguously draw species boundaries around particular populations. In other cases, such as gulls and baboons, species boundaries are fuzzy. None of this has any importance for the evolutionary changes going on within these populations. All life forms belong to one vast family, which has no abrupt discontinuities. Where clear species boundaries are found in living organisms, this is simply because the intermediate forms have all died out. Species boundaries are thus the *outcome* of evolutionary processes such as extinction, adaptation, and drift, and something that is the outcome of evolution cannot, of course, be necessary for evolution to occur.

In view of all this, it is perhaps unfortunate that Darwin chose to call his book *On the Origin of Species*. It might have been better if he had called it *On the Origin of Individuals that are Pretty Well Designed for their Environments and Look after their Own Interests*. Arguments about the interests, needs, or essences of species should always be rephrased in terms of individuals or, better still, alleles.

5.6.2 Directed mutation and directed heredity

Another common misunderstanding is to assume that the selective environment affects the probability of mutations with certain phenotypic effects. For example, when the population of black cats moves into a sandy environment where yellow colouring would be adaptive, people seem to believe that this makes it more likely that a mutation causing coats to be yellow would occur. This is a fallacy (review section 2.3). Mutation is governed only by biochemical processes within the cell. It is completely random with respect to the phenotypic effects that it produces. This is why most mutations are deleterious. A mutation producing a yellow coat could thus be no more likely where it might prove useful than where it might prove harmful. All selection can do is change the frequencies of alleles that do happen to arise; it cannot create new alleles.

A related misunderstanding is that the selective demands of the environment can somehow affect heritability. For example, people might believe that if a black cat and a yellow cat mate, they will produce more kittens with a yellow-producing genotype when yellow is adaptive in the environment and more black kittens with a black-producing genotype if black is adaptive. However, in general terms, this will not occur. The genotype frequencies of the kittens conceived follow from Mendel's laws (section 3.2) and nothing else.

5.6.3 Inheritance of acquired characteristics

Another misapprehension is the idea that phenotypic characteristics acquired during the lifetime of parents are transmissible to offspring. For example, people believe that if a herbivore spends its life craning its neck upwards to feed, its neck may become more flexible (which is true) and then its offspring will have more flexible necks than they otherwise would have done (which is false).

Such a belief is a fallacy because it contravenes the central dogma of genetics (section 2.2.1), which states that information flows from genotype to phenotype but not the other way. In the developing embryo, two things develop: a somatic line of cells, which go on to make necks and so forth, and a germ line, which is kept apart in sperm or ovaries. There is no way for what happens in a somatic cell to systematically alter the DNA sequence in that cell, still less the DNA sequence carried in germ line cells, and it is germ line cells which will determine the genotype of offspring. Thus, there is no mechanism for a characteristic acquired during life to be transmitted genotypically.

In this section and the previous one, I have argued that there is no way for the phenotype of one generation to affect the genotype of the next and this is true. However, there are a number of ways that the phenotype of one generation can affect the *phenotype* of the next, for example by altering the chemical environment *in utero*, or the nutrition they provide after birth, or behaving a certain way towards their young. Parents can also alter their offspring genotype frequencies by selective infanticide or spontaneous abortion of certain genotypes. None of these, however, induce changes in DNA sequence as a result of parents' phenotypic histories.

✔ Summary

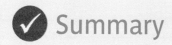

1. Natural selection can be purifying, stabilizing, or directional.

2. Whilst mutation increases genetic variation within populations, selection tends to decrease it. There are a number of mechanisms leading to the persistence of heritable variation even when selection is operative.

3. Directional selection can build phenotypic adaptations as long as there is a sequence of intermediate phenotypes between the start point and the adaptive end point, and each intermediate tends to have higher fitness than the last.

4. Natural selection will favour any proximate mechanism that is available and reliably produces the optimal phenotype. Thus, ultimate hypotheses can be agnostic about the proximate mechanisms producing the effect and proximate mechanisms cannot be deduced from ultimate explanations.

5. Phenotypes are not always optimally designed because of time lags, trade-offs, genetic correlations, and local peaks in the adaptive landscape.

6. Generating adaptive hypotheses is a useful device for finding out more about the characteristics under study. Adaptive hypotheses can be tested by making predictions about design features, by investigating the fitness of natural or experimentally produced mutants, and by using comparative evidence.

? Questions to consider

1. Two genes A and B affect body size in a haploid organism. The two loci act additively. Each has two alleles (a_1 and a_2, b_1 and b_2), such that the x_1 allele increases body size by one unit and the x_2 reduces it by one unit. The current population mean for size is the optimum, with reduced fitness as one moves away from the optimum in either direction. Show that if the frequency of the x_1 alleles is 50%, then both alleles of each gene have equal fitness. However, the x_1 allele of one of the genes and the x_2 of the other are likely to go to fixation in the end. Why is this, given that they are of equal fitness? Note that this is related to the reason that stabilizing selection reduces genetic diversity in the long term.

2. Kibbutzim are Israeli collective farms where all the children are brought up communally. Shepher (1971) showed that children brought up on the same kibbutz do not tend to be attracted to each other as adults, even if they are not biologically related. What might be the ultimate and proximate explanations of this finding? Is this behaviour the result of an adaptation and is it adaptive?

3. When a predator begins to chase them, gazelles may leap high into the air. This behaviour is known as 'stotting' and would seem costly, since energy and time spent stotting are not spent running away. Construct different adaptive hypotheses for stotting based on direct benefits and on kin selection. How would you test them?

Taking it further

Barton & Keightley (2002) provide a review of how genetic variation can be maintained in populations, for the reasons discussed here. Keller & Miller (2006) argue that the alleles underlying common mental disorders in humans (which are heritable) are maintained, despite selection against them, by force of mutation, whilst Nettle (2006) and Penke *et al*. (2007) argue that heritable personality differences in humans persist because selection on them is inconsistent.

Life history theory is clearly reviewed by Stearns (1992). Classic expositions of the power of selection to build complex adaptations are given by Dawkins (1986, 1996). For a review of the optimality approach, see Parker & Maynard-Smith (1990) and, for a user-friendly guide to actually doing the modelling, Kokko (2007). Many of the ingenious ways adaptive hypotheses are tested in non-human species are covered in Krebs & Davies (1993).

Sex

A fundamental feature of human life is that there are two sexes, with one member of each sex needed in order to reproduce. Pairing up is a complex business responsible for many of the scrapes that we get into and the emotional highs and lows that we experience. Movie-makers and writers would be short of material if there were no sex. The problem of sex is made even more difficult by the fact that men and women seem to differ in certain ways, including, often, in what they want. Why are humans like this?

This chapter is about sex and, in particular, why the sexual system which humans have has evolved. Why have any sex at all? Why have two *different* sexes, rather than anyone being able to have sex with anyone else? And why should the two sexes be different from each other in aspects of their phenotype other than just their gametes? The sexual system we humans have seems so normal, so inevitable, that these questions seem almost surreal, but in fact nature shows that a great variety of different ways of organizing reproduction are possible. For this reason, the chapter will focus mainly on examples from other organisms, until the final section, where we briefly apply some of the generalizations we have made to humans. However, you can probably identify other human parallels as we go along, and the Taking It Further section points you to avenues for exploring the human literature.

Figure 6.1 **Just some of the diversity of sexual systems in nature. Clockwise from top left: bdelloid rotifers have no sex at all; starfish can reproduce sexually or asexually; hamlet fish have both male and female genitalia; clownfish change sex from male to female during adulthood; pipefish males carry the eggs in a pouch or skin patch.** *Top left: © John Walsh/Science Photolibrary. All other photos: © Corel.*

6.1 The diversity of reproduction in nature

Nature contains an enormous diversity of systems of reproduction (Figure 6.1), the chief of which are described below.

6.1.1 Asexual reproduction

First, there are creatures who do without sex altogether. That is, reproduction simply involves the parent individual producing an offspring individual that is genetically identical to itself. This is asexual reproduction. It is common in many single-celled organisms, fungi, and plants, and there are even a few animals devoted to it. Bdelloid rotifers are small, plankton-like animals that live in water and in damp ground. There are several hundred closely related species, but all of them seem to reproduce exclusively asexually. Related groups of animals all have sexual reproduction, and so it looks like the bdelloids lost sexual function at some point in their evolutionary history and have survived for millions of years without it.

6.1.2 Obligate and facultative sex

Individuals of many species are capable of both sexual and asexual reproduction. Starfish, for example, can either release embryos, which do not need fertilization, into the water, or release gametes, which need fertilization by another starfish. Thus, they have both options available to them and their sex is said to be facultative. In facultatively sexual organisms, sexual behaviour is often most common when the population is crowded densely together or resources are scarce. These may be the times when sex is either least costly or most beneficial.

6.1.3 Isogamous sex and the number of sexes

In our system, one sex (the female) provides a large gamete (the egg) and the early parental care, whilst the other (the male) provides a small gamete (the sperm). This is not the only possibility. For example, in seahorses and pipefish, it is the male, not the female, who carries the fertilized eggs. (We still call him the male since he provides the smaller of the two gametes. This is the biological definition of maleness.) More radically, many algae and fungi display isogamy. This means that the two gametes are the same size and so there is no male or female. There is still a need to prevent gametes fusing with other gametes from the same parent, and so gametes in these organisms carry a marker and will only fuse with a gamete whose marker is different from their own. These markers define a mating type. The number of mating types is often two, but can also be much higher, and a gamete can fuse with a gamete of any mating type other than its own. In contrast to these systems, our system, featuring two strict sexes with gametes of different sizes, is called anisogamy.

6.1.4 Simultaneous hermaphroditism

Even if you have an anisogamous sexual system, there is no requirement for an individual to limit itself to just being one sex at a time. Most flowering plants are hermaphrodites, which means that they have both male and female parts. Amongst the animals, adult hamlet fish have both male and female genitalia. When they mate, they take turns in playing the male and female roles, so that each member of the pair provides both some eggs and some sperm. In the nematode worm, *Caenorhabditis elegans*, there are some individuals who are males and some who are hermaphrodites, but none who are full-time females.

6.1.5 Sex determination

In humans, which sex you are is determined genetically at conception, with males having one X and one Y chromosome, and females having two copies of the X chromosome. Even if you are going to have two anisogamous sexes and individuals can only be male or female at a given time, this is by no means the only way of arranging things. Amongst cockroaches, females are XX and males just have an X on its own. Amongst birds, it is the males who have two sex chromosomes the same (ZZ) and the females who have two different (ZW).

It is not even necessary to determine sex chromosomally. In many reptiles, the sex that an egg develops into depends on the temperature of incubation (in some species cooler temperatures producing females and in some species the opposite). Elsewhere, it depends on early experience. In the green spoon worm, *Bonellia viridis*, if the tiny larvae encounter a female in the first few days of their lives, they enter inside her and become male, thence producing sperm to fertilize her eggs for the rest of their lives. If they do not encounter a female in this early period, they grow and become female themselves. Still other organisms change sex as their lives progress. In the clown anemone fish, *Amphiprion ocellaris*, each sea anemone is home to a small group of individuals. The largest is the breeding female, the second largest the breeding male, and the rest juvenile males. When the female dies, the breeding male becomes female and the largest juvenile becomes a breeding male. In wrasse (a type of fish), the progression through the sexes is the other way around, with individuals female when they are small and male later.

6.1.6 How to think about the diversity of sex

It should be clear from this section that there are a great many viable ways of organizing reproduction. However, some kind of sex is very widespread. It looks like the ancestral condition for all eukaryotes is to have sex, with forms of asexual reproduction arising many times over evolutionary history, but rarely lasting very long or becoming very widespread. Sex roles and sex-determining mechanisms have also undergone numerous evolutionary changes through time. Thus, there is nothing inevitable about the reproductive system we happen to have.

In this chapter, we will be following the advice of Chapter 5 and taking the adaptationist stance on sex. That is, we will be asking why the alleles leading to sexual reproduction have out-competed alternative alleles for asexual reproduction and also why alleles for the particular form of sex one sees in mammals might have done better than their competitors. Where different organisms have come up with different reproductive systems, it must be because something about the ecology and history of those species has favoured the alleles making that system rather than another one.

6.2 Why have any sex at all?

The most fundamental question regarding sex is why there should be any at all, given that, as we have seen, asexual reproduction is possible. Section 6.2.1 will argue that there is always a cost to sex and that in anisogamous organisms that cost is quite large. Sex can only persist if it is providing some benefit that more than offsets the cost. Sections 6.2.2 and 6.2.3 review the two types of benefits that are best supported by evidence.

6.2.1 The cost of sex

Let us first consider the case where sex is isogamous (both parties provide gametes of the same size) and fertilization is external, so neither parent has to gestate the offspring in their body or carry them around. I have chosen this scenario because it probably represents the ancestral asexual condition from which sex arose.

Imagine that to produce a viable embryo requires 10 units of energy and an individual has 20 units of energy to spare. This individual has two choices. Strategy A would be to produce two offspring asexually. Strategy B would be to produce four gametes each of which contains 5 units of energy, in the hope that these will unite with four gametes from another individual, to produce four offspring in each of whom each parent has a half share.

Which is going to be better, strategy A or strategy B? Strategy A leads to a fairly certain reproductive success of two offspring. Strategy B could in principle also lead to a reproductive success of two offspring, since a 50% share in four offspring is equivalent to a 100% share in two offspring. However, strategy B is much more risky than strategy A. There might not be another individual around to mate with, the gametes released might fail to meet the other gametes, there might be genetic incompatibilities or malfunctions when the gametes fuse, and so on. All these factors will mean that a smaller fraction of energy invested in gametes will lead to offspring than would be the case for asexual reproduction. So all in all, since strategy B can at most be as good as strategy A, and is probably less good in practice, there would never be any reason to follow strategy B unless the offspring produced that way were in some way better than those produced asexually.

The situation is even worse when we come to consider the case of anisogamous sex. In anisogamous sex, the female pays essentially the full energy cost of setting up the offspring because the egg is large. Sperm, by contrast, are just DNA with a tail and so the male may pay almost nothing. Let us return to our example, where it costs 10 units of energy to make a viable embryo and an organism has 20 units of energy to spare. Strategy A would be to produce two offspring asexually. Strategy B would be to adopt the female role and reproduce sexually. As previously mentioned, an egg embodies all the energy required to set up an embryo and so costs 10 units just as an asexually produced embryo does. So strategy B is for the individual to produce two eggs, which then get fertilized by someone else's sperm. Thus, the choice is between a 100% share in two offspring, or a 50% share in two offspring. This means strategy A is giving twice the return of strategy B, even if all gametes are fertilized successfully.

This problem was described by the great evolutionary biologist John Maynard Smith (Figure 6.2) as the twofold cost of males (Maynard Smith 1978). In a sexually reproducing anisogamous

Figure 6.2 **John Maynard Smith (1920 – 2004), one of the greatest evolutionary theorists, applied himself to the problem of the evolution of sex.** © *Corbin O'Grady Studio/Science Photolibrary.*

population, any female carrying a mutant allele which made her give up males and start devoting her energies entirely to asexual reproduction would seem to double her reproductive success. Yet sex has persisted in the eukaryotes for hundreds of millions of years and has not, bdelloid rotifers aside, been outcompeted by asexual reproduction.

Sex must have evolved before anisogamy, so it did not immediately face the twofold cost of males. It must, however, have had some significant advantage in terms of offspring viability or success in order to have got going in the first place. However, once anisogamy had evolved, that advantage had to be very strong in order for sex to be maintained, for now the cost of sex to females had become much higher.

There are currently two main theories for why sex persists. The first focuses on genetic mutations and the second on the selection pressures brought about by parasites. Both of the hypotheses are similar in that the advantage of sex lies in creating genetic variation amongst one's offspring.

6.2.2 Mutation and the efficacy of selection

When an asexual organism reproduces, the offspring are identical to the parent, apart from the occasional new mutation. As we saw in Chapter 2, most mutations that have any phenotypic effect are deleterious. Other things being equal, then, the biological performance of organisms deteriorates as mutations accumulate (see section 2.3.3). What stops this happening is natural selection. Individuals carrying more deleterious mutations are less likely to reproduce and so the deleterious mutations are lost from the population.

First consider an asexual lineage in which a mutation occurs. The parent will pass that mutation on to all its offspring. If that mutation is deleterious, then *all* of the offspring will be at a selective disadvantage, as will all of the grand-offspring and all of the grand-grand-offspring. There is basically no way of getting rid of the mutation except by waiting for another mutation that exactly cancels the first, which is extremely unlikely to occur. Thus, if there is selection against the mutation, the likely outcome is that the whole lineage will lose out in competition with other lineages not carrying the mutation and go extinct. The lineage itself will not be able to evolve.

Now consider by contrast a deleterious mutation occurring in a sexually reproducing individual. The individual mates and, following the principles of inheritance, around 50% of the offspring will be carrying the mutation and 50% will not. Thus, even if the 50% carrying the mutation do extremely badly, the lineage has some descendants free of the mutation, who may go on to do well.

What sexual reproduction is doing, then, is creating genetic variance between lineage members, and this increases the efficacy of natural selection to remove the least fit alleles from the lineage, and also to fix the best ones, and to generate novel combinations of alleles. There is experimental evidence that sexual reproduction does increase the efficacy of selection. Goddard *et al.* (2005) studied two types of yeast. The first, the naturally occurring type, has facultative isogamous sexual reproduction. The second strain was genetically engineered to be incapable of sexual function and was thus restricted to asexual reproduction. The researchers allowed each strain to evolve for a number of generations in a novel environment (a Petri dish with limited glucose). They then compared the growth rate of the two experimental strains (the sexual and the asexual one) with that of the ancestral strain, which had never been exposed to limited glucose, in the glucose-limited environment.

The sexual strain got better at growing in the glucose-limited environment relative to its ancestral strain, at a faster rate than the asexual strain did (Figure 6.3). In other words, in the

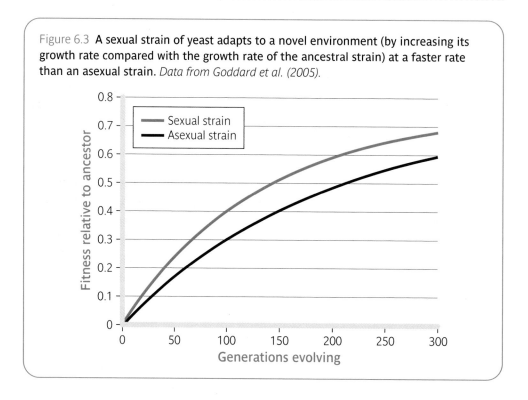

Figure 6.3 **A sexual strain of yeast adapts to a novel environment (by increasing its growth rate compared with the growth rate of the ancestral strain) at a faster rate than an asexual strain.** *Data from Goddard et al. (2005).*

sexual strain, selection was efficient at production adaptation to the glucose-limited environment, whereas in the asexual strain less evolutionary change occurred. This is because the variation produced by sex gives selection something to work with, eliminating the alleles that are deleterious under these conditions and fixing the beneficial ones.

The message of Chapters 4 and 5 was that arguments about selective advantage must always be expressed at the level of the allele ('Why would an allele causing sexual reproduction do better than its competitors?'). So how do we phrase the selective advantage of sex increasing the efficiency of selection in terms of alleles? An allele in a sexual versus an asexual individual would have a disadvantage in competition (because of the costs of sex, see section 6.2.1). However, on the other hand, it would in the long term be more likely to occur in some well-adapted individuals (and less likely to occur in only individuals so bad that they died out) exactly because sex is always shuffling the pack of which alleles co-occur with which others. This benefit could outweigh the cost of sex under certain assumptions about the frequency of mutations and the strength of selection.

6.2.3 The 'Red Queen' hypothesis

The second hypothesis centres on the idea that individuals benefit directly from being different from their parents. The most commonly discussed reason why this might be the case is because of parasites. Parasites—things like infectious diseases—are constantly evolving to be maximally efficient at infecting hosts of the most common type that they encounter. Thus, being of a different biochemical makeup from other individuals of one's population might make one less susceptible to infection. This provides a selective advantage to any mechanism that makes offspring different from their parents and sex does this. In fact, the selective advantage of sex is not just in the hosts, but also in the parasites; host immune systems get good at detecting the biochemical makeup of parasites, so sex will be favoured in the parasite too as a mechanism for staying ahead of the host's immune system. This hypothesis is called the 'Red Queen' hypothesis after the character in Lewis Carroll's *Through the Looking Glass* who has to run on the spot all the time just to stay still. Hosts and parasites need to change their biochemical makeup all the time just to maintain their existence level of immunity or ability to infect.

Evidence for the relationship of sex to parasites

Some of the best evidence for the Red Queen hypothesis comes from the study of a New Zealand snail, *Potamopyrgus antipodarum*. This creature exists in both sexual and asexual forms, often within the same habitat. Lively (1987) showed that the proportion of the sexual form in the population was correlated with the rate of infection by two parasites; the more parasites there were, the more common the sexual form was (Figure 6.4). In addition, Lively & Dybdahl (2000) have shown that the parasites are best at infecting whichever genotype of host is most common in their local area. This is a critical pre-condition for parasites to lead to the evolution of sexual reproduction.

Figure 6.4 **(a) In an experiment, snails (*Potamopyrgus antipodarum*) from either Lake Poerus or Lake Ianthe were exposed to parasites either from their own lake or from the other one. Parasites are best at infecting hosts they have been co-evolving with. Error bars represent 95% confidence intervals.** *From Lively & Dybdahl (2000).* **(b) Across different populations, the frequency of the sexual form of *Potamopyrgus antipodarum* is correlated with the frequency of parasites.** *Data from Lively (1987).*

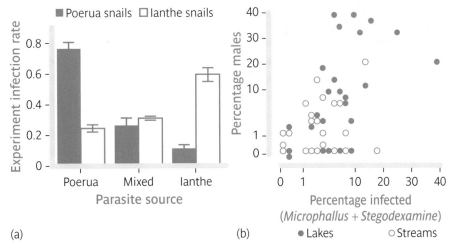

(a)   (b)

Major histocompatibility complex diversity and immunity

Another line of evidence that bears indirectly on the Red Queen hypothesis concerns a group of genes called the major histocompatibility complex (MHC). This group of genes is found in all vertebrates and is involved in how the immune system recognizes parasites to attack. There are often many alleles (e.g. in humans there are around 140 MHC genes and some of them have up to 500 alleles). A possible reason for all this diversity is that parasites become efficient at evading the immune systems made by the most common alleles and so rare types always have an advantage (this is an example of negative frequency-dependent selection, see section 5.2.2).

The relevance of this to sex is that when animals mate, they tend to prefer mates who are genetically dissimilar to them in terms of alleles at the MHC. Similarity is detected through odour. Preference for dissimilarity has been well documented in mice (Roberts & Gosling 2003). In humans, Wedekind & Furi (1997) showed that people prefer the scent of T-shirts that have been worn by people whose MHC alleles are unlike their own. By choosing a dissimilar mate, individuals are making their offspring more different from them at the MHC and thus giving them an advantage in parasite resistance.

The best evidence for such an advantage comes from Atlantic salmon, *Salmo salar*. These salmon are parasitized by small worms. Consuegra & Garcia de Leaniz (2008) compared the MHC genotypes and parasite loads of salmon in the wild from two groups. The first group's parents had mated in the wild. The second group's parents had been in hatcheries where they had no choice of mate and the young had then been released into the wild. The researchers found that the two copies of the MHC genes in the wild-mated salmon were less similar to each other than would be expected by chance. This was not true of the hatchery-mated group, which suggests that parental salmon were actively choosing mates unlike them at the MHC when they had an opportunity to do so. Moreover, the wild-mated salmon had lower loads of parasites than the hatchery-mated group (Figure 6.5), which turned out to be due to their greater MHC diversity.

This study does not compare sexual with asexual reproduction. However, it is relevant to the benefits of sex, in that it shows that combining genetic material with an individual different from oneself provides an advantage against parasites. An asexual salmon would be in an even worse position than the hatchery-mated group in the study since they would be genetically identical to their parent.

6.2.4 The pluralist approach

From the above discussion, you might be wondering what the difference is between the efficacy of selection hypothesis and the Red Queen hypothesis. They both seem to say that it is better to be different from your parents than be identical to them and so sexual reproduction evolved. However, there is a subtle difference. The efficacy of selection hypothesis does not require that sexual offspring have higher reproductive success than asexual offspring, only that the *variance* in their reproductive success must be greater. In the long term, this can lead to sexual reproduction winning out. The Red Queen hypothesis, by contrast, requires that the mean reproductive success be higher for sexual versus asexual offspring.

It is difficult to find evidence that conclusively dismisses the importance of either mechanism. For example, sex becomes more common the longer the generation time of the organism. It is ubiquitous in relatively long-lived organisms such as mammals, whereas many fast-reproducing single-celled organisms are asexual. Long generations increase the rate of mutation and so could make the role of sex in removing deleterious mutations more critical. On the other hand, long generations also means plenty of time for parasites to adapt to one's biochemical environment

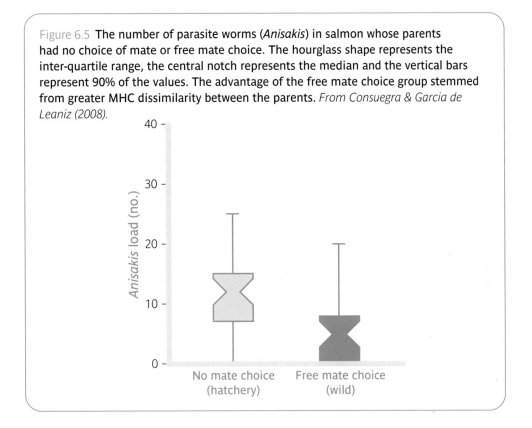

Figure 6.5 **The number of parasite worms (*Anisakis*) in salmon whose parents had no choice of mate or free mate choice. The hourglass shape represents the inter-quartile range, the central notch represents the median and the vertical bars represent 90% of the values. The advantage of the free mate choice group stemmed from greater MHC dissimilarity between the parents.** *From Consuegra & Garcia de Leaniz (2008).*

and thus makes the Red Queen advantage of sex all the greater. Thus, the relationship of sex to generation time is not decisive.

West *et al.* (1999) argue for a 'pluralist' approach, which stresses that both types of mechanism can operate. The efficacy of the selection hypothesis suggests that asexuals will experience greater likelihood of extinction in the very long term (as they fail to evolve), whereas the Red Queen gives them a handicap in the short term (as their offspring have higher rates of mortality due to infection). Thus, the reason that there is not more asexuality in the eukaryotes might be that when it occurs, its spread is slowed by greater vulnerability to parasites, and when it does spread, for example in environments where parasitism is rare, it eventually goes extinct because of reduced efficacy of selection.

This seems a plausible position. However, there are many questions unanswered about the origin of sex, including why sexual organisms tend to do it so often. Most models show that the advantage of sexual reproduction (particularly the mutational advantage, section 6.2.3) could be had by generations of asexual reproduction interspersed with occasional sexual events. There are facultatively sexual species, but in many others—all the mammals, for example—sexual reproduction is obligate in every generation. Why this should be the case is still a topic for investigation.

There are many fascinating further questions—beyond our scope here—concerning how to allocate energy to sexual functions, for example whether to be hermaphrodite or to specialize in one sex, whether to change sex during one's lifetime, and what is the best system (genetic versus environmental) for determining to which sex an individual belongs, all of which can be fruitfully approached from the adaptationist standpoint (see Questions to consider).

6.3 The evolution of anisogamy

The evolution of anisogamy is clearly a key development in the history of sex, since only once gametes are different sizes can there be said to be males and females. Both isogamous and anisogamous sexual reproduction are widespread and the exact conditions for one to be favoured over the other are still a matter of some debate (Randerson & Hurst 2001; Bulmer & Parker 2002). This is because anisogamous reproduction seems so obviously unfair. In isogamous reproduction, both parties get a half genetic share in the offspring and both pay equally for this. In anisogamous reproduction, the female gets a half share and pays almost all the cost (in cases where, as is common, there is no parental care from males). It is not difficult to see what the advantage is to the male, who gets the benefit of reproduction cheaply, but it is more difficult to see how females might have evolved to tolerate this apparent exploitation.

There are a number of models which suggest conditions under which anisogamy can evolve. Although they differ in details, they tend to share the following assumptions:

1. There are initially two mating types with equal gamete size.
2. Other things being equal, gametes that are larger than average have some advantage, for example they survive longer, are more likely to be found by the other gamete, or can build more viable offspring because they contain more raw materials.
3. Producing larger gametes means an individual produces fewer of them because more energy is required to produce each gamete if they are larger.
4. There is some random, heritable variation in gamete size to start the process off.

In such models, there is a trade-off between quantity of gametes produced and their size or the energy they contain, such that an individual producing many gametes has to reduce the size and energy of each one, and an individual maximizing their size or the energy contained can only produce a few. Crucially, the optimal compromise for one mating type in terms of size and number of gametes depends on what the other mating type is doing. Thus, if one mating type happens to have fewer, relatively larger gametes, the other is selected to specialize in producing more numerous smaller ones. Once this process of specialization has become established, it is very difficult to reverse. Any individual female in an anisogamous species who produces smaller gametes than other females will have reduced reproductive success, for example through her offspring having less energy to start their development, given that the sperm have become so small that they provide very little. Thus, once one mating type has a small gamete, the other is constrained to go on providing a large one. Similarly, any male who produces fewer, larger sperm will produce fewer than his male rivals and they will have no real advantage given the trivial energy contribution of the sperm to the fertilized egg.

Anisogamy is generally common in multicellular organisms and isogamy in single-celled ones. One possibility is that multicellular development, which requires that a lot of energy be present in the fertilized egg to power the process of embryonic development, selects strongly for increasing gamete size, and if one mating type happens to respond more quickly to this pressure, the other can become a quantity specialist or, in other words, male (Bulmer & Parker 2002).

6.4 Sex differences

In species with two sexes, the two sexes often look very different from each in ways that do not follow directly from their reproductive physiologies. For example, the male mandrill is twice the size of the female and has spectacular facial colouring, and in peacocks, males are brightly coloured with a splendid patterned train which is longer than their bodies, whereas females are more drab and lack the train (Figure 6.6). The difference between male and female forms of the same species is called sexual dimorphism. Why should sexual dimorphism exist? Darwin himself speculated that differences between males and females could often arise through what he called sexual selection. Sexual selection is natural selection on the ability to gain mates. If males with a particular trait, such as larger-than-average size or brighter-than-average colouring, can gain more mates than their rivals, then these traits will increase, even if they are detrimental to other aspects of fitness, such as the ability to avoid predators.

This section deals with sexual selection, examining exactly how it works, why it can lead to extravagant traits like the peacock's train, and why it so often (but not always) produces males that are larger and more showy than females. However, to understand sexual selection, it is essential first to clarify how the stakes in the game of reproduction differ between the sexes.

6.4.1 Bateman's principle

Recall that male gametes are very much smaller than female ones. For a male, producing some extra sperm takes little time and is not very costly in terms of energy. This means he could in principle father almost limitless numbers of offspring and his reproductive success will be limited by how many females he can persuade to mate with him. For a female, by contrast,

Figure 6.6 **Males and females are often different: left, male mandrills are twice the size of females and have spectacular facial colouring which the females lack; right, peacocks have bright coloration and an elaborate train, absent from the peahen.** *Left: © Corbis/Digital Stock; right: © Photodisc.*

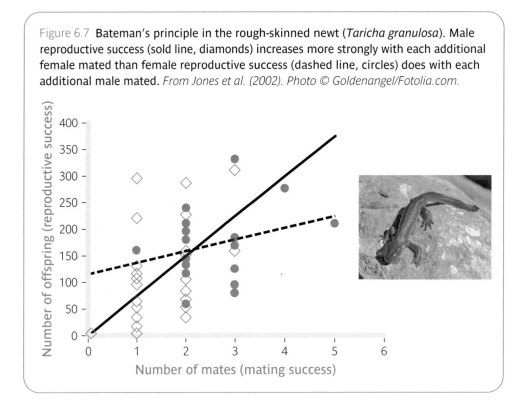

Figure 6.7 **Bateman's principle in the rough-skinned newt (*Taricha granulosa*). Male reproductive success (sold line, diamonds) increases more strongly with each additional female mated than female reproductive success (dashed line, circles) does with each additional male mated.** *From Jones et al. (2002). Photo © Goldenangel/Fotolia.com.*

producing extra eggs has a significant cost. She can only produce a certain number in her life-time. There is no reproductive success advantage for her in mating with more partners than are necessary to fertilize the number of eggs she can produce.

This leads to the prediction known as Bateman's principle (after Bateman 1948), which is the following: male reproductive success increases with each additional partner mated to a greater extent than is true for females. Bateman's principle can be tested empirically. For example, Jones *et al.* (2002) studied the rough-skinned newt, *Taricha granulosa*. They used DNA finger-printing techniques to establish the paternity of all the eggs produced in a pond. For males, the number of offspring sired increased sharply with the number of females mated (Figure 6.7). For females, by contrast, the increase in offspring production with increasing number of males mated was more modest.

A related principle is that the variance in reproductive success is greater for males than for females. A female will always find someone to mate with her, but has an upper limit on how many offspring she could ever produce. Males, on the other hand, have almost no upper limit, but they are faced with females that are choosy and will only mate with the best specimens. Thus, many males will manage no matings at all and a few high-quality individuals will manage to mate with a large number of partners and have very large numbers of offspring. For example, Clutton-Brock *et al.* (1988) studied lifetime reproductive success (number of surviving calves) for male and female red deer, *Cervus elephas*, on a Scottish island. Females who lived to breeding age had between 0 and 9 calves over their lives, with a mean of 5.03 and a variance of 9.09, whereas males who lived to breeding age had between 0 and 32 offspring (with many having 0), with a mean of 5.41 and a variance of 41.9 (Figure 6.8).

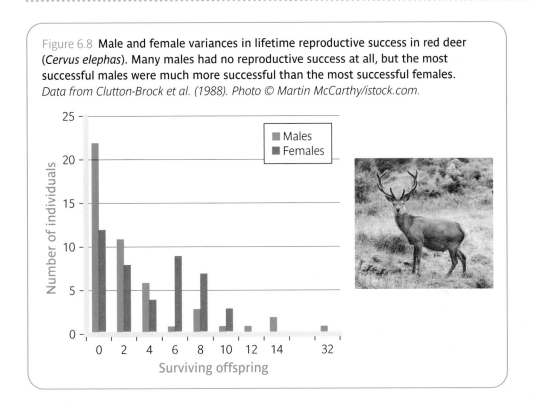

Figure 6.8 **Male and female variances in lifetime reproductive success in red deer** (*Cervus elephas*). **Many males had no reproductive success at all, but the most successful males were much more successful than the most successful females.** *Data from Clutton-Brock et al. (1988). Photo © Martin McCarthy/istock.com.*

Because of Bateman's principle, other things being equal, males should always seek and take any extra matings available (since the cost is low and the benefit relatively high), whereas females should be more choosy (since the cost is higher and the benefit smaller). This pattern is indeed common in nature. For example, grouse gather every breeding season on display grounds called leks. The males fight with each other for central positions, from whence they puff up their feathers, strut, and call. Females lurk, inspecting the talent on offer, and finally mate with one of the best specimens. Males, on the other hand, will accept all offers. Thus, females spend a long time choosing quality, whilst males maximize the quantity of their matings.

Since an extra mating is usually worth more to a male than to a female, there is generally more benefit for him than for her in investing in anything that might bring about an extra mating. This has consequences for how males compete with other males (section 6.4.2) and how they evolve to please females (section 6.4.3).

6.4.2 Intrasexual competition and sexual dimorphism in size

In many species, males fight with each other much more than females do. Fighting between males is an example of intrasexual competition. The reason males fight each other more than females (and intrasexual competition is generally more intense) follows directly from Bateman's principle. Fighting is a risky business and should only be undertaken to the extent to which the benefit is greater than the cost. The benefit of fighting will often be gaining access to more mates (by ousting or driving away one's rivals). Because of Bateman's principle, every extra mate gained will have a greater effect on reproductive success for males than for females. Thus, fighting is more often a worthwhile endeavour for males than for females.

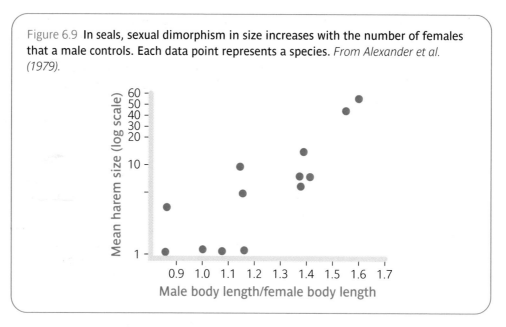

Figure 6.9 **In seals, sexual dimorphism in size increases with the number of females that a male controls. Each data point represents a species.** *From Alexander et al. (1979).*

This has a consequence for the evolution of body size. Having a larger body is advantageous in combat. However, growing large takes time and energy. A creature that carried on growing for its whole life would have a size advantage, but it would have no time or energy for reproduction. Thus, a trade-off point is reached where the returns to growing any more are not sufficient to outweigh the costs and organisms are selected instead to cease growth and begin reproducing. However, since for males the potential fitness benefits of successful combat are larger than they are for females, this trade-off point is reached later. In other words, to the extent that Bateman's principle is true, it is economic for males to invest in increasing their body size for longer than it is for females. This is why males are bigger than females in so many species.

This hypothesis can be tested directly because organisms differ in the extent to which the male reproductive variance is greater than the female one (for reasons we will discuss in Chapter 7). For example, male seals of some species defend large harems of females during the breeding season (and thus the male reproductive variance is large), whilst males of other species have smaller harems or do not defend harems at all, which means that the male variance is closer to the female variance. Alexander et al. (1979) showed that the greater the harem size, the greater the extent to which the male is bigger than the female (Figure 6.9). In elephant seals, with large harems, an adult male can be twice as long as a female and six times as heavy. Male–male fighting is very common and intense in this species in the run-up to the breeding season. In harbour or common seals by contrast, there are no harems, less male–male fighting, and the male and female are only modestly different in size. Thus, there is good comparative evidence that it is greater variance in male than female reproductive success that drives the evolution of sexual dimorphism in size and male–male aggression.

6.4.3 Female choice and ornamentation

A second way males may compete for extra matings is by their attractiveness to females. Darwin himself hypothesized that highly ornamented male traits such as the peacock's train might have evolved because it made the male bearing it more attractive to females. This hypothesis has been largely confirmed. However, it raises further questions. Why would females prefer ornamented males? And why would many of the sexual ornaments we see in nature, like the peacock's train, be so exaggerated? In this section, we first review some evidence that male ornaments are indeed the product of female choice, then look at the evolutionary mechanisms that lead to male ornaments becoming exaggerated.

Do male ornaments function to attract females?

The first step in showing that male ornaments exist because they attract females is to show that females actually prefer males with the ornaments. In a classic study, Møller (1988) examined tail length in the barn swallow, *Hirundo rustica*. Males of this species have elongated tail feathers compared with the females. Møller cut a section off the tail of some males and glued it onto the tails of others. Thus, he had birds with shortened tails and birds with artificially elongated tails. He compared the time it took from arrival at breeding grounds to find a mate for these two types of male, and also two control groups: males whose tails had not been manipulated and males who had had a section of tail cut off and glued back on again. Males with elongated tails paired up more quickly than the controls and males with shortened tails took a long time to find a mate (Figure 6.10). Thus, the long tail really does make males more attractive to females.

The tails of the elongated group were actually longer than any tail the females would ever have seen occurring naturally. This raises the question of why males do not just evolve tails of that length, given the mating advantage. The answer is that it is costly to do so. Males with the elongated tails caught less food and as a consequence were in poorer condition the next season. Thus, males have evolved tail elongation to the point where any extra mating success they gain from growing the tail any further will be outweighed by the costs to their flight and feeding.

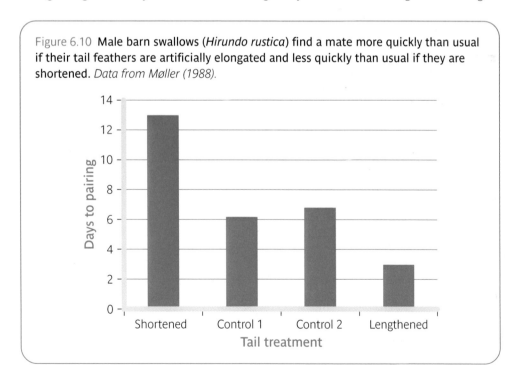

Figure 6.10 **Male barn swallows (*Hirundo rustica*) find a mate more quickly than usual if their tail feathers are artificially elongated and less quickly than usual if they are shortened.** *Data from Møller (1988).*

Results similar to those for the barn swallows have been found for many different types of male ornament in many different species. The ornament can be a behavioural as well as a bodily structure. For example, in a songbird, the European sedge warbler, Catchpole (1980) showed that males who sang the most elaborate songs were the ones who mated most quickly.

Why are male ornaments attractive to females?

The evidence that male ornaments are attractive to females is compelling, but why should they be so? What benefit is there to females in choosing the most ornamented males? Historically, there have been two approaches to this problem: Fisher's (1930) and Zahavi's (1975). These are known colloquially as the 'sexy son' hypothesis and the 'good genes' hypothesis. We will review each in turn and then conclude that they are not mutually exclusive and are usually likely to operate in tandem.

The sexy son hypothesis

The sexy son hypothesis is a great example of how evolutionary processes can be at once very simple and also extremely hard to understand. It basically states that if there is any initial slight preference amongst females for males with longer tails, then the preference for the long tail and the length of the tail itself co-evolve to both become ever greater over time.

The key to understanding this runaway is the following. Females who choose a mate with a longer than average tail will have sons who have longer than average tails (because the sons will inherit it from their fathers). Thus, there is a fitness advantage to choosing a long-tailed mate if only because one's sons will thereby be attractive. Fine, but how does this lead to tail length becoming exaggerated? In general, those females with the strongest preference for long tails will mate with those males who have the longest tails. This creates a genetic correlation between the preference for the trait and the trait itself. As we saw in section 5.4.3, when selection acts on one of a pair of genetically correlated traits, the other is changed too, so as selection favours a stronger preference (because females with a stronger preference have sexier sons), the length of the tail is dragged along as a correlated trait. However, as tail lengths get longer, the sons of the longest-tailed males get sexier and the fitness payoff for having a stronger preference becomes greater. The trait and the preference are in a positive feedback loop.

Fisher's hypothesis has nothing to say about where the initial slight preference for longer tails came from, but it only needs to be slight. It could be as simple as males with longer tails being easier to see or to identify as males. The point of the Fisher process is that the slight initial difference can be amplified by selection into a strong female preference and an exaggerated male trait without having to postulate any further functional effects of long tails.

The good genes hypothesis

The good genes hypothesis starts from a slightly different point of view. What male ornaments such as long tails, bright coloration, or energetic singing all share is that they are costly, that is it would be difficult to allocate enough energy to doing them well unless one had energy to spare, meaning that one was feeding effectively, not too infested with disease, not carrying too many deleterious mutations, and so on. Thus, females choose males with the largest ornaments because those males are proving that they have the quality to do well in the current environment (hence 'good genes'). Lower-quality males simply cannot produce signals as elaborate as those produced by higher-quality males. Males invest in ornaments as much as they can to signal their quality to females. The reason that the ornaments tend to be so exaggerated is that if they were not costly, all males could produce them and they would not discriminate the high- from the low-quality ones.

In the good genes model, the female is choosing males of high quality because quality is heritable. This makes a subtly different prediction from the Fisher hypothesis. Under Fisher's hypothesis, the (male) offspring of long-tailed males need to also have long tails and this needs to make them more attractive to females. Under the good genes model, all the offspring of long-tailed males need to be of better quality in general, that is you would expect them not just to be more attractive, but to survive better and have lower parasite loads.

Evidence for good genes and sexy sons

There is evidence for sexy son effects and for good genes effects on female mate choice. One of the key predictions of the Fisher model is that there should be a genetic correlation between preference for the ornament and the ornament itself. Bakker (1993) studied sticklebacks, amongst whom red coloration is a sexually selected male ornament. He was able to measure experimentally the strength of the preference that females have for red coloration in males. He showed that the stronger a male's red coloration, the stronger his sister's preference for red coloration in males (Figure 6.11). This shows that the alleles for the preference and the trait are indeed assorting together, as Fisher's model requires.

As for good genes, a number of studies have shown that the offspring of highly ornamented males survive better or have lower parasite loads than the offspring of males with small ornaments. Amongst the barn swallows, the offspring of long-tailed males' resistance to parasites is heritable. Møller (1990) showed this by an adoption study in which he moved chicks into different nests (to eliminate effects of shared environments). He found that the parasite of the biological parent predicted that of the offspring and moreover that chicks whose fathers had

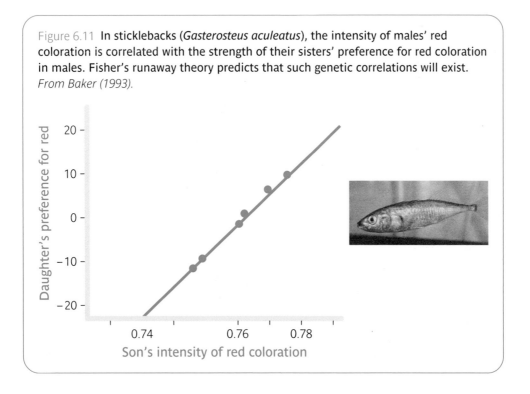

Figure 6.11 **In sticklebacks (*Gasterosteus aculeatus*), the intensity of males' red coloration is correlated with the strength of their sisters' preference for red coloration in males. Fisher's runaway theory predicts that such genetic correlations will exist.** *From Baker (1993).*

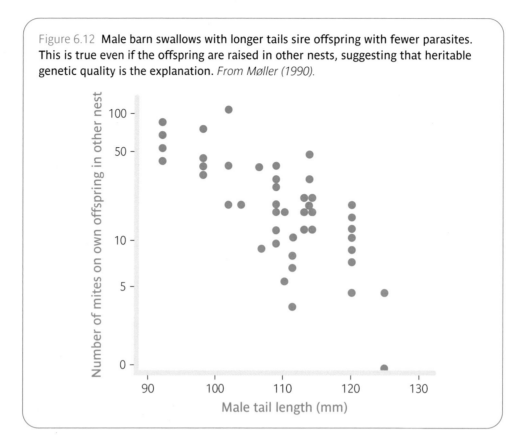

Figure 6.12 **Male barn swallows with longer tails sire offspring with fewer parasites.** This is true even if the offspring are raised in other nests, suggesting that heritable genetic quality is the explanation. *From Møller (1990).*

longer tails grew up with fewer parasites (Figure 6.12). These findings are consistent with the idea that males, through their long tail feathers, are signalling their heritable genetic qualities such as parasite resistance.

Complementarity between good genes and sexy sons

The good genes and sexy sons models are often framed as alternatives, but in fact they are not mutually exclusive. Moreover, they are both likely to operate. Choosing a male who is signalling good genes will benefit a female both through making her offspring more healthy in general and making her sons more attractive to females in particular. Also, as a trait becomes more and more exaggerated through the Fisher process, it becomes more and more costly to produce it and males will vary in their ability to do so. A trait under Fisherian selection is therefore likely to become revealing of male quality because some males will be able to allocate enough energy to displaying it and some will not. Theoreticians have thus realized that the two processes are tightly connected and may often occur in tandem (Kokko *et al.* 2002).

6.4.4 Sex-role reversal

Is Bateman's principle always true, producing choosy females and males that compete for mates in every anisogamous species? Our discussion of Bateman's principle focused exclusively on the costs of producing the gamete. However, these are not always the only costs of reproduction. Parents of many species feed their young after conception or even develop them inside their bodies. Trivers (1972) pointed out that, in considering the costs and benefits of mating, what matters is the *total* cost to each sex of a reproductive episode, not just the cost of the gamete.

In mammals, it is the female that gestates the offspring inside her body, the female that lactates, and often the female that provides other care too. Mostly, males do nothing for the offspring apart from mating. Thus, the difference between the costs for the two sexes is made even greater by considering the costs of post-conception parental investment as well as those of the gamete and Bateman's principle holds true. However, there are organisms—many penguins, for example—where mating pairs share the raising of offspring more or less equally. This means that the difference in cost of a reproductive episode for a male and a female is slight, and, accordingly, males tend to be neither larger nor more ornamented than females in these species.

There are also species where the male does all the post-fertilization care. This more than offsets the initially lower gamete cost and means that the cost asymmetry between the sexes is exactly reversed. Accordingly, we should predict that in these species it will be females who have the larger variance in reproductive success, females who will be larger and more ornamented, and males who will be more choosy.

Jones *et al.* (2001) studied mating success in a pipefish, *Syngnathus scovelli*, off the coast of Florida. In these animals, the female transfers the fertilized eggs to a patch or pouch on the male, who carries them around until they hatch, and so the male invests more time and energy in a clutch than the female does. Using DNA fingerprinting, Jones and colleagues showed that the variation in reproductive success was also the reverse of the usual pattern. All males held just one brood (from one female), whilst the most successful females had broods with four males and many had none (Figure 6.13). The variance in female mating success was around seven times the variance in male mating success, which is an asymmetry as strong as that seen

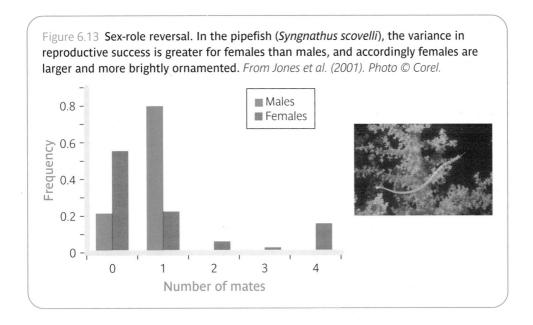

Figure 6.13 **Sex-role reversal. In the pipefish (*Syngnathus scovelli*), the variance in reproductive success is greater for females than males, and accordingly females are larger and more brightly ornamented.** *From Jones et al. (2001). Photo © Corel.*

the other way around in sexually dimorphic mammals. Satisfyingly, in this species of pipefish, females are larger than males and have brilliant stripes, which the males lack. The researchers were also able to show that, amongst the females they captured, those that had managed to mate were larger and more brightly ornamented than those who had not, which demonstrates sexual selection in action.

Pipefishes and other sex-role-reversed species are examples of 'the exception which proves the rule'. That is, typical sex differences are reversed, but this is because the difference in costs of reproduction is reversed and so theory predicts that the females should be larger and more ornamented. Their existence does mean, however, that we need to formulate Bateman's original principle slightly more carefully. Rather than saying that the male will always have a larger variance in reproductive success than the female and will therefore compete for mates, we need to say that whichever sex invests less per episode of reproduction will have the larger variance in reproductive success and will therefore compete for mates.

6.5 Pluralism in sexual strategies

Our discussion so far has stressed how the sex with the lower parental investment grows as large as possible and competes for mates, whilst the sex with the higher parental investment is choosy and only seeks one mate. This picture is a slight simplification for two reasons. First, females do often seek more mates than necessary to fertilize all their eggs (section 6.5.1) and, second, males often have alternative tactics for gaining mates (section 6.5.2).

6.5.1 Female multiple mating and extra-pair copulation

Most birds and some mammals form long-term pair bonds and biologists used to believe that females in these species only mated with one male. DNA fingerprinting has allowed us to understand that this is not so; females often mate with multiple males, even if they have a social pairing with just one. Matings that take place with a male other than the social partner are called extra-pair matings. Extra-pair matings are quite common in many pair-bonding species, and female multiple mating in general more common still. Some people have argued that the existence of multiple mating by females is a challenge to Bateman's principle, but this is not strictly correct. Bateman's principle only states that the benefit of an extra mating is greater for a male than for a female; it does not state that there is no benefit for the female.

Females mate with multiple males for several reasons. Where there are pair bonds, the best males may already be paired up and so a female has to settle for a male that is available and prepared to choose her. It might still be worthwhile for her to accept even quite a low-quality male as a social partner because having a social partner may aid in parenting or territory defence. However, in terms of genetic makeup of her young, she will do better to be fertilized by the highest-quality male (or the one most dissimilar to her at the MHC) than by her social partner.

Evidence for effects of this kind comes from the blue tit, *Cyanistes caeruleus*. Kempenaers *et al.* (1997) showed that between one-third and one-half of all nests contained chicks fathered by an individual other than the social partner of the female. It was rarely the case that all the chicks were the result of extra-pair matings. Rather, the females were obviously mating with both the partner and another male in close succession and raising mixed broods. The researchers

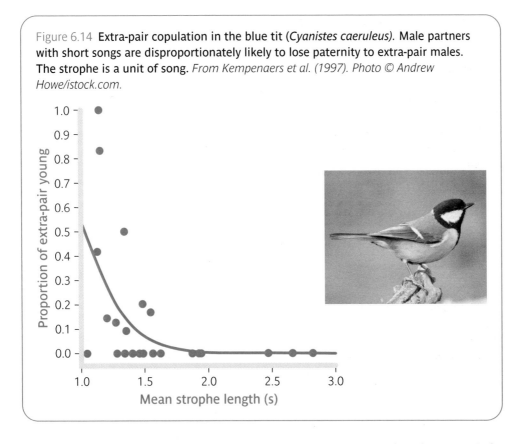

Figure 6.14 **Extra-pair copulation in the blue tit (***Cyanistes caeruleus).* **Male partners with short songs are disproportionately likely to lose paternity to extra-pair males. The strophe is a unit of song.** *From Kempenaers et al. (1997). Photo © Andrew Howe/istock.com.*

showed that it was low-quality males who lost paternity to an extra-pair male. For example, males who sang long songs fathered all the offspring in their nests, whereas males with short songs had extra-pair offspring in theirs (Figure 6.14). The mechanism for this seems to be that females with a neighbour who is more attractive than their social mate seek extra-pair matings. The reasons for them doing so are also clear: in this species, the offspring resulting from extra-pair copulations are more likely to survive than their nest mates fathered by the social partner. Thus, females are using extra-pair mating to choose good genes.

The dynamics of these situations are very interesting. It is still worth low-quality social partners investing in their partnership because at least some of the offspring may be theirs. However, higher-quality males gain a disproportionate fraction of all extra-pair matings, which amplifies the variance in male reproductive success and strengthens the operation of sexual selection.

6.5.2 Alternative male reproductive strategies

The best outcome for a male seeking to maximize his reproductive success is to be the largest male, the most dominant in contests with other males, and the most ornamented. However, by definition, not every male can be these things. Males often have secondary tactics for gaining some matings if they are not doing well in the primary competition. For example, in a colonial bird called the great-tailed grackle, the largest males hold a territory and sire offspring via both social partners and extra-pair matings. Smaller males are unable to defend a territory successfully. They either remain in one colony seeking the occasional extra-pair mating and waiting for a territory to become available or else they become transient, roaming from colony to colony picking up occasional extra-pair matings and providing no further investment. The reproductive success of these transient males is very much less than that of the successful territory holders (Johnson *et al.* 2000). However, they are making the best of a bad job and it is at least greater than zero. Such alternative male tactics are very widespread.

6.6 Sexual selection and mate choice in humans

You have probably been wondering through sections 6.3 and 6.4 to what extent principles like Bateman's, and the concomitant sex differences in mating strategy, apply to humans. In this final section, we establish how humans fit in to the pattern described for other organisms.

6.6.1 Bateman's principle in humans

A first question to ask is whether the variance in male reproductive success in humans is greater than that in female reproductive success. Table 6.1 shows estimates of male and female variance in reproductive success (number of offspring) for several different societies. For contemporary Britain (at the age of 45 years), the male variance is only around 6% higher than the female variance, whereas for the Kipsigis, a group of Kenyan farmers amongst whom rich men have many wives and poor men tend not to have any, the male variance is around 15 times the female variance. Two groups of hunter-gatherers are intermediate between the British sample and the Kipsigis. Thus, such cross-cultural evidence as there is suggests that the variance in male reproductive success does tend to be higher than the variance in female reproductive success, but

Table 6.1 **Variance in human male and female reproductive success in four very different societies. In contemporary Britain, male variance is only slightly higher, although these data are from 45 years of age, and the difference may become more marked as the men get older, whereas for the Kipsigis, an African society where rich men have many wives, the disparity is vast.** *Data from Daly & Wilson (1983), Borgerhoff Mulder (1988), and Nettle (2008).*

Society	Description	Male variance	Female variance	Ratio male:female variance
!Kung	African hunter-gatherers	9.27	6.52	1.42
Xavante	Brazilian hunter-gatherers	12.1	3.6	3.36
Kipsigis (Nyongi cohort)	Kenyan farmers	54.03	3.62	14.93
Contemporary Britain	Industrial and service economy	1.74	1.64	1.06

that the local economic and social situation modifies the extent of the difference (see Chapter 8 for why different societies might have different marriage systems). However, the difference is always in the direction of red deer, not of pipefish, and thus we should expect humans to have evolved the corresponding pattern of larger male size and greater female choosiness.

6.6.2 Big males and choosy females

Within the primates, as in the seals and their relatives, the degree of sexual dimorphism in size covaries with the intensity of male–male competition. Gorilla males, who defend harems, are about twice the size of females, whereas amongst gibbons, who are extremely monogamous, males and females are the same size. Humans are intermediate and rather closer to the gibbon pattern, with males around 10% bigger than females. This suggests an evolutionary history of men experiencing slightly larger variances in reproductive success than women, as corroborated by the hunter-gatherer data described in section 6.6.1.

Human males are much more likely to be involved in violence than human females, and both the victims and the perpetrators of homicide are overwhelmingly unmarried young men (Wilson & Daly 1985). Men compete with each other more riskily than women do, leading to greater male than female death rates from accidents and violence. This is particularly pronounced during the period of peak reproductive competition, from about 16–25 years (Figure 6.15).

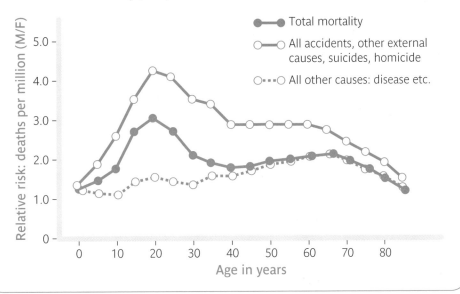

Figure 6.15 **The male death rate relative to the female death rate in the USA, broken down by age. Amongst the very young and very old, there is no sex difference in death rate, but in the peak years of reproduction, men are three times more likely to die than women. Breaking the causes of death down into deaths attributable to disease versus those attributable to violence and accidents shows clearly that the sex difference is due to greater risk-taking and aggression through the reproductive years.** *Data from Wilson and Daly (1985).*

Choosy females

Bateman's principle suggests that human females should be choosier when it comes to mating, whereas males should be more interested in gaining additional mating partners. There is cross-cultural data showing this pattern. Schmitt *et al.* (2003) showed that, in 52 countries from across the globe, men express a desire for a greater number of sexual partners in the future than women, whilst women report requiring a longer period of acquaintance with a man before consenting to sex than men do for sex with a woman (Figure 6.16). There are very interesting local differences in these attitudes and in the size of the sex difference, but the direction of the sex difference is never reversed. Clark & Hatfield (1989) illustrated the difference vividly in a famous study on a university campus, where men or women were approached by an attractive stranger of the opposite sex and asked if they would like to go to bed with them that night. None of the women, but a sizeable fraction of the men, said that they would.

6.6.3 Parental investment

There is a slight paradox in data showing that men desire partner variety and are not choosy. This is because humans pair for long periods and human males make large investments in their offspring. Thus, mating represents a very considerable investment for men, almost as much as for women, and thus one might expect that both sexes would be choosy in humans.

In fact, survey evidence suggests that both sexes are choosy, particularly when selecting a long-term partner, and both sexes place kindness and reliability high on their lists of desired

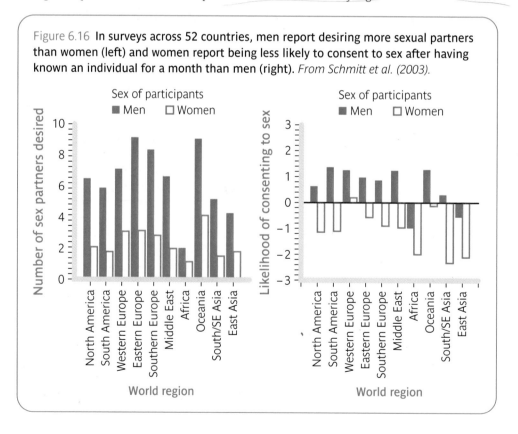

Figure 6.16 **In surveys across 52 countries, men report desiring more sexual partners than women (left) and women report being less likely to consent to sex after having known an individual for a month than men (right).** *From Schmitt et al. (2003).*

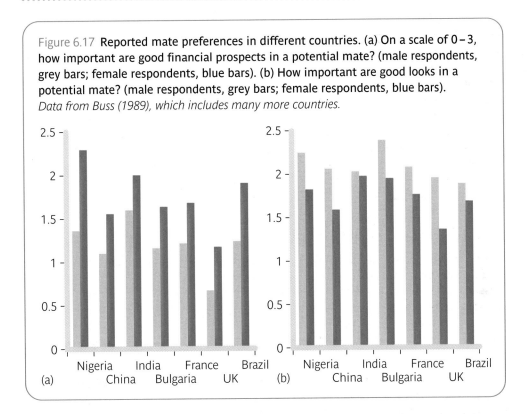

Figure 6.17 **Reported mate preferences in different countries. (a) On a scale of 0 – 3, how important are good financial prospects in a potential mate? (male respondents, grey bars; female respondents, blue bars). (b) How important are good looks in a potential mate? (male respondents, grey bars; female respondents, blue bars).** *Data from Buss (1989), which includes many more countries.*

characteristics. However, the two sexes may be choosy in subtly different ways. Men need to know that the women they are investing in will be fertile and produce healthy offspring, and they place a relatively high value on physical appearance, particularly such characteristics as symmetry and body fat distribution, which signal health and fertility. Women have different priorities. Human children are extremely costly and require a long period of material investment, and women in many cultures prefer men with material resources (income or wealth) to offer. Buss (1989), in another cross-cultural survey study, showed that, across 37 countries, women put a higher value on income in a potential mate and men a higher value on physical appearance (Figure 6.17). Once again, the precise attitudes and the magnitude of the sex difference vary considerably with local ecology, but the sex difference is never reversed.

These patterns may explain why ornamentation takes the forms it does in humans. Unlike peacocks and mandrills, men do not have brightly coloured or ornamented bodies, but they do exploit cultural opportunities to signal their wealth and status, whilst women's fashions often emphasize their body composition and cues of youth and health.

6.6.4 Strategic pluralism in human mating

How do we reconcile the data showing that men desire partner variety and consent to sex at low acquaintance with the argument of section 6.6.3 that human males are high investors who ought to be choosy? The answer lies in recognizing that there are alternative mating strategies available to both males and females.

Male short-term mating

Mating can be an expensive decision for a man, if he is going to devote himself to the relation-ship and the offspring it might produce, but it is not *necessary* that it be costly. Unlike the woman, he can produce an offspring with no cost beyond that of copulation itself. Thus, although men might be choosy for a long-term relationship in which they plan to invest, they also have available to them a short-term mating strategy with no post-conception investment, which they might resort to if the context favours it. The male reports of desire for sexual variety and for willingness to consent to sex at low acquaintance may reflect the operation of this short-term alternative strategy.

Female extra-pair mating

Just as men have an alternative strategy available to them, so do women. In choosing a social partner, women have to accept someone who is also available and prepared to invest, and this may not represent the maximal genetic quality they could achieve. Surveys show that a size-able fraction of women at least sometimes have sex outside of their established relationship (Koehler & Chisholm 2007) and genetic studies reveal that a small fraction of children are not fathered by their mother's long-term partner.

Like the blue tits, then, women sometimes seek genetic quality by extra-pair mating. Further evidence that this is the case comes from studies showing that highly symmetrical (i.e. high-quality) men report having more often been a partner in an extra-pair mating (Gangestad & Thornhill 1997). Moreover, at the point of their menstrual cycle where women are most fertile, they show more interest in and fantasize more about men other than their current partner (Gangestad *et al.* 2002), as well as being more likely to actually commit an extra-pair mating (Bellis & Baker 1990).

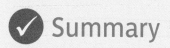 # Summary

1. Sex of some kind is extremely widespread in nature, but there are many different types of sexual system.

2. Sex is adaptive because of some combination of increasing the viability of offspring (e.g. through their resistance to disease) and increasing the variance amongst them, which makes natural selec-tion more efficient.

3. Because of anisogamy and differential parental investment, the two sexes often benefit to differ-ent extents from gaining an additional mate. This leads to the evolution of one sex that is larger, more aggressive, and more highly ornamented than the other.

4. Male ornamentation is attractive to females in many species because the offspring of such males do better through some combination of overall higher quality and greater attractiveness.

5. There are often multiple mating strategies available to both sexes, such as territory-holding versus mobility or in-pair versus extra-pair mating.

6. Many of the principles seen elsewhere in nature, such as greater male than female variance in reproductive success, greater male aggression, female choice, and extra-pair mating, also characterize humans.

? Questions to consider

1. What factors do you think might determine whether an organism evolves to be simultaneously hermaphrodite like the hamlet fish rather than exclusively male or female like humans?

2. In many organisms, the optimum size in terms of reproductive success for males is larger than the optimum size for females. For this reason, males are often larger than females. However, there are at least three different ways of bringing this about. One is the pattern seen in many mammals, including humans, where sex is determined genetically and males grow for longer than females. The second is seen in a shrimp-like creature, *Gammarus duebeni*, where day length when the eggs are laid determines which sex the egg develops into. This results in eggs laid early in the year becoming males, who are then larger when the breeding season comes around. Another is the pattern seen in wrasse, where all individuals grow on the same trajectory, but are female when they are small and then change their sex to male as they get bigger. What kinds of factors would favour one of these systems over the others?

3. In Møller's barn swallow experiments, why do you think he had a group of birds where a section of the tail had been cut off and stuck back on again?

4. In Britain, a man is only allowed to be married to one woman and yet the variance in male reproductive success is higher than the variance in female reproductive success. How can this be?

→ Taking it further

The sheer diversity of sexual systems in nature is outlined entertainingly by Judson (2002) and the diversity of male and female mating strategies laid out by Birkhead (2000). A readable book-length treatment of the issue of the evolution of sex is Ridley (1993), whilst a recent review article is West *et al.* (1999). On current theory relating to mate choice and the evolution of ornamentation, see Kokko *et al.* (2002, 2003). Important contributions to the psychology of mating in humans include Buss (1989), Gangestad & Simpson (2000), and, with a focus on the way mate preferences vary according to ecology, Gangestad *et al.* (2006).

Life histories

The lives of living things have a cyclical quality. Individuals are born, mature, reproduce, and die, for the next generation to do the same in their turn. This is true of humans just as much as any other organism. There are many questions we can ask about life cycles. What determines how long the cycle lasts? How long should one grow before maturing and beginning to reproduce? How soon should one stop caring for an offspring after it is born and have another one? These questions are addressed by a branch of biology called life history theory. Life history theory takes the adaptationist stance, that is it assumes that alleles producing the life cycles we see in nature have out-competed alternative alleles for longer, shorter, or otherwise different life cycles. Thus, the life cycles we see must be ones that make sense in terms of long-term reproductive success, and our job is to understand why the particular life cycles that we see might be better than the alternatives.

The central concept of life history theory is that organisms have finite resources. They can only capture energy from the environment—in the form of food—at a certain rate. They thus face a problem of allocation. How much of the energy they manage to capture should they expend on improving their own bodies (growth, repair of tissues), how much should they expend on reproduction (attracting a mate, producing gametes or embryos), and how much should they expend on aiding their existing descendants or kin? The general answer to the allocation problem is that it depends on their current state, and on their likely life expectancy and that of their offspring. These in turn depend on the ecology in which they are operating. We thus see big differences in life cycle between related species, or even different populations of the same species, if their ecological regimes are different.

Section 7.1 addresses the question of why organisms should have the lifespans that they do and section 7.2 looks at the issue of why they reproduce when (and as much or little) as they do. Section 7.3 looks at parental care for offspring and the conflicts between parents and off-spring over when this care should cease. Finally, section 7.4 examines the role of grandparental investment, particularly in humans.

7.1 When to die: the evolution of lifespan

Darwinian evolution produces organisms that seek to preserve their own lives because one's reproductive success is usually higher alive than dead. This accounts for a great deal of our motivated behaviour, from avoiding predators and diseases to seeking food and water. It is thus paradoxical that even where there are no external causes of death, such as predators, diseases, and accidents, organisms do eventually die. Domestic cats and dogs in affluent countries, for example, are kept well shielded from external risks of death, and yet they do not live much longer than the oldest wild cats or wolves. Instead, they reach an age where multiple systems of their bodies begin to deteriorate until one or other fails completely. The situation with humans is not much different. After around 80 years, if not earlier, multiple physiological processes deteriorate markedly. This is known as senescence, and because of it most people die by the age of 100 years even with no external causes. This means that, although the proportion of people reaching a ripe old age has increased dramatically with better living standards and medicine, the age of the oldest people has not increased by much. Why should this be? Why should we not simply go on living?

This may seem like asking the impossible, but think about it another way. The processes of tissue repair and maintenance in humans are good enough that they can keep our bodies working just as well after 40 years as they work after 1 year. In 39 years of living, every cell in the body will have had to be replaced or repaired many times. However, if the body can repair itself with essentially complete effectiveness early in life, why should it not go on doing so indefinitely?

The disposable soma

To understand the evolution of senescence, we need to take the gene's eye view. A gene resides in a particular body at a particular instant, but that body could die and the gene endure as long as that body had reproduced (or caused close kin to reproduce) before it died. Thus, the gene is, in an important sense, potentially immortal, whereas the individual body is disposable. Senescence could not possibly be advantageous to the individual senescing, who would always prefer an extra year of healthy life to death, but it could be advantageous to the immortal gene lines currently residing in that body. This section explains how.

We first need to introduce the idea of **extrinsic mortality**. The extrinsic mortality risk is the probability of an individual dying in a given time period through events in the environment that they could do essentially nothing about, such as predators, fluctuations in the food supply, accidents, weather fluctuations, and so on. Extrinsic mortality risk might vary, but it is never completely reduced to zero. Even a well-nourished organism taking full precautions against disease and predators could be struck by lightning. Extrinsic mortality is contrasted with intrinsic mortality, which is the mortality due to the organism's own ageing.

Organisms reproduce in part because it allows their genes to buffer themselves against extrinsic mortality. From the gene's point of view, residing inside a single individual, when that individual could get struck by lightning at any moment, is not a safe route to persistence. Far better to cause that individual to invest some of its energy in making other bodies that will be alternative vehicles for the genetic cargo. Thus, all else being equal, an allele which caused its bearer to devote a little more of its time and energy to making copies of itself could have higher fitness than its competitors. However, it is also in the interests of genes that whichever body they happen to be in at this moment continues to function, so an allele which caused bodies to repair themselves more effectively could also out-compete its rivals.

The organism with the highest reproductive rate would be one that allocated all its energy to reproduction and none to self-repair, whilst that which repaired itself best would be the one that allocated all its energy to repair and none to reproduction. Between these two extremes, there is a **trade-off** between self-repair and reproduction, whereby doing more of one means doing less of the other. We will now analyse how this trade-off leads to the evolution of senescence, following what is known as the 'disposable soma' theory (Kirkwood 2008).

The disposable soma theory assumes that ageing, and the associated spontaneous death, arise from the accumulation of unrepaired damage to body cells. Energy devoted to self-repair can slow this accumulation of damage or even abolish it completely. However, the energy that is needed to do this is energy that could also be spent on reproduction instead and, according to the disposable soma theory, reproductive success will be higher for an organism that does not repair itself perfectly, but instead uses the energy for reproduction. Why is this?

Evolution of senescence: a simple model

To answer this question we have to consider the proportion of individuals that are alive at different ages, given that extrinsic mortality is not zero. Let us say, for example, that the rate of extrinsic mortality for some organism is 1% per year even if the organism repairs its cells with total effectiveness. Figure 7.1 shows, for 100 individuals who are born, how many are left alive at different ages. By 100 years, only about 36 are still alive, and after 150 years, only about 22 are. For the sake of argument, we will assume that these organisms can reproduce every year from as soon as they are born and as long as they are alive.

Let us compare three hypothetical types of organism. Type A organisms repair themselves with complete effectiveness and so they never senesce. However, this costs them energy, so they can only produce one offspring per year. Type B organisms repair themselves moderately effectively, leaving enough damage unrepaired that by the time they are 100 years old the damage accumulates to a critical point, and they senesce and die. The payoff for repairing less is that they can produce two offspring per year. Finally, type C organisms allocate even less to repair, with the result that damage accumulates to a critical level much earlier and they senesce and die at 50 years. However, they can produce three offspring per year.

Figure 7.2 shows the number of offspring produced by 100 individuals of each type at different ages, and in total across the lifespan. Type A continues to produce offspring long after

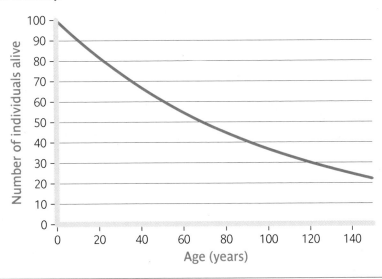

Figure 7.1 The number of individuals from an initial population of 100 expected still to be alive at different ages, given an extrinsic mortality rate of 1% per year and no intrinsic mortality.

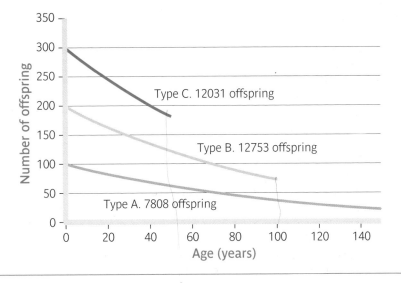

Figure 7.2 Comparison of the numbers of offspring left at different ages and in total for initial populations of 100 of three types of organism. Type A do not senesce but produce one offspring per year. Type B senesce after 100 years but produce three offspring per year. Type C senesce after 50 years but produce three offspring per year. Type B have the highest reproductive success overall. The extrinsic mortality rate is 1% per year.

the other two types have died. However, by this point, most individuals have been eliminated by extrinsic factors anyway and so the number of extra offspring created by this lengthened lifespan is small. It is not enough to offset the lower reproductive output earlier in life and the total number of descendants created by the type A organisms is much lower than the other types. Type C reproduce at the fastest rate whilst alive, but their short viable lifespan means that they leave fewer offspring than the type B organisms, who invest enough in repair to remain viable for as long as most of them will be around anyway, but not more.

This example shows that an allele that caused its bearer to leave some somatic damage unrepaired and instead used the energy to reproduce could out-compete an alternative allele that caused longer life. This is the best explanation we have for why we senesce. The gains in reproductive success from living longer would not be enough to offset the cost, in terms of reduced energy allocated to reproduction earlier in life.

The disposable soma theory predicts that the higher the rate of extrinsic mortality, the earlier senescence should occur, and the more energy early in life should be devoted to reproduction. For example, if we increase the rate of extrinsic mortality to 2% per year, organisms of type C in Figure 7.2 now have a huge advantage over type B. Does this explain why some animals reproduce much more quickly and live less long than others? There is a great deal of comparative evidence bearing on this question, as we shall see in the next section.

7.2 When to breed: the evolution of reproductive strategies

Looking across different organisms, there is a great deal of variation in length of life. For example, cats rarely live beyond 15 years, whereas brown bears can live for up to 50 years. It is difficult to say how much of this variation is due to extrinsic mortality and how much is intrinsic. However, Figure 7.2 suggests strongly that the greater the mortality rate, the greater the extent to which higher reproductive rate should be favoured over long-term maintenance of the soma. In other words, the higher the extrinsic mortality rate, the more it pays to live fast and die young, whereas the lower the extrinsic mortality rate, the more it pays to invest for the long term.

If this prediction is correct, then organisms with high mortality rates ought to show a suite of other characteristics: they should start reproducing earlier, have more (and/or more frequent) offspring, and devote more effort per year to reproduction. Promislow & Harvey (1990) examined how aspects of the life cycle correlated with mortality rates in the wild across 48 species of mammal. Some of the results are shown in Figure 7.3. The higher the rate of mortality, the sooner the animal matures, the larger the litter it has, and the shorter the interval until the next litter. However, this comes at a cost: the higher the mortality rate, the smaller the relative body weight of the offspring, the less time it can be gestated, and the sooner the mother stops nursing it. Thus, we can array all mammalian species along a continuum of slow to fast life history. At the slow end is the African elephant, not maturing until it is 15 years and having one calf which suckles for nearly 3 years, whilst at the fast end are bank voles, maturing in around a month and having around five pups per litter which suckle for less than 3 weeks.

Body size is associated with life history strategy, with larger bodies associated with slower life histories. Growing large takes time and so, other things being equal, large size delays age at first reproduction, and increases gestation and lactation times. It also slows down reproductive

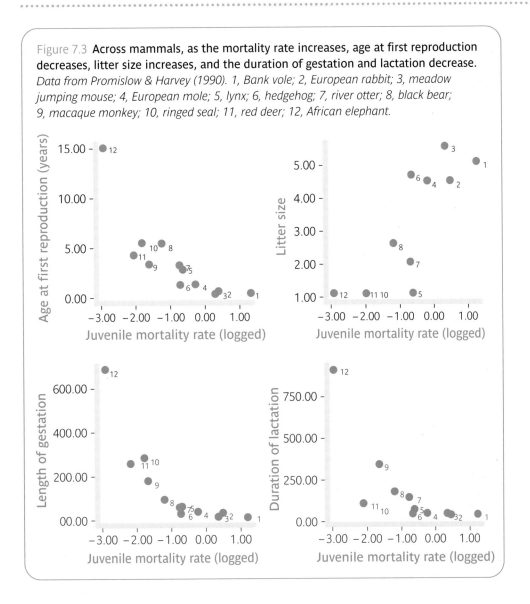

Figure 7.3 **Across mammals, as the mortality rate increases, age at first reproduction decreases, litter size increases, and the duration of gestation and lactation decrease.** *Data from Promislow & Harvey (1990). 1, Bank vole; 2, European rabbit; 3, meadow jumping mouse; 4, European mole; 5, lynx; 6, hedgehog; 7, river otter; 8, black bear; 9, macaque monkey; 10, ringed seal; 11, red deer; 12, African elephant.*

output, since large organisms metabolize less energy per unit mass and thus can allocate relatively less energy to reproduction. However, the patterns in Figure 7.3 cannot be reduced to differences in body size. Even when body size is statistically controlled for, there are significant associations between mortality rates and age at first reproduction, litter size, gestation length, and so on. In fact, there are interesting differences in life history between mammals of the same size. Bats have extremely slow life histories compared with ground-living mammals of similar size, and this may reflect the fact that their lifestyle allows them to escape a lot of the predation that would affect them were they on the ground, so they live slowly and for a long time, but their body size is kept small by the demands of flying. Marine mammals such as seals have fast reproductive rates for their size and this may reflect the abundant nutritious resources that their marine adaptations allow them to access (Sibly & Brown 2007).

Nonetheless, if other ecological factors remain constant, the trade-off between energy allocated to growth and energy allocated to reproduction means that an increased mortality rate tends to result in the evolution of smaller size at maturity and earlier reproduction, and selection for larger bodies tends to slow down life history.

Human life history

Where do humans fit into all this? The first point to make is that human life history is slow in the mammalian scheme of things, with large body size, sexual maturity not occurring for at least 15 years, 9 months of gestation, and a usual litter size of one. This suggests an evolutionary history in which extrinsic mortality has been rather low and indeed estimates suggest that the probability of survival to the age of 15 years for a human baby in a hunter-gatherer society without modern medicine is about 0.6, whereas the figure for a chimpanzee is around 0.35 (Kaplan *et al.* 2000). Accordingly, the maximum age for humans is about twice the maximum age of chimpanzees, and chimpanzees begin to reproduce by around 14 years, compared with around 20 years for hunter-gatherer women.

However, human life history is puzzling in other ways. We achieve a rather high reproductive rate given our body size. Chimpanzees give birth around every 5 years, whereas women in natural fertility populations do so every 3.5 years. On the other hand, the childhood period of dependence is prolonged in humans. The combination of increased birth rate and prolonged dependency is only possible because humans reproduce cooperatively. In human societies, fathers, grandparents, and other kin and friends help provide resources for the dependent offspring, which allows women a faster reproductive rate than they could otherwise achieve. Chimpanzee fathers and grandparents provide no care or resources. We return to the significance of fathers and grandparents for human offspring in later sections.

Humans living fast and dying young: pygmy populations

The trade-off between continuing to grow and starting to reproduce is manifest not just between species, but also within them. For example, there are several dozen known human populations with adult male heights of less than 155 cm. These populations are known as pygmies, and pygmy stature seems to have developed independently many different times: there are pygmy populations in Africa, Malaysia, Thailand, New Guinea, Brazil, Bolivia, and Papua New Guinea, amongst other places. One interpretation of pygmy stature is that these populations are nutritionally stressed, but this cannot be the whole story. East African herders such as the Masai and Turkana, for example, are just as nutritionally challenged and grow no faster than pygmy children in the first 10 years of life, but the Masai and Turkana continue growing and are amongst the tallest of the world's populations, whilst growth flattens out earlier for pygmy populations.

Life history theory suggests instead that the pygmy populations might have a fast life history as an adaptation to high rates of mortality. Migliano *et al.* (2007) showed that this explanation is indeed correct. Pygmy populations live in high-disease environments such as tropical forests. Data from several different pygmy populations revealed much harsher mortality rates than non-pygmy populations such as the Turkana and, correspondingly, the pygmy populations stopped growing sooner and reproduced earlier in life (Figure 7.4). The high risk of death in these populations has made them live fast and die (or at least start to reproduce) young.

This means that the pygmy pattern of earlier cessation of growth and onset of reproduction is adaptive given the ecology that these populations experience, but note that there are two

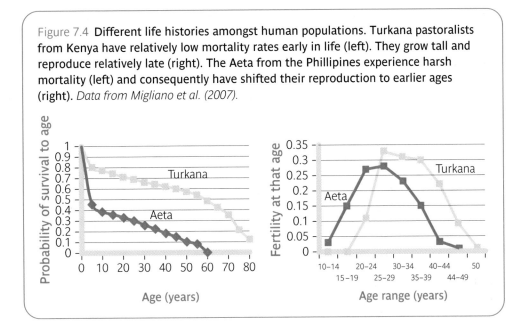

Figure 7.4 **Different life histories amongst human populations. Turkana pastoralists from Kenya have relatively low mortality rates early in life (left). They grow tall and reproduce relatively late (right). The Aeta from the Phillipines experience harsh mortality (left) and consequently have shifted their reproduction to earlier ages (right).** *Data from Migliano et al. (2007).*

quite different proximate mechanisms that could bring this adaptation about. One is genetic differentiation, with natural selection favouring different alleles for timing of growth and puberty in pygmy and non-pygmy populations. The other is phenotypic plasticity. Phenotypic plasticity is the evolved capacity of the phenotype to alter its development in response to cues during development that the world is in one state or another. We will discuss this further in Chapter 9. On this account, there need be no genetic differences between pygmy and non-pygmy populations. Rather, all humans have evolved the capacity to switch resources away from growth and into reproduction if they receive cues in their lifetime that the mortality rate is high, and in pygmy populations they receive such cues.

Living fast and dying young in Chicago

The idea that humans might have evolved phenotypic plasticity to respond to harsh environmental conditions by speeding up their life histories is supported by a number of studies in Western populations. Wilson & Daly (1997) examined life expectancy and reproductive behaviour across 77 neighbourhoods of the US city of Chicago. Life expectancy varied dramatically across neighbourhoods (for men, mid-50s in the worst neighbourhoods and high 70s in the best). The biggest differences between the best and worst neighbourhoods were in external causes of death such as homicides (the rate varied more than 100-fold across neighbourhoods) and accidents. Wilson & Daly compared the age profile of mothers between the ten neighbourhoods with the lowest life expectancy and the ten neighbourhoods with the highest (Figure 7.5). As you can see, where life expectancy is low, women have more babies overall and in particular carry out much more reproduction early in life (before the age of 25 years). This looks like adaptive behaviour given the risk of dying young.

The proximate mechanism for these adaptive differences between neighbourhoods cannot possibly be genetic differences. There is far too much mixing and moving between neighbourhoods

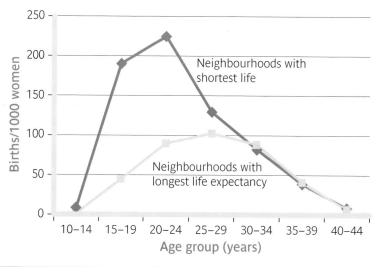

Figure 7.5 **The number of babies born per 1000 women at different ages in the ten Chicago neighbourhoods with the lowest life expectancy and the ten with the highest life expectancy. In the lowest life-expectancy neighbourhoods, women are shifted to doing more reproduction overall, especially more early in life.** *Data from Wilson & Daly (1997).*

for any genetic differences to be maintained. Instead, people must be responding to cues in their neighbourhood, such as seeing family members or friends dying young, by changing their life history strategy. Other types of early life stress can also induce changes in life history strategy. Pesonen *et al.* (2008) studied a cohort of Finnish people born during the Second World War. Because of the fighting between Finland and the Soviet Union, a large number of children at this time were separated from their parents and sent away to Sweden and Denmark. Pesonen *et al.* found that those women who had had this traumatic separation in childhood had earlier onset of menstruation and more children overall than those who had stayed at home. Amongst the men, the former evacuees had an earlier age at first parenthood and a shorter interval between successive children.

7.2.1 Quantity and quality of offspring

The previous section suggested that following a faster life history strategy (more or earlier offspring) always came at a cost (those offspring must be smaller or receive less investment). This section considers further the tension between number of offspring and the investment available for each one. The basic theory in this area was set by David Lack, a British biologist who worked on clutch size (how many eggs an individual lays in a breeding season) in birds. A bird's reproductive success is obviously higher the more chicks it has, other things being equal. However, Lack argued that the rate of mortality of chicks would increase with the number of chicks in the nest, since the food that parents could supply would be divided amongst more mouths. This increased mortality is intrinsic, as opposed to extrinsic, since it follows from

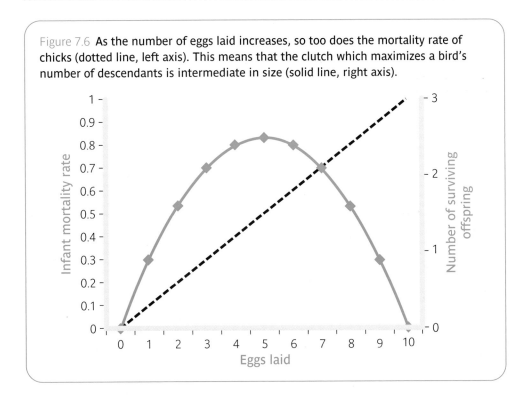

Figure 7.6 **As the number of eggs laid increases, so too does the mortality rate of chicks (dotted line, left axis). This means that the clutch which maximizes a bird's number of descendants is intermediate in size (solid line, right axis).**

the behaviour of the parents rather than being independent of what the parents do. If intrinsic mortality increases with clutch size, there will be a trade-off: increasing offspring *quantity* will tend to decrease their *quality* (meaning in this instance their probability of survival). This means that there is an ideal clutch size which maximizes the number of surviving descendants (Figure 7.6).

The ideal clutch size is generally well below the maximum number of eggs that the bird could lay. Thus, we should not expect organisms to have as many offspring as they can, but rather the number which maximizes surviving descendants in future generations, and this number is a compromise between quantity and quality.

Subsequent research has shown that the optimal clutch size in birds is probably smaller than that which maximizes the number of surviving chicks in the brood. This is because the parent might have an opportunity to reproduce again the next year and so should keep some energy back for maintaining itself in good condition for future years. Thus, the optimal brood size is the one which maximizes the parent's *lifetime* number of surviving descendants, not just the surviving descendants from this brood.

Evidence for optimal clutch size

The key assumptions of optimal clutch size theory have all been well tested. For example, Daan *et al.* (1990) manipulated brood size in the kestrel, *Falco tinnunculus*, by taking chicks from some nests (the reduced clutches) and adding them to other nests (the enlarged clutches). This is possible because birds do not commonly recognize their own young. They compared survivorship of chicks to fledging and of parents to the next breeding season for the reduced and

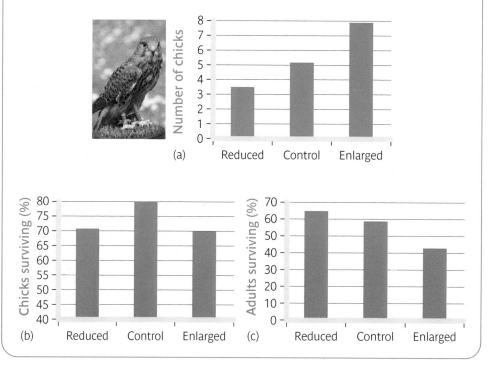

Figure 7.7 **Effects of enlarging or reducing clutch size in the kestrel** *Falco tinnunculus.* **(a) Number of chicks in reduced, control, and enlarged clutches. (b) Where the clutch is enlarged, there is a small drop in the percentage of chicks surviving, but in this case the percentage is not improved by reducing the clutch. (c) The percentage of parents surviving to the next breeding season is reduced by enlarging the clutch and increased by reducing it.** *Data from Daan et al. (1990). Photo © Paul Heasman/Fotolia.com.*

enlarged clutches compared with unmanipulated controls (Figure 7.7). The chances of survival of a chick were slightly worse in enlarged clutches compared with controls, but the main effect was that parents with enlarged broods had a much smaller chance of surviving to the next year compared with controls and reducing the clutch increased the survival rate of parents relative to controls. Thus, kestrel parents are having smaller clutches than they could rear in order to retain resources for their own maintenance and survival to the next breeding season.

Selection on brood size in humans

The vast majority of human pregnancies result in only one baby, with twins the relatively rare exception. Following the reasoning above, it seems likely that human brood size has been subject to natural selection, with mothers bearing two or more babies subject to increased mortality of either themselves or their offspring. There is good evidence for both of these effects. In a study of a rural Gambian population with relatively little access to modern medicine, Sear *et al.* (2001) showed that only 17% of twins survived to the age of 15 years, compared with 47% of singletons

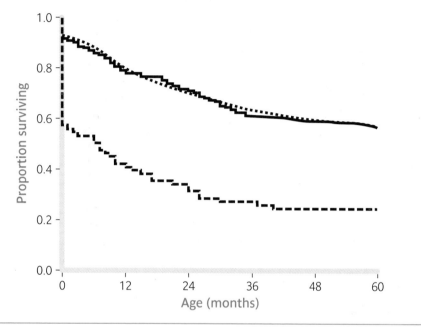

Figure 7.8 **Proportion of babies surviving over time for a rural Gambian population. The lower line represents twins. The solid line represents singletons, whilst the upper dotted line represents the singleton siblings of twins. This is an important comparison since it shows that the higher mortality of twins is not to do with the families into which they are born, but from being a twin per se.** *From Sear et al. (2001).*

(Figure 7.8). They also estimated that the rate of maternal mortality for twin mothers was two or three times that of mothers of singletons.

These are large costs to twinning. However, in Sear *et al.*'s study they were not enough to offset the increased offspring numbers of twin mothers. Sear *et al.* note that mothers of twins tend to be taller and heavier than mothers of singletons. Thus, they suggest that twinning might persist in humans as an alternative reproductive strategy that is sometimes advantageous for women in good phenotypic condition. However, Helle *et al.* (2004) showed for a Finnish population that having had twins reduces the length of a woman's lifespan even if she survives the immediate reproductive period. Thus, the increased reproductive effort has consequences in the long term. The fact that most human births are singletons suggests that the decrease in quality and maternal survival of multiple births more often than not outweighs the advantage of additional offspring.

Quality–quantity trade-offs in sequential reproduction

For organisms whose young disperse soon after birth, the main quality–quantity trade-off concerns how many offspring to have in a single reproductive event. However, where there is a prolonged period of juvenile dependency, as in the human case, there is a trade-off between quality and quantity of successive offspring, since an older child will still be dependent on parental input at the time the next child comes along, and that parental input will be reduced by having to be shared. There is quite a lot of evidence for such trade-offs. In a Finnish population born between 1946 and 1958, Helle (2008) showed that every additional sibling reduced a person's adult height by nearly half a centimetre, suggesting reduced access to resources. For contemporary Britain, Lawson & Mace (2009) showed that each additional sibling reduced the amount of parental care an individual received, and in the long term this means that children from large families have lower IQ scores and are less upwardly socially mobile than children with one sibling (Nettle 2008). These conflicts are exacerbated by the inheritance of wealth, which in many societies is needed for offspring to marry and reproduce in their turn. Borgerhoff Mulder (2000) showed that Kipsigis women from Kenya who had the largest numbers of children had fewer grandchildren than those with an intermediate number of children. This is at least partly because of their inability to set their sons up with the wealth that in this society is a strong prerequisite for getting wives and reproducing. Thus, having the greatest possible number of children is not the way to maximize one's long-term number of descendants.

7.2.2 Sex of offspring

The previous section dealt with parental decisions about how many offspring to have, but there are also decisions about what type of offspring, specifically male versus female, to have. There are conditions, reviewed in this section, under which it would be advantageous for parents to vary the sex of the offspring they have according to their current circumstances. You might be slightly puzzled by this idea, at least for mammals, since the sex of offspring is something set genetically (e.g. by the presence or absence of a Y chromosome), and therefore would not seem to be something that parents could adjust. However, there are a number of possible mechanisms that could allow adjustment. Females, for example can promote differential success at fertilization by X- and Y-bearing sperm, or differential implantation and survival of male and female embryos. Thus, it could be possible for parents to vary the sex of their offspring according to current circumstances even in organisms where sex determination is genetic. However, what would be the advantage of doing this?

The Trivers–Willard hypothesis

An influential idea in this field is from Trivers & Willard (1973). Their argument can best be expressed in the following way. In many species, females are choosy, whereas males are relatively indiscriminate. This means that females will basically always be able to find mates, whereas males in good phenotypic condition can have many mates and thus, because of Bateman's principle, very high reproductive success, whilst males in poor phenotypic condition will often find no mates at all and have zero reproductive success. Thus, the slope of the relationship of phenotypic condition to reproductive success will be steeper for males than females (Figure 7.9).

Phenotypic condition of offspring is likely to correlate with phenotypic condition of the mother, since a mother in good condition can provide more resources and better care for her young. Thus, if the mother is in poor condition, and therefore liable to produce offspring in poor condition, her offspring's expected reproductive success is going to be higher if those offspring are daughters rather than sons. You can see this in Figure 7.9: at the low-quality end, the female line is above the male one. On the other hand, if the mother is in good enough condition to produce high-quality offspring, then she will do better if those offspring are male (on Figure 7.9, at the high-quality end, the male line is above the female). Thus, the Trivers–Willard hypothesis is that, in populations where the relationships between condition and reproductive success are

Figure 7.9 **For many organisms, the relationship between phenotypic condition and reproductive success is stronger for males than for females. This means that it can be adaptive for parents liable to produce low-quality offspring to produce more daughters and parents liable to produce high-quality offspring to produce more sons, the so-called Trivers – Willard effect.**

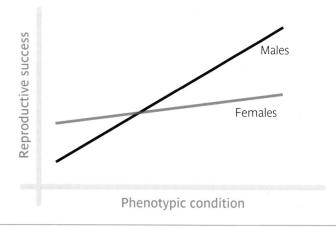

as shown in Figure 7.9, mothers able to produce high-quality offspring will produce more sons and those unable to do so will produce more daughters.

The Trivers–Willard hypothesis has spawned a huge number of empirical investigations and there are some compelling findings consistent with it. Clutton-Brock *et al.* (1986) studied the red deer, *Cervus elephas*. Female deer can be assigned a dominance rank based on their ability to displace other herd members from resources. Dominance is a behavioural measure, but it correlates with physiological condition, with more dominant females having better body condition and calf survival. The researchers first established that the conditions for the Trivers–Willard hypothesis were met, that is sons of high-ranking females had higher lifetime reproductive success than daughters of high-ranking females, but daughters of low-ranking females did better than sons of low-ranking females (Figure 7.10a). They also found that the higher a hind's dominance rank, the greater the proportion of males amongst her offspring (Figure 7.10b). This is exactly as the Trivers–Willard hypothesis would predict. However, it should be noted that many studies in many species have failed to find Trivers–Willard effects and so, although they do occur, they are by no means ubiquitous.

Trivers–Willard effects in humans

Do Trivers–Willard effects occur in humans? The first question to ask is whether we should expect them. Remember, they are only predicted where the pattern shown in Figure 7.9 is obtained. There are some parental characteristics for which the pattern does seem to be obtained quite often. For example, good maternal nutritional status will produce larger offspring and being larger has a more beneficial effect on male than female reproductive success in at least some human populations (Nettle 2002). Thus, we might expect women in better nutritional condition to produce more sons. Material and financial resources also benefit the reproductive

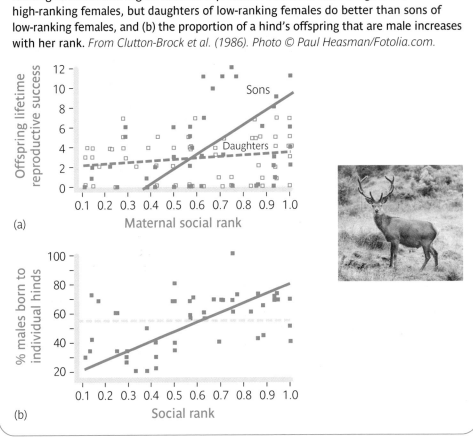

Figure 7.10 **For red deer (*Cervus elephas*) on the isle of Rhum: (a) sons of high-ranking females have higher lifetime reproductive success than daughters of high-ranking females, but daughters of low-ranking females do better than sons of low-ranking females, and (b) the proportion of a hind's offspring that are male increases with her rank.** *From Clutton-Brock et al. (1986). Photo © Paul Heasman/Fotolia.com.*

success of men more than women in a number of populations (Nettle & Pollet 2008), and so maternal access to these might be expected to correlate with proportion of sons.

For maternal nutritional status, one of the clearest results is from Gibson & Mace (2003), who studied Oromo agro-pastoralists in rural Ethiopia. They assessed nutritional condition by a measure of muscle mass on the arm. As Figure 7.11 shows, women with the most arm muscle mass were more than twice as likely to have had a son at their most recent birth than the women with least muscle mass. These are very large effects and studies in other populations have not found comparable ones. This may be because the Oromo at the time were living under high nutritional stress and some of the women were very malnourished.

In more affluent populations, the range of body condition amongst women may not be broad enough to observe significant correlations with sex ratio. However, dieting may mimic the effects of nutritional stress in the short term. Mathews *et al.* (2008) found for British women that mothers of sons ate more around the time of conception than mothers of daughters (food intake was around 100 calories more, including significantly greater quantities of carbohydrate, protein, and fat). Neither body mass nor food intake later in pregnancy were significantly

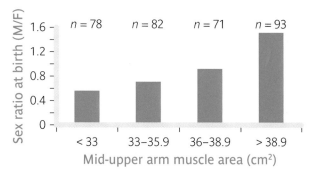

Figure 7.11 **In a rural Ethiopian population, women with the lowest arm muscle mass have many more daughters than sons, whilst women with the highest muscle mass have more sons than daughters.** *From Gibson & Mace (2003).*

different, suggesting that food intake during pregnancy is acting—misleadingly—as an immediate cue that affects implantation of male embryos. Mathews and colleagues suggest that the gradual decline in the proportion of boys born in Western countries in recent decades may be linked to dieting and changes in women's eating habits.

What about access to resources? A number of studies have found increased proportions of sons in women who are married versus unmarried, more educated compared with less educated, or where family status or wealth is higher (Hopcroft 2005; Almond & Edlund 2007). However, the effects are often small and a number of studies have failed to find them, suggesting that, although adaptive sex ratio variation in humans is real, the effects are sometimes weak and not ubiquitous.

7.3 Parental care

Having decided to reproduce and produced offspring, the next set of life history decisions concerns how much parental care to provide. As humans, we tend to assume that it is normal for both parents to provide prolonged parental care, but actually this is quite a rare situation in the natural world. In this section, we review the selective pressures affecting parental care and some of the behaviours that result.

7.3.1 To care or not to care?

In most organisms, there is no parental input after the production of gametes. In many fish, for example, eggs and sperm are released into the water and left to fend for themselves. However, in a minority, one parent stays around and protects the developing eggs or fans water over them. When should we expect such behaviour to evolve?

Time and energy allocated to caring is time and energy that the parent could also be using elsewhere, to gather food for itself or to reproduce again. The *cost* of caring is thus the opportunities for these other things which are forgone in order to care. The *benefit* will be

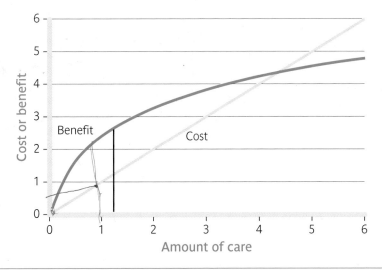

Figure 7.12 **The costs of providing parental care (light blue line) accumulate linearly with care provided, whereas the benefits (dark blue line) flatten off. Organisms are selected to care up to the point where the cost of one more unit of care is greater than the extra benefit it would bring. This is the point where the two curves are furthest apart, marked by the vertical line. Note that if the benefit of the first unit of care is less than its cost, no parental care will evolve.**

the improvement of offspring survival or condition, which is brought about by acts of care. There will come a point where there are diminishing returns to any more care (e.g. once an offspring is feeding and swimming about for itself, the benefit of providing more care is likely to be close to zero), and so the benefits of caring will flatten off as the amount of care provided increases (Figure 7.12). On the other hand, the costs of caring continue to add up at at least a constant rate. Given this scenario, we should expect parents to be selected to care until the point where the additional benefit of providing one extra unit of care does not exceed the cost of so doing (the point marked on Figure 7.12).

If the benefit of the first unit of care does not exceed its cost, no care will be provided at all. As mentioned, this scenario is common amongst fish. However, some kind of parental care is universal amongst mammals and birds, in mammals because young are fed milk, and in birds because eggs must be incubated and chicks fed. Mammals and birds would have zero reproductive success if they did not provide some parental care, and parental care is thus said to be obligate (as opposed to facultative, which would mean that care is beneficial but not absolutely necessary). What this amounts to in terms of Figure 7.12 is that for birds and mammals the benefit of the first few units of care is so large as to always outweigh the costs. However, for mammals and birds, providing care also has diminishing returns and selection will favour stopping at the point where the marginal costs start to outweigh the marginal benefits.

Figure 7.13 **Most organisms provide no parental care, but there are exceptions, with uniparental care by females common in mammals, uniparental care by males widespread in fish, and biparental care common in birds.** *Left: © Jeff Chiasson/ istock.com; right: © Toos van den Dikkenberg/istock.com.*

7.3.2 Who cares?

The previous section discussed when care behaviour should be expected to evolve, but said nothing about which parent should provide it. Care can be either uniparental (one parent provides) or biparental (both parents; Figure 7.13). What conditions will favour one or other of these outcomes?

One way to think about this problem is to think about costs and benefits for the second parent, given that the first parent is already caring. For the second parent, it is not the benefit of an additional unit of care in general, but the benefit of an additional unit of care *given that the other parent is already caring*, that has to exceed the cost of providing it. This criterion is harder to meet than the criterion for a first parent to care and, in many types of organisms, the first parent cares, the second deserts, and care is uniparental.

You might be forgiven for thinking that the first, caring parent is always the mother, and the second, deserting parent the father. For mammals and for fish with internal fertilization, this is indeed the case, for the obvious reason that it is much easier for the male to desert than it is for the female. However, in fish with external fertilization, the female deposits the eggs before the male releases his sperm onto them. Thus, she has the earlier opportunity to desert, and where there is any care, it is more often the male who provides it.

Biparental care in birds

The great exception to the prevalence of uniparental care is found amongst the birds, most of whom show biparental care in the form of sitting on eggs and bringing food to the nest. The evolution of biparental care is related to the fact that bird eggs must be incubated. Given that birds carry little stored fat and thus the female must frequently feed, and that even brief periods away are a danger to eggs, the benefit to the second parent for remaining involved are large.

7.3.3 Care by male mammals

In mammals, the females, being the sex that gestates and lactates, are obligate carers, whilst males are facultative carers and usually do not provide any care at all. The males are the sex with the opportunity to desert, but they also have other reasons for not investing. Whereas a female is certain that the offspring she produces are her own, a male may or may not be the sire of his mate's offspring. In other words, females are certain of maternity, whereas males are always faced with some degree of paternity uncertainty.

Paternity uncertainty means that the hurdle to be overcome before it is worthwhile investing in offspring is even higher than discussed before: the additional benefit in terms of offspring survival or quality, multiplied by the level of paternity certainty, and given that the female is already investing, must exceed the cost of providing the care. That this threshold is usually not met is suggested by the fact that unambiguous paternal care for offspring is reported in fewer than 10% of mammal species. Some of the clearest exceptions occur where the social system means that mated females encounter few males other than their mates. In the owl or night monkeys of South America (genus *Aotus*), there are long-term pair bonds, and the social group consists of just a breeding pair and their offspring, so males have high paternity certainty and they, rather than the females, carry the offspring around.

However, even where there are social groups containing many males, some kinds of paternal care can be found. Savannah baboons (*Papio cynocephalus*) live in large groups containing multiple adults of both sexes, and females mate with multiple males. Buchan *et al.* (2003) showed that in disputes between juveniles, adult males intervened on the side of their offspring more often than would be expected by chance. It is not clear whether they identify their offspring by directly recognizing some aspect of the phenotype or by tracking the amount of time spent consorting with the mother before the offspring was born, but clearly some kind of paternal input has evolved. Moreover, having a father present in the social group seems to improve offspring growth rates (Charpentier *et al.* 2008).

The other side of the coin: infanticide

Males favouring their own offspring has a flip-side, namely disfavouring the offspring of other males. One of the commonest causes of death of infants in mammalian groups is killing by males other than the father. By doing this, males bring females into reproductive condition sooner than they otherwise would, which gives them an opportunity to mate with them themselves. Having mated with a particular female reduces males' propensity to kill her offspring (e.g. in mice, Cicirello & Wolff 1990), and the most compelling argument for why females in multi-male groups often mate with several males is to confuse and discourage infanticide by giving more males the possibility of being the father (Wolff & Macdonald 2004).

Paternal investment in humans

In humans, it is universally women who provide more care for children than men, which is unsurprising given prolonged lactation. However, paternal investment in humans is considerable and widespread. Given that humans live in multi-male groups where there is always the opportunity for extra-pair copulation, we might thus expect human males to have evolved sensitivity to likely genetic relatedness in order to prevent them allocating investment to infants who are not their own.

There is evidence that men are sensitive in this way. Anderson *et al.* (2007) found, for a sample of men in New Mexico, that feeling uncertain of being a child's biological father increased the probability of the father divorcing the mother and decreased his involvement in the child's development. Platek *et al.* (2004) conducted an experiment where men had to choose a photograph of a child from an array of five in response to questions like 'Which one of these children would you adopt?' and 'Which one of these children would you resent least paying child support for?' The researchers had taken a photograph of the participant beforehand and, unbeknown to him, they had used computer morphing to make the face of one of the five children resemble him. Men chose the child that looked like them for hypothetical investment more often than would be expected by chance. Interestingly, mothers of newborns are overwhelmingly likely to claim that the baby resembles the father, especially in the father's presence (Daly & Wilson 1982).

Effects of paternal investment

Men would only be selected to provide paternal investment if it had some benefit. However, cross-culturally, the evidence for the benefits of fathers is somewhat equivocal. Sear & Mace (2008) reviewed 45 studies of infant mortality from pre-industrial human populations across the globe, including historical European populations. They showed that having a mother who is alive and resident improved a child's chances of survival in all societies studied. Having a father alive and resident only made a difference in a minority of societies. Positive effects of fathers in keeping children alive were less widespread than positive effects of maternal grandmothers (see section 7.4). Why then is human paternal care so widespread?

There are several possible answers. One is that paternal investment is actually a form of mating effort or mate guarding. In other words, by spending time with a woman and her children, and providing resources, men are increasing their chances of being the father of her future children. Another possibility is that fathering represents facultative investment that may not be essential for survival in all ecologies, but can improve offspring quality. There is some evidence for this position from contemporary Western societies. British adults whose fathers played a significant role in their upbringing have higher IQs and more upward social mobility than those lacking paternal input (Nettle 2008). A third, more disturbing, possibility is that fathers are protecting their children from other males.

Infanticide in humans

Just as unrelated males wishing to consort with the mother can pose a danger to infants in other mammal species, so it is with humans. Amongst the Aché of Paraguay, the mother remarrying someone other than the father doubles a child's risk of dying before the age of 15 years. In Western nations, the risk of child injury and death is greatly increased by the presence of a stepfather (Daly & Wilson 1988). Indeed, the rate of child homicide increases around 100-fold, making this the strongest predicting factor for violence against children that has ever been discovered. The age pattern of these homicides (Figure 7.14) shows that the big risks are when the child is under 5 years and particularly under 2 years. This is the age when children are most physically vulnerable and also when men have most to gain in terms of diverting the woman's energies into their children rather than those from a previous relationship. Children in households containing stepfathers are, perhaps unsurprisingly in view of the above, more stressed than those living with biological fathers (Flinn & England 1995).

Set against these rather disturbing findings is the fact that many men provide good stepparental care and nurturance. From an evolutionary perspective, this is perhaps best interpreted as a form of mating effort. By investing in a woman's children, a man strengthens his relationship with her and increases the chance of being the father of her future offspring. Evidence consistent with this view comes from the fact that step-parental care is sharply reduced if the man is not in an ongoing relationship with the child's mother (much more sharply than his investment in his own genetic children; Anderson *et al.* 1999).

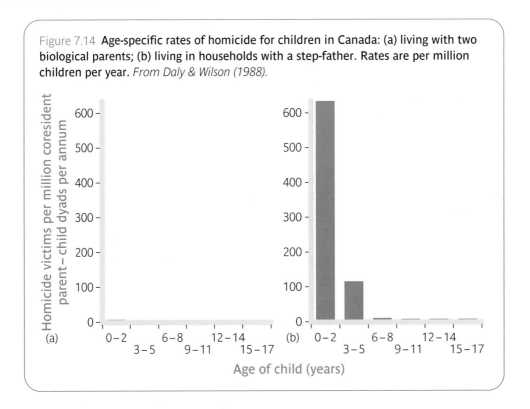

Figure 7.14 **Age-specific rates of homicide for children in Canada: (a) living with two biological parents; (b) living in households with a step-father. Rates are per million children per year.** *From Daly & Wilson (1988).*

7.3.4 Parent–offspring conflict

Both parents and offspring have an interest in the offspring doing well, the offspring for obvious self-interest reasons and the parents because the offspring are their genetic future. However, the interests of parents and offspring are not perfectly aligned. Instead, evolutionary theory predicts that there will be a degree of parent–offspring conflict over the amount of parental care provided.

The fundamental insight into why this is the case comes from Robert Trivers (Trivers 1974). Let us take the case of a mammal with uniparental care by females. We saw in section 7.3.1 that such a mother faces a trade-off in how much care to provide. Energy allocated to caring for the current offspring is energy that she could spend on garnering resources to have future offspring. Figure 7.12 showed mothers are selected to provide care up until the point where the benefit to the offspring of the next unit of care is less than the cost of providing that unit (i.e. $b < c$, where b is the benefit and c is the cost).

The mother is equally related to her current and her future offspring. This is not the case for the offspring. The offspring is perfectly related to itself, but its coefficient of relatedness to its mother's future offspring is $\frac{1}{2}$ (less if the fathers are different, but we will ignore this for now). Thus, for the offspring, energy allocated to mother's future offspring has only half as much value in terms of inclusive fitness as energy allocated to itself. This means the offspring's fitness interests are best served by the mother continuing to invest in it until the point where $b < \frac{1}{2}c$. This means, effectively, that the offspring is selected to want more resources from the mother than the mother is selected to want to give it, and in the region where $\frac{1}{2}c < b < c$, the two parties are in conflict about the allocation of care. Figure 7.15 illustrates the situation graphically.

Figure 7.15 **The logic of parent–offspring conflict. Parents are selected to provide care to the point where** $c < b$ **(solid vertical line, in this case 1.3 units of care), whereas offspring are selected to seek care to the point where** $\frac{1}{2}c < b$ **(dashed vertical line, in this case 2.8 units of care). Between the two vertical lines, conflict exists between parents and offspring over the allocation of care.**

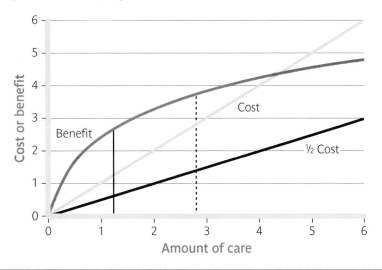

The theory of parent–offspring conflict has proven informative in many respects. For one thing, there appear to be hormonal and physiological battles between embryo and mother during mammalian gestation, in which the embryo tries to extract more resources from the mother for growth than she is prepared to supply. Conflict also occurs over weaning in mammals, and this can be easily observed in domestic or farm animals (Figure 7.16). Mothers

Figure 7.16 **Parent–offspring conflict over weaning. When offspring are small, mothers provide continuous access to milk, but as they grow, they reach a point where offspring continue to seek access but mothers start to block it.** *Left: © Sportlibrary/Foltolia.com; right: © Marilyn Barbone/Fotolia.com.*

of newborn offspring provide continuous access to teats. However, as the offspring grow, a point is reached at which they increasingly block off access and may drive away offspring trying to suckle. The offspring continues to try to suckle when it can, long beyond this point. As for humans, you can reliably observe parent–offspring conflict in your local shops, where the timeless drama of small children wanting more toys or treats to be put into the basket and mothers grim-facedly explaining that they cost too much can be observed any day of the week.

7.4 Grandparental care

In most mammals, reproduction goes on until general senescence. This means that death closely follows the cessation of fertility. Human females are unusual in this regard, experiencing menopause, the cessation of fertility, several decades before general senescence. Why has this long post-reproductive period evolved?

Linked to this question is the fact that grandparental care is ubiquitous in humans. Indeed, grandmother presence has been shown to improve child survival in more populations than father presence (Sear & Mace 2008). This makes humans unusual: in most mammals, there is no grandparental care of grandoffspring. Thus, it could be that menopause in women is selected for because women after the age of around 50 years can better enhance their lifetime reproductive success by helping their children to reproduce rather than having more children of their own (Hawkes *et al.* 1997).

This is a plausible argument and the evidence on the magnitude of benefits to grandoffspring of grandparental involvement suggests that it may be correct (Shanley *et al.* 2007). However, the evolution of a switch from mothering to grandmothering is only likely where some other unique conditions are met. Mortality during childbirth has to increase strongly with maternal age and offspring have to be sufficiently expensive to rear that having care from one individual makes a substantial difference. Both of these assumptions are met in humans, but may not be in other female mammals, accounting for the rarity of menopause.

There is no evidence that men have been selected to switch from fathering to grandfathering, since the presence of grandfathers less often has positive effects on offspring than the presence of grandmothers (Lahdenperä *et al.* 2007). However, male fertility does not cease so abruptly or so long before senescence as female fertility does and men in many societies continue to sire children at much later ages than women, so there is less of a need for an adaptationist explanation of male post-reproductive lifespan.

Lineage differences

Grandparents, and associated relatives such as uncles and aunts, can be divided into two lineages: the matriline (mother's parents and their relatives) and the patriline (father's parents and their relatives). Matrilineal relatives are connected to an individual via the certainty of maternity and patrilineal relatives via the uncertainty of paternity. Thus, other things being equal, the fitness return on investment in matrilineal kin will be higher than that in patrilineal kin. One consequence of this is that the maternal grandmother may be more disposed to provide care than the paternal grandmother. A number of studies have found that this pattern does indeed occur (see e.g. Pollet *et al.* 2007), with similar effects for uncles and aunts (McBurney *et al.* 2002).

Slightly at odds with the tendency of grandmothers to invest more time in their daughters' children than their sons' children is the fact that in most existing human societies the overall cultural significance of the patriline is higher than that of the matriline. For example, names often pass down with the father and with them the family land or herd. Why would this be, given that matrilineal relatives are more surely related?

The most likely answer is related to the Trivers–Willard hypothesis (section 7.2.2). Inherited wealth or status will often produce greater benefit when given to sons than daughters because the slope of the relationship of reproductive success to wealth or status is higher for men than for women. This can be enough to make it adaptive for parents to pass all to their sons, even at the risk that their sons then pass it to individuals who are not biologically related (Holden et al. 2003). Holden and colleagues' model predicts that where either there is little by way of inherited wealth to pass on, and/or where paternity uncertainty is high, the matriline will become more important than the patriline.

This prediction is met. A number of matrilineal societies, where mother–daughter relationships are more culturally important than father–son relationships, are found across the world. They tend to be associated with social practices that lead to paternity being uncertain and also with the absence of substantial transmissible wealth assets such as herds of livestock (Holden et al. 2003). Since hunter-gatherers do not have substantial transmissible assets, it is possible that the preponderance of patrilineal social organization is a development that has arisen since the origin of agriculture and herding within the last 10,000 years.

 # Summary

1. Life history theory is the branch of evolutionary theory dealing with how natural selection acts on the allocation of resources to reproduction over the life course.

2. The disposable soma theory shows how, because of extrinsic mortality, it can be adaptive not to perfectly repair the soma, but instead to divert energy into reproduction, even though this eventually leads to senescence and death.

3. Mammalian life histories can be arrayed along a general continuum from fast (maturing young, high reproductive effort, and low per offspring investment) to slow (late maturing, low reproductive effort, and high per offspring investment). High extrinsic mortality rates select for fast life histories.

4. Humans have a generally slow life history pattern, but have the flexibility to respond to high-mortality conditions by speeding it up.

5. Organisms have to trade off the quality of their offspring against the quantity. This generally leads to the evolution of a clutch or litter size considerably smaller than the largest they could produce.

6. It can be adaptive for females in good condition to bear more sons and females in poor condition to bear more daughters.

7. Parents provide care to the extent that the benefit exceeds the cost. This extent may differ for the two parents.

8. Parents and offspring are in conflict over the amount of energy allocated to care.

9. Human females may have evolved menopause as an adaptive switch from reproductive effort to increasing their fitness via grandmothering later in life.

❓ Questions to consider

1. We reviewed the argument that organisms keep their clutches or litters smaller than they could be, in order to retain energy for their own survival. What predictions would you make about how an individual's last reproductive event might differ from its first?

2. In the past couple of hundred years, developed nations have gone through what is known as the demographic transition. This is the transition to having much smaller families (in Britain, from five to six children per woman 200 years ago to fewer than two now). How might we interpret this change in behaviour from an evolutionary perspective?

3. A number of human populations, particularly in lowland South America, have social customs known as partible paternity. This means that several men are thought of as fathers of a particular child, often to differing extents. Partible paternity is associated with multiple mating by women, and children with more than one father seem to have better outcomes than those with just one. Why do you think this custom may have evolved?

4. If a woman with a young child loses her husband, when is it in her interests to marry again and when is it in her child's interests for her to do so? Would you expect family conflict over her remarriage?

➡ Taking it further

The account of life history theory described here is quite simplified, for example the expected effects of increased mortality on life history depend on when the mortality occurs and how variable it is. For more detailed, if mathematically challenging, treatments of the theory, see Stearns (1992) and Roff (1992). Also, the idea of a single slow–fast continuum is a simplification; see Bielby et al. (2007) for recent work. For the rest of the material in this chapter, the original studies cited are readable and easy to get hold of, and I recommend the reader to go directly to them and the other references they cite. A key review of recent ideas about human life history is given by Kaplan et al. (2000).

Social life

The last chapter concerned relationships between parents and offspring, and grandparents and grandoffspring. We also have important relationships with others who are not our descendants or ancestors, such as our mates, our colleagues, and our friends. This chapter concerns those social relationships and the evolutionary pressures which affect them. In section 8.1 we consider the question of why (some) animals live in groups at all and section 8.2 describes the various types of social group that occur amongst non-human mammals. In section 8.3 we look at the consequences of group living and section 8.4 asks how human groups fit into this scheme. Section 8.5 then takes on a key question for understanding human behaviour, namely how does social cooperation between unrelated individuals evolve?

8.1 Why live in groups?

We humans are such strongly social creatures that we forget that not all animals belong to social groups. Leopards, for example, live solitary lives, each one patrolling a discrete territory

Figure 8.1 **Leopards are solitary, whereas lions live in prides. Both sociality and non-sociality have evolved multiple times in animals.** *Photos © Digital Vision.*

(Figure 8.1). Male–female interaction is limited to brief mating, and male–male interaction to fighting each other in territorial disputes. Cubs remain with their mothers for up to 2 years, but then disperse to find territories of their own. In contrast to leopards, lions live in prides comprising several adult females, one or two adult males, and their cubs. The adult females usually sleep, move, hunt, and feed as a group.

The leopard/lion contrast is mirrored across many groups of animals. There are solitary rodents and social rodents, solitary whales and social whales, solitary bats and social bats, and solitary fish and social fish. Thus, it is clear that both non-social and social forms of living can evolve, and have done so many times. The question is, then, under what circumstances sociality versus non-sociality is favoured. As ever, the general evolutionary expectation is that we should expect sociality to be found wherever the benefits exceed the costs and to be absent wherever the costs exceed the benefits. However, what are the benefits and costs of sociality?

8.1.1 Benefits of group living

A number of different benefits of sociality have been proposed and tested. Which ones are important for a particular animal—and there might be several—will obviously depend on its ecological niche. Here, we review some of the most important.

Predation

A key benefit of group living for many social species is a reduction in the risk from predators. Being in a group reduces the risk of predation for two main reasons. One is simple dilution. A sparrowhawk can only carry off one sparrow at a time. If a sparrow lives in a flock of 100 individuals, then if the sparrowhawk strikes, the chance of being the victim is just 1% on average, whereas for a solitary sparrow the chance of being the victim in a strike is 100%. The advantage of flocking together might be diminished by the large flock being more visible to sparrowhawks and thus suffering more attacks, but the number of attacks is unlikely to be 100 times greater and so the risk of predation is still lower overall in the flock.

In addition to this effect, predators may be less effective when they attack grouped prey animals than when they attack solitary ones. One reason for this is that the larger the group, the greater the chance of someone spotting the predator early and beginning evasive action. In fact, the

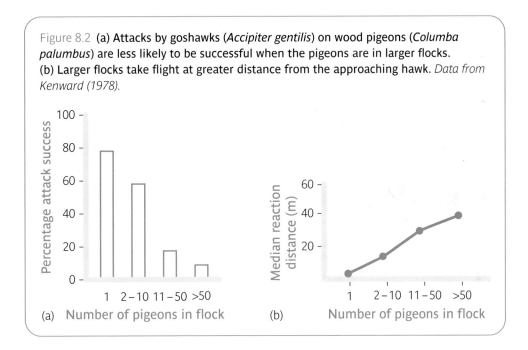

Figure 8.2 **(a)** Attacks by goshawks (*Accipiter gentilis*) on wood pigeons (*Columba palumbus*) are less likely to be successful when the pigeons are in larger flocks. **(b)** Larger flocks take flight at greater distance from the approaching hawk. *Data from Kenward (1978).*

larger the group, the smaller the proportion of time any individual needs to spend scanning around for predators in order for constant vigilance to be maintained at the group level.

Kenward (1978) demonstrated this effect in action in attacks by goshawks on wood pigeons (Figure 8.2). The larger the flock, the smaller the proportion of attacks that ended in a kill. This was because larger flocks tended to take flight when the predator was further away, presumably because with so many eyes the probability of at least one individual detecting the hawk early was increased.

In addition to fleeing predators, larger groups of animals may also have advantages in other types of anti-predator action, such as attacking back and confusing predators with massed movement.

Joint foraging

Animals in groups may also enjoy advantages of taking on joint ventures, for example joint hunting. Pack hunters can take larger prey together than any one individual could and predatory fish derive advantages from being able to surround a shoal of prey. Note that these kinds of joint ventures are favoured not when the total return from the hunt is greater in a pack than alone, but when the return *per individual* is greater in a pack than alone. This means that social hunting has to be very much more fruitful than solitary hunting to offset the greater number of mouths at the kill. For example, hunting in pairs is only favoured if it yields overall more than twice the calories of hunting alone, hunting in groups of six if it yields six times the calories, and so on. In addition, social hunting raises problems of cooperation. It is always better for each individual to let someone else do the dangerous bit of actually killing the prey and just feed from the results. We return to these problems of cooperation in section 8.5.

Note that animals hunting in a pack does not mean that the pack has formed because of its advantages in hunting. In African lions, *Panthera leo*, for example, Packer *et al.* (1990) showed

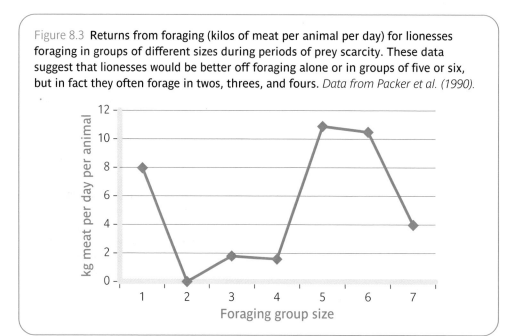

Figure 8.3 **Returns from foraging (kilos of meat per animal per day) for lionesses foraging in groups of different sizes during periods of prey scarcity. These data suggest that lionesses would be better off foraging alone or in groups of five or six, but in fact they often forage in twos, threes, and fours.** *Data from Packer et al. (1990).*

that when food was scarce the return of meat per day was high for females hunting alone or for females hunting as a group of five or more, but low when hunting in twos, threes, or fours (Figure 8.3). This is because twos, threes, and fours seek the same prey species as a solo huntress would but are less successful and have to share it out when they do succeed. Groups of five or more can target larger prey (or larger carcasses, since lions actually scavenge a significant part of their diet).

Despite these clear advantages of foraging either alone or in a large group, the most common lioness party sizes observed were twos, threes, and fours. This suggests that the efficiency of group hunting is not the benefit driving the social aggregation of these animals, and some other factors must be relevant.

Defence of territory

A different type of joint venture is the defence of a territory. It can be advantageous for an individual to control a territory containing sources of food and water, but this means that it will have to be able to repel others of the same species trying to come in. A pair or group of animals may be able to defend a territory where a single animal could not. Indeed, if some members of a population have become social, this exerts pressure on others to follow suit, on the basis that a group can only be repelled by another group of at least equal size. A limit on such defensive group formation is imposed by the number of mouths that the territory can feed, and thus territory-defence sociality is particularly likely where there are resource-rich patches such as fruit trees or waterholes which are defendable but can support more than one animal.

Resource defence appears to be important for lionesses. Packer *et al.* (1990) observed a number of inter-group encounters between lionesses and these usually ended with an intense chase, which the larger group won. Several prides of one or two females lost their territories, which did not happen to larger prides.

Care of offspring

Another type of joint venture that may make grouping beneficial is joint care of offspring. If a parent is to provide care, then it must remain with the offspring for a period of time, and if both parents are to provide care they must remain with each other. Thus, where there is biparental care, species have to be social, at least for the reproductive period, with a minimal group of two parents and the young. Where the parental relationship extends over several reproductive episodes, a form of social group centred around the monogamous couple and their current and recent offspring is found. This kind of social organization is found, for example, in the night monkey (see section 7.3.3).

Biparental care is not the only kind of joint care. To return to the African savannah, lionesses of the same pride pool their cubs together into a 'crèche' from the age of about 6 weeks to 2 years. The greater the number of females who have cubs in the crèche, the higher the proportion of cubs that survive. The main source of this benefit appears to be defence of the cubs against infanticidal males, who accounted for 27% of cub deaths in Packer and colleagues' 1990 study. Where there were aggressive encounters between the pride and male strangers, all the cubs died in five of six cases where there was just one mother with cubs, but some cubs always survived where there was a crèche of more than one female. Thus, lionesses have good reasons for sticking together.

Information transfer

Another potential benefit of living in a group is that seeing the behaviour of others allows access to useful information. This idea has been best developed for colony-living animals that forage separately and return to a base, such as birds or bats. There is good evidence that individuals can identify others who have foraged successfully and follow them to find resources the next time. More generally, observing other members of the same species can provide cheap information about behaviours that are suitable for the current environment (see Chapter 9).

8.1.2 Costs of group living

We have examined a number of benefits to group living. However, if there were only benefits then we should expect all animals to live socially, which they do not, so there must be costs as well. Although there are a number of likely costs to sociality, for example the increased transmission of infectious disease in more social animals, the best documented cost is feeding competition. The more mouths there are to feed, the quicker local resources will run out.

There would seem to be two responses to feeding competition within a group. Some animals disperse temporarily to forage independently, but return to group together the rest of the time. Colony-roosting bats or birds would be good examples. The anti-predator advantages of being in a group are thus enjoyed whilst the animals are at their most vulnerable, such as when they are resting, and the competition for food is attenuated by their temporary dispersal. However, the advantages of being in a group are lost while foraging.

The second strategy is to retain the integrity of the group the whole time. This has the effect of depleting food sources rapidly and so, the larger the group, the more quickly it has to move from patch to patch in order to get enough to eat. Thus, a cost of being in a larger group is increasing travel around in order to acquire enough food.

The costs and benefits of sociality are well demonstrated in a recent study of a number of groups of blackbuck antelope, *Antilope cervicapra*, in India (Isvaran 2007). The larger the group, the more quickly an approaching 'predator' (the researcher himself) was detected (Figure 8.4a),

Figure 8.4 **In blackbuck antelope (*Antilope cervicapra*), (a) the distance at which an approach is detected increases with group size, (b) the proportion of time individuals devote to vigilance decreases with group size, and (c) the distance travelled per hour increases. Data such as these shed light on the benefits and costs of living in a group.** *From Isvaran (2007). Photo © tephen Bonnau/istock.com.*

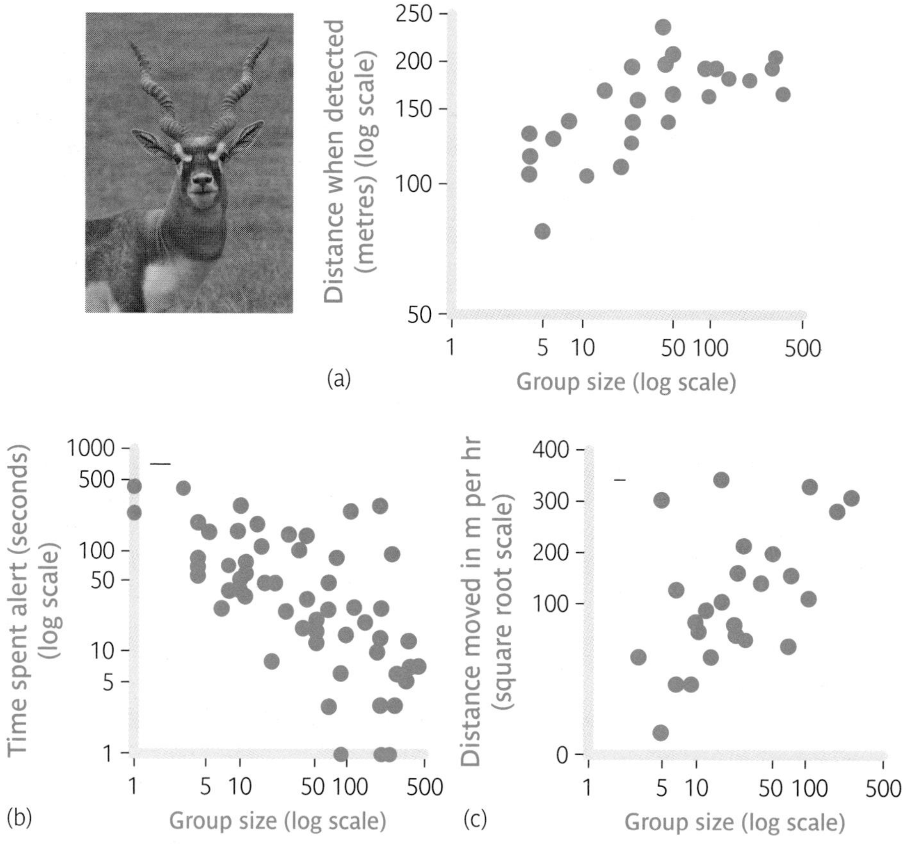

this despite the fact that in larger groups individuals spent less time on vigilance (Figure 8.4b). However, as groups got larger, the animals had to move further per hour in order to forage (Figure 8.4c). There will come a point where this increased travel is energetically uneconomic and at this point the costs of being in such a large group outweigh the benefits.

8.1.3 Optimal group size

As group size increases, the benefits of extra members may begin to level off, whilst the costs in terms of feeding competition may begin to become prohibitive. Thus, there will be an intermediate level of group size that reflects the optimal compromise. There is good evidence of selection for intermediate-sized groups and against extremes. For example, Armitage & Schwartz

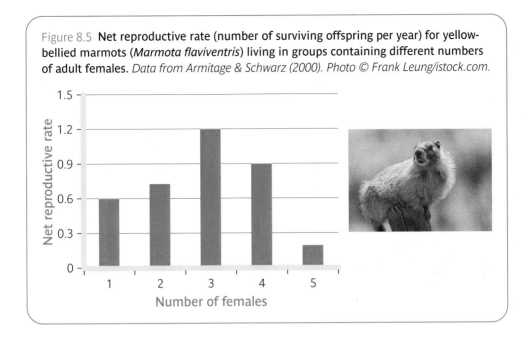

Figure 8.5 **Net reproductive rate (number of surviving offspring per year) for yellow-bellied marmots (*Marmota flaviventris*) living in groups containing different numbers of adult females.** *Data from Armitage & Schwarz (2000). Photo © Frank Leung/istock.com.*

(2000) studied female reproduction in yellow-bellied marmots, *Marmota flaviventris*, in Colorado, USA. These ground-living squirrels form groups of related females ranging from one to five adults. Female reproductive success was highest in groups of three and significantly lower in lone females and in larger groups (Figure 8.5).

This raises the question of why not all the marmots lived in groups of three. Population dynamics constantly affect group size. When a female dies in a group of three, it becomes a group of two. When a female disperses, she cannot instantly recruit two group-mates and she may have to disperse if local resources are insufficient to feed three mouths. As for groups being larger than optimal, this is a common finding. The reason is that, although reproductive success is lower in a four than a three, an extra female still does better joining an existing group of three than she would by living alone. The residents may try to exclude her, but this will be costly for them. Thus, groups slightly larger than optimal will be common. Only where the group is so big that all parties would gain from its subdivision will the group definitely fission.

These considerations suggest that social animals should have evolved to be sensitive to the size and dynamics of their current group and be able to alter their social behaviour as strategically appropriate. A number of studies show that this is the case. Mantled howler monkeys, *Alouatta palliata*, have a social system similar to that of the yellow-bellied marmots, with similar optimal group size considerations. Half of all females born into groups with two adult females remain in their natal group, whereas only 10% of females born into groups of three, and none of the females born into groups of four, remain (Pope 1998). This shows that females are able to assess local group conditions in their decisions about dispersal.

Ring-tailed lemurs, *Lemur catta*, live in groups of up to 25 females. Pride (2005) showed that levels of cortisol (a stress hormone) in these animals are lowest when they are in intermediate-sized groups and significantly elevated if the group is very small or abnormally large (Figure 8.6). Again, this suggests that the lemurs' cognitive and emotional mechanisms are attuned to the current state of the social group.

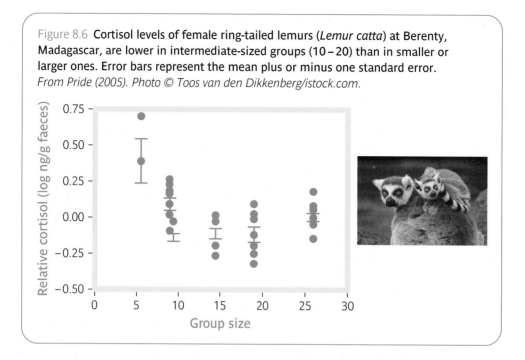

Figure 8.6 **Cortisol levels of female ring-tailed lemurs (*Lemur catta*) at Berenty, Madagascar, are lower in intermediate-sized groups (10 – 20) than in smaller or larger ones. Error bars represent the mean plus or minus one standard error.** *From Pride (2005). Photo © Toos van den Dikkenberg/istock.com.*

8.2 Types of group

Animal groups vary not just in size but also in their composition: how many of each sex, with what relationship to one another. In this section, we review some of the main types of social organization that can be found. The social system of a species is often related to its mating system, so before turning to social systems in section 8.2.2, we investigate the different types of mating systems that occur.

8.2.1 Mating systems

Among sexual species there are a number of ways that matings can be distributed between the two sexes. These are as follows:

1. monogamy: both males and females have one mate
2. polygyny: a male mates with several females, but each female only has one mate
3. polyandry: a female mates with several males, but each male only has one mate
4. promiscuity: any female may mate with any male.

Monogamy, polygyny, and polyandry can be further divided into cases where the mating bonds last for just one breeding event and where there is longer-term association between males and females (e.g. lifetime monogamy versus monogamy for one breeding season). Promiscuity is associated with a lack of long-term bonds between mates, although males and females may have more transient consortships.

Where only one (or neither) sex makes any post-reproductive investment in caring for the offspring, the mating system is basically determined by ecology. For example, in mammals where there is little paternal care, females distribute themselves so as to optimize their access to resources and males distribute themselves to optimize their access to females. Thus, if females are dispersed singly around the environment, monogamy may arise simply because a male can only monopolize one female. This happens, for example, in some monogamous deer. If females forage in groups, a male may be able to exclude other males from mating a whole group of females, giving a polygynous harem system, as seen in gorillas or elephant seals. Polygynous harem systems are associated with strong sexual dimorphism in size (see section 6.4.2). If females are aggregated or mobile, but it is uneconomic for any one male to exclude others from access to a female, the system becomes promiscuous. Females in promiscuous systems may benefit from mating multiple males in order to confuse paternity and thus reduce the likelihood of infanticide (section 7.3.3). Males in promiscuous systems, such as that of the chimpanzee, tend to have large testicles, since female mating with multiple males generates competition between the sperm of different males, leading to selection for larger ejaculates.

Polygyny is very common across the animal kingdom, due to Bateman's principle (see section 6.4.1) making it more often beneficial for males to expend energy recruiting multiple mates than it is for females to do so. Polyandry is rarer and often associated with sex-role reversal, as in the case of pipefish (section 6.4.4). Monogamy is much more common in cases where both sexes provide parental care than where just one does. Many more birds than mammals are monogamous, for example, and this is associated with males playing a role in the incubation of eggs and provisioning of chicks (however, see section 6.5.1 on extra-pair copulation).

8.2.2 Social systems

Just as we categorized mating systems in the previous section, we can roughly categorize the social systems most commonly found in animals as: solitary; one male, one female; one male, multiple females; and multiple males, multiple females.

Solitary systems
Solitary systems are characterized by a lack of mating pair bonds and a similar lack of same-sex relationships in adulthood. Note that one sex can be solitary whilst the other is not. In the cheetah, *Acinonyx jubatus*, for example, females are solitary, whereas males form coalitions with their brothers, apparently in order to be able to defend territories from other males. Solitary species may also aggregate temporarily, for example to choose mates.

One male, one female
This type of social organization is strongly associated with mating monogamy (but monogamy can also be found in other types of group, see below). Often, the young will delay dispersal and so the reproducing male and female will form the nucleus of a small family group consisting of themselves, their current infants, and not-yet-dispersed juveniles who may act as helpers at the nest. This is the social system of a large lemur called the indri, for example (Figure 8.7).

One male, multiple females
A polygynous harem mating system often gives rise to a one male, multiple female social system. In gorillas, for example (Figure 8.7), there is only a single mature male, the silverback,

Figure 8.7 Indris (*Indri indri*, left) live in monogamous families where males carry the offspring. Gorillas (*Gorilla gorilla*, centre) live in polygynous groups with a dominant male. Chimpanzees (*Pan troglodytes*, right) live in multi-male, multi-female groups with promiscuous mating and no paternal care of offspring. *Left to right: © Wolfgang Kaehler/Alamy; Photodisc; Corel.*

in a troop. Other males are tolerated until they reach sexual maturity, at which point they disperse and found or take over groups for themselves. One male, multiple female social systems, because of the sex-ratio imbalance within the groups, generate a cadre of lone or wandering males who have not successfully recruited or captured a group.

The females in such groups may be closely related, either as sisters or mothers and daughters. This comes about by females remaining in their natal group. This is called female philopatry. Where there is female philopatry, there tends to be male dispersal. Female philopatry enhances the benefits of group living, since kin selection gives every female a positive genetic interest in all the young, and collective care and even nursing of offspring may result.

A variant of the one male, multiple female social system is seen in lions, where each pride is associated with a coalition of males rather than just one. The coalition often consists of two males, but can be up to four, and they tend to be brothers. Coalitions are probably favoured for their ability to exclude rivals, whilst larger coalitions, especially of non-relatives, are disfavoured because of the loss in reproductive success with every additional male in the pride.

Multiple males, multiple females

Many animal groups contain multiple adult males and multiple adult females. In the primate literature, these are often described as 'multi-male' groups, to contrast them with harem-based single-male groups. Multiple male, multiple female groups are often based on promiscuous mating, as in the chimpanzee (Figure 8.7). However, this is not a necessary connection. Many colonial birds, for example, maintain monogamous pairs within a wider multiple male, multiple female roost, whilst hamadryas baboons (*Papio hamadryas*) maintain small harems within a larger multiple male, multiple female society.

Unless the social group is vast, one sex or the other of maturing young will tend to disperse. Baboons have female philopatry and male dispersal, and as a consequence matrilineal kin relationships bind many female group members. Chimpanzees, by contrast, have male philopatry and female dispersal, and so patrilineal, father, and brother kin relationships become an important part of group structure.

8.3 Consequences of group living

Having briefly surveyed some of the reasons social groups may form, and the types of group that are found, we now turn to the consequences of living in a group. In particular, we focus on two issues: the emergence of dominance hierarchies and the cognitive demands of social life.

8.3.1 Dominance hierarchies

Although the individuals within a group may all share an interest in the group continuing to exist, there will also be conflicts between them over food and access to mates. Members of the same group encounter each other sufficiently often that the outcomes of such contests can be established without a lengthy confrontation each time. Thus, in many mammalian social groups, dominance hierarchies tend to emerge, with some individuals consistently able to displace others from a resource. Ranks in the hierarchy will depend on such factors as size, strength, age, and coalitional support within the group. As these factors change, rank orders are challenged by fighting, and may be reversed. Ranks can be relatively dynamic, as in male chimpanzees, or remarkably stable, as in baboons, where daughters take a place adjacent to their mothers in the hierarchy, which is stable for many years.

It is important to stress that just because subordinate individuals accept low-ranking positions does not mean that being of low rank is adaptive. Reproductive success tends to be higher for high- than low-ranking individuals, and thus other things being equal it is always in an individual's interest to be of high rank rather than low. However, for low-ranking individuals, the expected cost of contesting their rank may be higher than the expected return (or success may be impossible), meaning that by accepting a low rank such individuals make the best they can of a bad situation. Where dominance hierarchies are steep and stable, low-ranking individuals often show physiological evidence of increased stress and have worse health in the long run as a consequence. However, when the hierarchies are unstable, or need to be constantly maintained through fighting, it is the high-ranking individuals who suffer the most stress (Sapolsky 2005).

8.3.2 The cognitive demands of social life

Social relationships can be psychologically demanding. For example, maintaining a pair bond with a mate means recognizing that individual, keeping track of where they are and what they are doing, and coordinating one's behaviour with theirs. Maintaining a coalition of female relatives is even more demanding, since there might be several individuals to keep track of. These observations lie at the heart of what is known as the social brain hypothesis. In its broadest sense, this is the idea that maintaining social relationships requires devoted brain mechanisms. One of the predictions of the social brain hypothesis is that social species will tend to have relatively larger brains than non-social ones.

Some of the most compelling evidence for the social brain hypothesis is the finding that, separately in carnivores, ungulates, birds, and bats, species that form long-term pair bonds, particularly monogamous ones, have relatively larger brains than those with either social or solitary promiscuous mating systems (Shultz & Dunbar 2007; Figure 8.8). Shultz and Dunbar interpret this as the consequence of selection for greater cognitive capacities to manage the dynamics of a close and long-lasting social relationship.

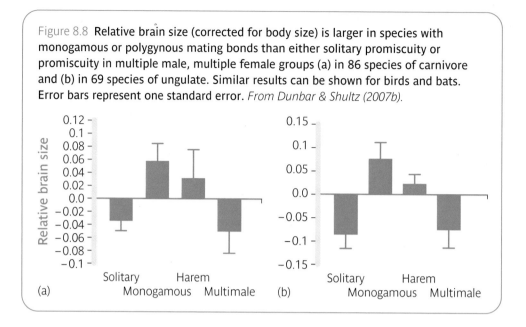

Figure 8.8 **Relative brain size (corrected for body size) is larger in species with monogamous or polygynous mating bonds than either solitary promiscuity or promiscuity in multiple male, multiple female groups (a) in 86 species of carnivore and (b) in 69 species of ungulate. Similar results can be shown for birds and bats. Error bars represent one standard error.** *From Dunbar & Shultz (2007b).*

Primates provide a slightly different picture. Solitary and monogamous primates have the smallest relative brain size, whilst primates in multiple male, multiple female groups have the largest relative brain size. In fact, across the monkeys and apes, there is a linear positive correlation between relative brain size and social group size (Figure 8.9). Shultz & Dunbar (2007) suggest that what has happened in primates is that the kind of closely coordinated social relationship which in other taxa is reserved for bonded mates is generalized in primates to all other group members. There is good evidence that primate social relationships are particularly intense; individuals maintain long-term reciprocal grooming relationships with several others, form coalitions, and track the kinship, mating, and coalitional status of all the others in their group. This suggests that the cognitive demands of group living will rise directly with total group size, consistent with what the brain data show.

However, correlation (of brain size and group size) does not prove causation and it is important to stress that there are a number of pathways by which the two could come to covary. Grouping reduces predation, for example, and this would allow a slower life history (see Chapter 7). Slow life history favours large brains, since growing brain tissue represents a costly allocation of resources to the soma, which only pays off over the length of the lifespan. Selection for fast life history, such as where predation is high, favours the diversion of these resources to become reproductively active earlier instead. Similarly, the rich resources that allow large groups to form might also fund the energy costs of growing large brains. Testing between these alternative pathways is still going on (Dunbar & Shultz 2007b), but the comparative evidence for a special relationship between social relationships and brain evolution seems relatively strong.

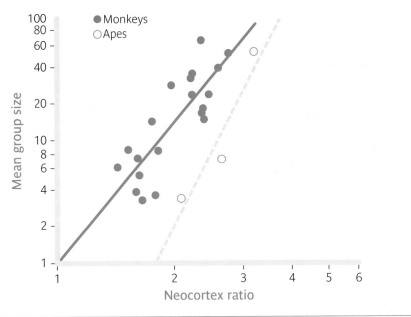

Figure 8.9 **Relationship between neocortex ratio (a measure of the relative size of the 'higher' parts of the brain) and group size for monkeys and apes. There is a positive linear relationship overall and a suggestion that apes have relatively larger brains for their group size compared with monkeys.** *From Dunbar & Shultz (2007a).*

8.4 Human groups in comparative perspective

Now that we have examined the main forms and consequences of sociality in our relatives, it is time to look briefly at human groups from a comparative, evolutionary perspective. The first point to make is that humans are intensely social. Although human forms of life are very variable from one society to the next, they are always social. It is very rare to find humans living alone. Lack of social connections is something that most people find deeply unpleasant, and is a risk factor for stress-related health problems such as depression.

Second, human groups are typically large by primate standards, and always of the multiple male, multiple female type. From the smallest bands of hunter-gatherers to the largest nation-states, it is normal for several adult males and several adult females to live in close proximity and have intense ongoing social relationships. However, embedded within these broader social networks there are enduring mating/marriage bonds, as we shall see below. Beyond these obvious points, what can we say more specifically about human groups?

8.4.1 Group size and the social brain hypothesis

An interesting approach to human group size is to extend the regression line of Figure 8.9 to the neocortex ratio of humans (which is higher than that of any other primate). Reading off the predicted group size on the vertical axis gives a figure of about 150 (Dunbar 1993). Thus, given that neocortex ratio is generally correlated with group size in primates, and given the neocortex ratio that humans have, we should expect humans to live in groups of around 150 individuals.

At first sight this prediction for human group size is immediately contradicted by experience. People can live in any social unit from an Inuit wintering band of a few families, to a mega-city of several million people. However, the latter does not exactly contradict the hypothesis. The social brain idea is that brain capacities constrain the number of other individuals with whom it is possible to have intense, ongoing personal relationships. In a city of millions, not all the inhabitants know each other. Instead, their interactions, to the extent they interact at all, are governed by formal systems of law and policing, monetary exchange, and so on. If a person had to name all those others with whom he or she habitually interacted and could describe the current ebb or flow of their social relationship, the number might be closer to the 150.

A fairer test of the social brain hypothesis for humans is the idea that human groups that are based on purely informal social mechanisms (i.e. do not have any police force or defined hierarchical structure) cannot generally exceed 150 individuals. Dunbar (1993) reviewed a number of documented hunter-gatherer societies and showed that the mean size of their bands was indeed around 150. At any one time, these bands might be subdivided into smaller camps, for ecological reasons, but several camps would be united by mutual friendships and interchanges. Another example comes from the Hutterites, a branch of the Christian Anabaptist movement, who live in Canada and the USA and are committed to living an egalitarian and peaceful life in isolated rural communities. Hutterite communities have a maximum size of 150. When a village approaches this size, land is sought elsewhere for a daughter colony and the group divided into two, often by lots, with one half leaving immediately for the new settlement. The reason the Hutterites do this is that they feel it would be impossible to maintain cohesion and community mindedness within the colony by informal peer pressure alone and so rather than create a police force they prefer to split.

Thus, there is some evidence for the idea that around 150 represents an upper limit on the size of social group whose cohesion can be maintained by informal social communication alone. This means that larger social groups tend to have two important properties. First, they tend to have formal mechanisms of power and social control, like kings, law courts, policing, and so on, that are not so obviously required in very small groups. Second, they tend to have what social scientists call a segmentary organization, which means the whole subdivides into many duplicate parts. An army division, for example, may consist of 20,000 troops, but they will be subdivided into several brigades, who are in turn subdivided into several battalions, who are subdivided into several companies each consisting of several platoons. Each company within a battalion has the same structure and function. The point of the subdivision is to create entities of a manageable size for cohesive and coordinated behaviour. Interestingly, the company, which is the smallest independent unit of an army and the largest within which all the soldiers might all know each other personally, tends to consist of around 150 individuals (Dunbar 1993).

8.4.2 Dominance hierarchies

To what extent are human societies characterized by dominance hierarchies, like social groups of many other species? Sometimes clear hierarchies can be generated, as with military ranks, chiefs, or other grade systems. However, these seem to be most salient in large social structures where not everyone knows everyone else, such as armies, empires, or universities. Within small social groupings, like hunter-gatherer bands, Hutterite villages, and individual university departments, there tends to be a strong preference at least in principle for egalitarianism and collective consensus.

There is quite a lot of evidence that being at the bottom of a social hierarchy has negative effects on health and well-being in humans. For example, a famous series of studies of the British civil service showed large differences in self-perceived well-being according to the occupational grade reached (Marmot *et al.* 1991). Moreover, these perceived differences panned out into large differences in the incidence of heart disease, stroke, and other problems, and in life expectancy. Part of the gradient may be explicable in terms of material resources, but this is not a population living in poverty and even the lowest grades have stable and adequate incomes. Lifestyle factors such as smoking and exercise also account for part of the difference. However, at least part of the effect appears to be due to the adverse psychological and stress effects of being at the bottom of a stable social hierarchy.

8.4.3 Mating system

We have noted that within the multiple male, multiple female groups which humans form are nested long-term mating bonds. What, then, is the mating system characteristic of humans? There is clearly a great deal of flexibility, since monogamy, polygyny, and polyandry are all found in human societies. However, they are not found with anything like equal frequency. An influential cross-cultural survey (Murdock 1967) estimated that 82% of human societies were polygynous, about 17% monogamous, and only about 1% polyandrous. However, there are a number of caveats to note. First, in monogamous societies, marriage bonds may be dissolved and if men are more likely than women to marry again, then the system is effectively polygynous via serial monogamy. Second, although most *societies* are polygynous, most *marriages* are monogamous, since in many societies that allow polygyny, only a few men actually have more than one wife. Thus, we can conclude that human societies are mostly mildly polygynous, which is what the degree of sexual dimorphism in size among humans would predict (see section 6.4.2), but with the flexibility to alter the system according to ecological contingencies.

The degree of polygyny in human societies is conditioned by two main factors (Marlowe 2000). First, the more contribution that males make to their offspring (direct paternal care and resources into the family), the more monogamous the system becomes, other things being equal. This is because the greater the post-reproductive involvement of males, the more beneficial it becomes to have the undivided attention of one. By contrast, where males make a lesser post-reproductive contribution, there is less reason to need one for oneself, as it were, and women might do better choosing a male of higher genetic quality even if he is already married. Going with this, marriages are more stable (divorce is harder and rarer), and women's extramarital affairs are less tolerated, in societies where men make a larger contribution to subsistence.

Second, the greater the gap between the richest and the poorest males, then, other things being equal, the greater the degree of polygyny. For example, some African farming and cattle-herding societies feature very large disparities between the farm or the herd sizes of the richest and poorest men, with rich men having several wives and poor men having none. It is easy to understand why this might be, from an evolutionary perspective. A woman choosing between being a first wife of a poor man and a second wife of a rich man might maximize her reproductive success by choosing the latter if the disparity between them was sufficiently great (however, on the lack of direct evidence for the benefits of polygynous marriages to women, see Gibson & Mace 2007).

8.4.4 Philopatry

We saw that baboons have female philopatry and male dispersal, whilst chimpanzees have male philopatry and female dispersal. What about humans? In most traditional human societies, neither sex disperses until marriage and then there is variation in whether the wife moves to the husband's group or vice versa. Wives moving to be near their husbands' families is called virilocality, whilst husbands moving to be near their wives' families is called uxorilocality. Among hunter-gatherer societies, there is no consistent bias towards either virilocality or uxorilocality (Marlowe 2000). Among subsistence farmers and herders, however, virilocality is the most common pattern. This shift seems to arise because of the bias, already discussed in section 7.4, towards passing accumulated resources to sons. This is easier, especially if the resource is land, if the son is nearby. Thus, preferential patrilineal inheritance creates a demand for sons to be near their fathers. Hunter-gatherers have little by way of transmissible wealth and so there is no particular pressure towards virilocality.

Contemporary developed societies have been described as neolocal, in that couples often establish households in new locations that are away from both sets of parents. With greater mobility and the transfer of wealth holdings from land towards portable assets such as money, there is not such a strong pressure to remain geographically close to kin.

8.4.5 Joint ventures

It is worth making one more observation about human groups. Through all their great diversity, one feature shines through, namely the high effort devoted to joint ventures. Humans often defend their groups collectively, hunt, forage, or farm collectively, build shelter or irrigation collectively, and form all kinds of associations such as age sets, armies, companies, and sects to undertake joint activities. This key aspect of the human way of life raises issues of cooperation, a subject to which we now turn.

8.5 Cooperation

A key issue for evolutionary theory is how cooperation is maintained within social groups. By cooperation, we mean behaviours that provide benefits to individuals other than the actor and have been selected because they do so. For example, alarm calling by prairie dogs when a predator is spotted (section 4.4.2) is a cooperative behaviour.

Although cooperation is discussed in many different ways (and using many different terminologies) in the literature, there are really only two classes of situation that we need to consider. Where the behaviour positively affects the recipient's lifetime reproductive success but negatively affects the actor's lifetime reproductive success, then the behaviour is true altruism. True biological altruism can only evolve through kin selection (section 4.4) or some similar mechanism that directs the benefits to individuals disproportionately likely to be also carrying the alleles that code for the behaviour. Parental care, for example, is true altruism of this kind and is, of course, maintained by kin selection.

The second class of situation is where the behaviour positively affects the lifetime reproductive success of the recipient and also positively affects the lifetime reproductive success of the actor. Such behaviours are often misdescribed as altruism, but are better designated mutual-benefit behaviours. There are interesting issues surrounding the evolution of mutual-benefit behaviours because they often raise issues of cheating and enforcement. For example, let us say that every individual does better in terms of calories gained relative to those expended by going on a group hunt rather than hunting alone. Group hunting is therefore straightforwardly advantageous and should be expected to evolve at the expense of individual hunting. However, an individual who goes along on the group hunt but does not do any actual killing, instead waiting for others to do the kill and then joining the feed, is going to do even better than a normal group hunter. This cheater gets all the benefits of the cooperative venture without paying the costs and cheaters will therefore have high reproductive success and become more numerous, until the point where there are only cheaters in the population. Mutual-benefit behaviours can only be maintained if cheating strategies can be prevented from prospering.

A number of mechanisms have been discussed by which mutual-benefit behaviours can be maintained, which we now review.

8.5.1 By-product benefits

We must first remind ourselves that not all behaviours that provide benefits to others are cooperation nor are prone to the cheating problem. For example, consider deer that live in a group to dilute the risk of predation. Each deer, by being in the group, provides a benefit to all the others (by diluting the predation risk by a certain amount). However, each deer is also doing the best thing for itself (its own risk of predation is lower in the group than outside it). Thus, the mutual benefit arises simply from every individual following its own immediate self-interest. This situation is sometimes described as a by-product benefit because the benefits to others arise simply as a side-effect of the benefits to the actor. Note that there is no cheating problem in situations like this; you cannot take the benefit of being in the group without bearing the cost of being in the group.

8.5.2 Direct reciprocity

A possible mechanism for the evolution of mutual-benefit behaviour was discussed by Robert Trivers (1971). He called the mechanism reciprocal altruism, which is unfortunate since altruism is not involved. It is better named direct reciprocity. The idea is very simple. Individual A helps individual B in some way, and individual B returns the favour to individual A at some later point. Both can end up better off.

Theoretical work shows that direct reciprocity can evolve only if certain conditions are met. The benefit of the behaviour to the recipient must be greater than its cost to the actor. For example, direct reciprocity in food sharing could only evolve if there were times where a calorie of food was more valuable to individual B than to individual A, and times where the reverse was true. (This is quite reasonable; food becomes more valuable the closer to starvation one moves.) Interacting individuals must re-encounter each other multiple times and always have a substantial probability of repeat interaction. Finally, there must be some mechanism for allocating cooperation differentially to individuals who have been cooperative in the past. This is to prevent cheaters—who receive cooperative benefits but never return them—from proliferating. The most obvious mechanism is being able to recognize and remember the individual concerned.

Figure 8.10 **In the ring-tailed coati (*Nasua nasua*), individuals intervene in support of those who have supported them in the past.** *Photo © Fabio Liverani/Nature Picture Library.*

These conditions can be summed up in a stability condition for direct reciprocity, which is that $c < wb$, where c is the cost of the behaviour to the actor, b is the benefit of the behaviour to the recipient, and w is the probability of the recipient reciprocating in the future.

There are some interesting cases of mutual benefit through direct reciprocity in nature. The ring-tailed coati, *Nasua nasua*, is a social carnivore from the Americas (Figure 8.10). When there is a fight between two coatis, a third individual may intervene on the side of one or the other. Romero & Aureli (2008) studied a group in a zoo and showed that the more individual A intervened in support of individual B in disputes, the more individual B would intervene in support of individual A when it was A who was in a fight. In addition, the more individual A groomed the fur of individual B, the more likely individual B was to intervene on behalf of individual A. The researchers did not know all the kin relations obtained within the group and so were not able to decisively rule out kin selection as the mechanism maintaining the behaviour, but it looks very much as if there is reciprocity of support in these animals.

Direct reciprocity is also important in humans. In the famous 'live and let live' system in the trenches of the 1914–18 European war, soldiers on one side would allow the other side to go out to repair their defences and collect their casualties, with the other side returning the

Figure 8.11 **The Christmas truce of 1914 in France and Belgium, during which soldiers of opposing armies moved about safely, repaired trenches, played football, and even roasted some pigs together, is reputed to have begun with German troops raising signs saying, 'You no shoot, we no shoot', a clear grasp of the principle of direct reciprocity.** *Photo © 2000 Credit:Topham Picturepoint/TopFoto.co.uk.*

favour (Figure 8.11). The conditions for direct reciprocity were favourable, since the same units would face each other across the line for many months and thus have the opportunity for repeat interaction, the costs of not shooting at the enemy outside of a direct assault were low, the benefits were potentially high, and cheating could be immediately detected. Direct reciprocity would be much more difficult to establish in mobile warfare or where the same units did not interact for a prolonged period.

8.5.3 Indirect reciprocity

Direct reciprocity is the idea that we help individuals who have helped us in the past. By contrast, indirect reciprocity is the idea that it might be advantageous to help individuals who we have seen helping others in the past, even if that help was not specifically directed to us. In indirect reciprocity, then, the payback for being helpful to individual A is not that individual A will necessarily return the favour, but rather that by helping A, I will gain a good reputation, in virtue of which others might bestow benefits on me. The disincentive against cheating in an indirect reciprocity system is that cheaters get bad reputations and thereby are denied participation in future joint ventures.

Indirect reciprocity can be an evolutionarily stable strategy as long as, once again, the cost of the cooperative behaviour to the actor is smaller than its benefit to the recipient and, crucially, where the availability of accurate information about every individual's reputation is high. Formally, indirect reciprocity is advantageous where $c < qb$, where q is the probability of having correct information about the partner's prior behaviour. For q to be high, all behaviour of all social group members needs to be visible to everyone else or else there needs to be an accurate system for sharing information about the prior behaviour of others. Human language provides such a mechanism (see below).

Indirect reciprocity has been empirically demonstrated in humans in laboratory settings. Wedekind & Milinski (2000) had participants interact repeatedly in a group where, in turn, they could choose to transfer money to another player. Every Swiss franc given was doubled or quadrupled by the experimenters so that the benefit to the recipient was always greater than the cost to the actor. The participants were made anonymous from each other (they interacted via electronic key pads). However, at the point of making a decision about whether to give to someone, the actor could view how often that person had given to other group members in the past.

The prediction of the indirect reciprocity model is that those who give to others should themselves be given to and this is exactly what happened. Participants more generous in bestowing benefits also received more benefit, and not especially from those to whom they had given (this was veiled in the experiment by the use of keypads), but from everyone.

Indirect reciprocity is especially significant because humans are often generous towards causes where there is no expectation or even possibility of direct reciprocation, such as giving to charity or doing good works. One explanation of such behaviour is that it is beneficial through its reputational effects. For example, Milinski *et al.* (2006) showed in a laboratory task that people contributed more money to a climate-change fund when other group members would see their decision than when their decision was private. Moreover, people who had been seen to contribute to the fund were more likely to be chosen by others for a different, mutually beneficial cooperation game played later on.

Thus, it seems we humans are concerned about maintaining our reputations and this prompts us to behave cooperatively, even where direct reciprocation is impossible, wherever our actions are visible to others. In fact, it may not be necessary for our actions to actually be visible to others as long as we feel that they are visible. Bateson *et al.* (2006) studied contributions to an honesty box, which pays for supplies in a staff coffee room at Newcastle University. On alternate weeks, above the instructions for payment, they placed either a picture of human eyes or of flowers. Contributions were significantly higher in the weeks when the eyes were displayed (Figure 8.12). Since the eyes were only a picture, there were no actual reputational consequences of their being there, but they were obviously sufficient to make people feel that their behaviour was being seen and therefore known to others.

Indirect reciprocity may be a particularly potent mechanism in humans because human language is an effective means for sharing information about people's past behaviour. It is not necessary for humans to actually see how individual A behaves in joint ventures. Instead, if A is a cheater, someone will talk about him and everyone will come to know. Gossip—who did what with whom and why—is a central conversational activity in all cultures and always of great interest. Sommerfeld *et al.* (2007) studied a similar game to that of Wedekind & Milinski (2000), but this time participants could also write a piece of 'gossip' about the individuals they had interacted with, which would be shared with other participants. The researchers found that gossip information was more positive the more cooperative the person was, and was used by other players in subsequent cooperation decisions.

Figure 8.12 **Contributions to an honesty box in a coffee room are higher in weeks where there is a picture of eyes on the wall than in weeks where there is a picture of flowers.** *From Bateson et al. (2006).*

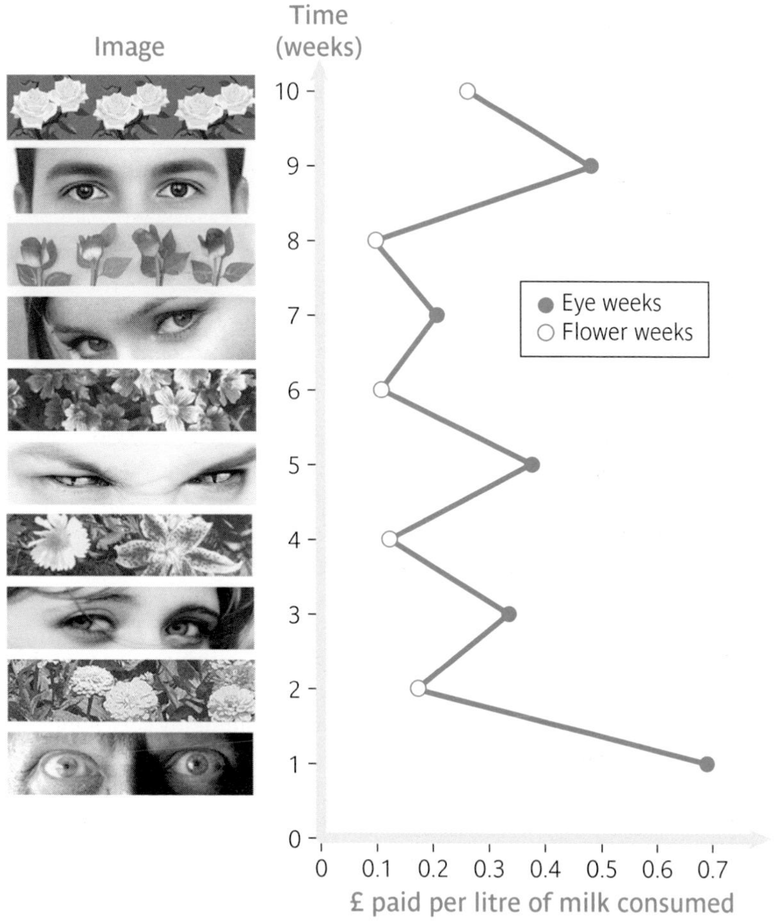

8.5.4 Punishment

Another mechanism that can make cooperation stable even in the absence of direct reciprocity is punishment. If cheating individuals are punished, cheating becomes an economic strategy. In laboratory experiments on humans, individuals will punish non-cooperators, even at cost to themselves and even if the cheating was not directed at them. The presence of punishment for non-cooperation sustains cooperation at much higher levels than where there is no possibility of punishment (Fehr & Gächter 2002; Figure 8.13).

Figure 8.13 **Results from a laboratory experiment where group members (who were kept anonymous from each other) could repeatedly choose to contribute money to a joint venture, but benefited from it regardless of whether they contributed or not. Left: The normal result in such a game is that cooperation gradually declines over time to the point where no one is contributing. Right: When people are given the opportunity to punish others who don't contribute, even at cost to themselves, cooperation is high and remains so over time. Upper and lower lines represent the 95% confidence interval for the mean.** *From Fehr & Gächter (2002).*

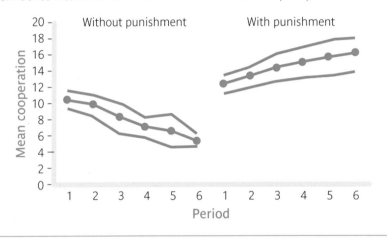

This is a very interesting finding, but it does raise further issues, principally, why should people punish? Given that punishment might be dangerous and costly, at first glance it seems like I would always be better off leaving others to do the punishment in my social group. Thus, providing punishment is itself a form of cooperation and is equally subject to problems of cheating. The most promising solutions to this issue focus on the idea that punishing non-cooperators brings reputational benefits (Barclay 2006) and thus links punishment to indirect reciprocity.

8.5.5 Issues surrounding human cooperation

There is a vigorous ongoing literature on human cooperation. This is because the level of cooperation between non-relatives that is found in humans seems to be much higher than that found in other animals. Cooperation in joint ventures must be part of the human success in conquering so much of the planet.

We will not review the entire literature on human cooperation here, but merely make a few observations for the reader who wishes to explore it for him- or herself. The first is that terminology is used rather variably in this literature. Humans are often described as altruistic because they are often prepared to make a short-term sacrifice of resources or time for others. However, it is not clear that such behaviours, like contributing to charities or joint ventures, are really altruistic in the long run. The reputational and social benefits I might derive from doing so could easily outweigh the costs over my lifetime. Theories of 'altruism' discussed in the human literature are usually not theories of altruism at all, but theories of benefits for cooperation.

A second point to make is that what makes humans so special is not that they have directly reciprocal one-on-one social relationships (like friendships, marriages, and so on). As we have seen, these kinds of relationship exist in some form in coatis and other animals. Rather, what makes humans unique is large-scale collective actions, such as a group of dozens or hundreds of men all going to war together. In these behaviours, the difference to the outcome made by one extra man is very small, but the costs to that man of taking part are potentially very large. Thus, there are strong individual disincentives for participation. These collective actions seem to be sustained by some combination of reputational benefits of participation, the danger of ostracism for not taking part, and punishment, which is in turn sustained by reputational benefits to the punishers (Milinski *et al.* 2002; Panchanathan & Boyd 2004; Barclay 2006). The human capacity for language facilitates all this by allowing reputational information to be shared.

However, the issues are not all solved. Contribution to joint ventures is evolutionarily stable in populations where non-contributors are shunned, but where does the shunning of non-contributors to joint ventures come from? Wearing a blue hat is evolutionarily stable in a population where individuals who do not wear blue hats are shunned, but this does not mean that groups of blue hat wearers will necessarily evolve. Thus, there must be some reason why humans evolved a psychology inclined to reward people for contribution to joint ventures and shun those who do not contribute, rather than a psychology inclined to reward people for some other arbitrary characteristic like wearing a blue hat. Boyd (2006) suggests competition between groups as a key factor here. With a norm of taking part in joint ventures stable in one social group, and a norm of wearing blue hats stable in another one, it is likely that the group performing joint ventures would proliferate at the expense of the other group.

 Summary

1. Animals live in groups when the benefits of so doing outweigh the costs. Benefits of group living include reduction in predation risk, joint foraging, territory defence, care of young, and information transfer. A key cost is increased feeding competition.

2. Mating systems can be polygamous, polyandrous, monogamous, or promiscuous.

3. Social groups can be based around a single monogamous pair, one male and multiple females, or multiple males and multiple females.

4. According to the social brain hypothesis, close social bonds are cognitively demanding and require specialized brain mechanisms.

5. Human social groups are based on multiple males and multiple females, with a flexible mating system that is most commonly mildly polygynous. They often feature segmentary organization.

6. Cooperation within social groups can be sustained by some combination of kin selection, direct reciprocity, indirect reciprocity, and punishment.

7. Humans are notable for the extent of their cooperation, particularly joint ventures undertaken by large groups.

Questions to consider

1.   Why do you think herbivores often live in vast social groups, whilst carnivores seldom do so?

2.   In multiple male, multiple female social systems in primates, why does either one sex or the other disperse, rather than neither or both?

3.   Why are the seas and public lakes over-fished, whilst private fishing lakes are well stocked?

4.   You may have noticed that there is an uncanny similarity between Hamilton's rule (kin altruism can evolve where $c < rb$), the condition for direct reciprocity to be evolutionarily stable ($c < wb$), and the condition for indirect reciprocity to be evolutionarily stable ($c < qb$). Why do you think these three expressions are so similar to each other?

 # Taking it further

A good account of the benefits of social relationships is Silk (2007). On mating systems and the social brain, see Shultz & Dunbar (2007). For a review of social rank and its relation to stress in humans and other primates, see Sapolsky (2005). The literature on cooperation, especially in humans, grows all the time. For a good overall review, see West *et al*. (2007). For a few of the many different approaches, see Roberts (2005), Milinski *et al*. (2006), Choi & Bowles (2007), and Barclay & Willer (2007).

Plasticity and learning

We now come to one of the most fascinating and often misunderstood areas of evolutionary biology, namely the evolution of plasticity. Plasticity is the ability of the phenotype to alter in response to experience, and it is an important feature of more or less every living organism.

Misunderstanding is so common in this area because of a widespread—and erroneous—tendency to assign behaviours to one of two apparently mutually exclusive causes: 'nature' vs. 'nurture', 'genes' vs. 'the environment', 'innate' vs. 'learned' (Bateson & Mameli 2007). There tends to be a linkage in people's minds between the 'nature' side of the dichotomy and evolution, so that 'innate' behaviours are argued to be the product of evolution. This implies that behaviours that are learned are not the result of evolution, but are instead caused by some other process, a view that makes no real sense on closer examination. Nonetheless, people continue to oppose 'evolved' to 'learned', and those who want to refute a Darwinian account of a particular behaviour often do this by showing that learning is involved, as if this was evidence against it being an adaptation.

This chapter seeks to clarify this muddle by making a number of key claims and fleshing them out with examples. First, I will argue that it is not feasible to separate behaviours into those that are 'learned' and those that are 'innate'. Learned behaviours depend on the existence of evolved cognitive mechanisms that guide and cause the learning, whilst unlearned behaviours can only develop with the appropriate environmental inputs. Instead, we can sometimes distinguish between behaviours that are relatively robust (they develop in much the same way across a wide variety of environmental contexts) and those that are more plastic (their form is altered in specific ways by environmental input).

Second, 'learning' as we think of it is not a single process. There are many different evolved mechanisms that cause phenotypic plasticity, and each of these has different characteristic features such as the type of environmental inputs that are important, the timing of those inputs, the reversibility of the phenotypic change, and so forth. If a child learns from her early experience that the world is a hostile place and she should not trust others, she is not using the same mechanisms as she is when she learns to ride a bicycle, learns what the word 'apple' means, or mounts an immune response to the polio virus. Each of these phenotypic changes will be the product of different evolved mechanisms being activated by different classes of environmental input.

Third, learned behaviours are just as much a product of evolution as unlearned ones are. To see why this is the case, consider an example such as hunting behaviour in cats. To hunt competently, cats must go through a series of learning experiences, including pouncing on leaves and balls of wool, experimenting with pursuit of prey of different types, trial and error, and possibly observation of their mother. Without these inputs, the fully functional adult hunting repertoire would not appear and, moreover, cats developing in different environments (e.g. where there are birds rather than rodents to hunt) have different learning experiences and end up with different hunting behaviours. Thus, hunting is a learned behaviour. However, it would be extremely odd to claim that the ability to hunt was not an evolved characteristic of cats. Hunting develops reliably in all cats and is a central part of their evolutionary niche. Cats that cannot hunt have lower fitness than those that can. Thus, there is strong selection on hunting ability and hunting is an evolved ability. What natural selection does is to fix traits like interest in hunting-like play and chasing things with certain properties, ability to improve one's skill by trial and error, and a motivation to observe and experiment with hunting. Genotypic variation in any of these traits will affect the hunting phenotype, even though hunting itself is learned, and selection can work on such genotypic variation. Thus, hunting is both learned and the outcome of an evolutionary process of adaptation.

Fourth, plasticity exists because it has selective advantages, that is the ability to learn is favoured by selection because it provides a solution to certain types of adaptive problem. Far from being the opposite of evolution, learning is a mechanism that evolution produces as a way of maintaining adaptive behaviour under certain conditions. We will discuss what those conditions are in the next section.

The structure of the chapter is as follows. In section 9.1, we look at a simple example of plasticity of the phenotype and consider under what circumstances plasticity should generally be expected to evolve. The next four sections deal with some specific types of plasticity that are relevant to behaviour: developmental induction (section 9.2), imprinting (section 9.3), associative learning (section 9.4), and social learning (section 9.5). This is not an exhaustive or indeed mutually exclusive list of types of plasticity, but it allows a review of some of the main evolutionary issues involved. Section 9.6 then examines the relationship between learning and evolution by natural selection, including how learning can influence the course of genetic evolution.

9.1 Conditions for the evolution of phenotypic plasticity

When we expose our skins to sunlight, we produce more pigment and our skins become darker. This is the mechanism of tanning. Tanning is a good example of phenotypic plasticity. As we saw in section 5.5, the coloration of human skin represents a trade-off between the advantages of making more vitamin D, which favours lighter skins, and the advantages of protecting cells from ultraviolet damage, which favours darker skins. If there is very strong sunlight, the optimal skin colour is darker than if the sunlight is less strong. Human beings have, over evolutionary time, lived in many different places, where the sunlight has sometimes been strong and sometimes not (and, of course, sunlight changes with the season). How does natural selection solve the problem of maintaining the right skin colour, given this variation in the environment?

One solution is to make skin colour not fixed, but plastic, becoming darker when the sun is strong, and paler when it is not. This is what tanning does. This mechanism does not account for all of the variation in human skin coloration—there are differences in baseline pigmentation between populations too—but it does seem to be an important evolved ability. It also provides us with a chance to specify the conditions under which a mechanism for plasticity, like tanning, could be adaptively favoured. Those conditions are the following:

1. The population must have encountered a range of environmental variation over its evolutionary history. If all humans who had ever lived had experienced the same level of sunlight, the tanning mechanism could not have evolved. Environmental variation can arise through change over time or through movement of the population across space.

2. There must be different optimal phenotypes in different environments. As we have seen, this is true in the case of sunlight and skin colour. If it were not true, the ability to tan would provide no advantage.

3. The mapping between the environment and the optimal phenotype must be consistent. That is, it is *always* optimal to have a darker skin where the sunlight is stronger and a paler one where it is less strong. If we imagine a hypothetical world where stronger sunlight sometimes favoured those with darker skins and sometimes favoured those with lighter skins, then the ability to tan would not evolve, since becoming darker when the sun was stronger would sometimes be an advantageous thing to do and sometimes a disadvantageous thing to do, and would provide no net benefit overall.

4. There must be reliable cues available of what state the environment is in (what the level of sunlight is going to be) for the organism to exploit in triggering the phenotypic change. In the case of tanning, the cue that the sun is going to be strong in the immediate future is that it is strong at the moment (i.e. tanning is triggered by strong sunlight). Note that the evolution of plasticity requires a certain degree of regularity in the environment over the short term. If the sun is very strong today, then on average it is likely to be pretty

strong tomorrow too. Tanning exploits this regularity. Imagine instead a strange hypothetical world in which the sun being very strong today meant that it was going to be very weak tomorrow. It is unclear that the tanning mechanism would be useful in such a world, since the darkening produced by 1 day's sunlight would make the individual *less* well suited, rather than better suited, to tomorrow's environment.

Note the contrast between condition (1) and condition (4). Plasticity is useful where the environmental is variable over the long term (e.g. over generations), but regular over the short term (e.g. from day to day). If the environment is invariant over even very long timescales, then plasticity has no advantage, and if the environment is completely irregular even over the short term, then it is impossible to learn effectively since present experience is no guide to future experience anyway.

Even where the four conditions listed above are met, it is not necessarily the case that plasticity will evolve. This is because plasticity has costs as well as benefits. The mechanisms needed to produce plasticity are more complex than those required to produce a fixed phenotype. For example, the ability to tan requires the production of tissues that are sensitive to sunlight and alter their production of pigment in response to it. These tissues require more energy than a simpler system producing a fixed amount of pigment. Being more complex, they may also be more likely to go wrong. A plastic mechanism may also take longer to produce the appropriate phenotype than a fixed one would because there are more stages involved and because it depends on environmental cues that might or might not come along at the appropriate time. Finally, plastic mechanisms can be triggered inappropriately or cues can prove unreliable (e.g. there is a freak sunny day), meaning that the wrong phenotype ends up being activated.

In view of all this, we should only expect plasticity to emerge where conditions (1)–(4) are all met, *and* the benefits of being plastic outweigh the costs of the extra complexity and energy expenditure, the slower development, unreliability of cues, and so on. Given that plasticity is so widespread amongst living things, this situation must often prevail. However, it does serve to remind us that evolution will not produce infinite plasticity and will tend to reduce plasticity where it is not providing some fitness benefit.

9.2 Developmental induction

In this section, we review examples of plasticity where exposure to a cue or cues early in life induces a permanent change in the phenotype. Such effects are known as developmental induction. An example comes from the freshwater snail *Physella virgata*. These snails face a variety of adaptive challenges in their lives. Sometimes they share their habitat with freshwater crayfish, which eat snails. In this case, their optimal response is to grow big as fast as possible and for their shell to be relatively elongated. This is because crayfish only take small individuals and do so by entering the aperture of the shell to extract the body within. A more elongated shell makes this form of predation more difficult.

Sometimes, however, the snails share a habitat with fish that feed on molluscs, such as the pumpkinseed, *Lepomis gibbosus*. These fish crush the whole shell in their jaws. The optimal response to the presence of fish is to grow a more rotund shell shape, which is more crush resistant, and also to crawl out of the water as often as possible. This is a dilemma because making the shell as rotund as possible precludes making it as elongated as possible and, since

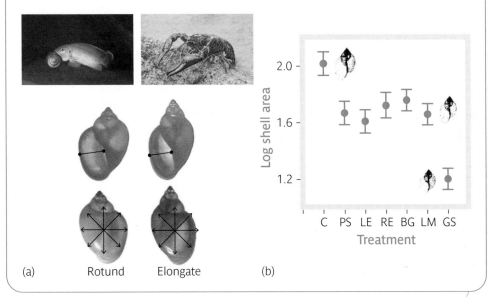

Figure 9.1 **Freshwater snails (*Physella virgata*) have different developmentally induced phenotypes. (a) Snails raised in water containing mollusc-eating fish have more rotund and less elongated shells than their siblings raised in water containing crayfish. (b) Snails raised in water containing fish are smaller than controls for their age. C represents 'control' conditions, and the other six initials represent different species of fish. Error bars represent one standard error.** *From DeWitt (1998) and Langerhans & DeWitt (2002). Photos: left: © Oxford Scientific (OSF)/Photolibrary.com; right: © John Barber/Fotolia.com..*

the snails only feed whilst in the water, crawling out is incompatible with growing as fast as possible. The optimal trade-off balance of feeding and crawling out, shell rotundity and shell elongation, will thus depend on which type of predator is present.

The evolutionary response to this situation has been to create a system of developmental induction. Snails that are raised in water containing pumpkinseed fish grow more rotund shells than those raised with crayfish or no predators (Figure 9.1a). This is done by sensing the chemical cues from fish or crayfish in the water and changing growth pattern accordingly. Thus, the snail achieves the best phenotype for its situation whichever predator is present.

This system illustrates many of the advantages but also many of the limitations of systems of developmental induction. The snails are able to develop the best phenotype for whichever of several different environments they happen to be born into. On the other hand, there are costs. The chemical cue makes the individual commit to an irreversible pattern of growth. This is fine as long as the cue received early in life accurately predicts the adult environment. However, imagine a situation where there are crayfish in proximity early in development, but later in life pumpkinseeds move into the habitat. Here, there is a mismatch between the cue and the reality of the adult environment, and developmental induction is worse than useless since the snails develop elongated shells that the fish can crush. Thus, the system is only useful to the extent that crayfish around early in life predicts crayfish around later in life.

Moreover, there may be limits to how well the system can discriminate the appropriate cues. Langerhans & DeWitt (2002) raised snails with one of six different types of fish: five related species of sunfish and a type of bass. All six of the fish species induced the rotund shells and the crawling out behaviour to develop (Figure 9.1b). However, only two of these fish species actually eat snails! The problem is that the chemical signature of mollusc-eating fish is probably extremely similar to that of related non-mollusc-eating fish. Thus, having the system of developmental induction, as well as allowing the 'correct' calibration to predator presence, is prone to produce 'incorrect' calibration by exposure to similar but actually harmless fish. This adds to the potential costs of plastic development.

In humans, there has been much recent interest in the idea that aspects of the adult phenotype are developmentally induced by the availability of resources to the foetus in the prenatal period. Female babies that are born light for their length, which suggests restriction of resources *in utero*, reach puberty earlier than other babies (Adair 2001; Figure 9.2). This is a striking finding, since nutritional restriction after birth has precisely the opposite effect, delaying puberty, as the child needs longer to grow. Small babies also develop into adults with other features, such as a greater tendency to lay down fat around the middle of the body, and a greater likelihood of diabetes and high blood pressure later in adulthood. The pattern can be interpreted in the following way: restriction of resources in the womb is a cue that the environment the baby will be born into will be characterized by nutritional scarcity. The individual is thus developmentally induced to have a fast life history strategy (get on with reproduction early, since life expectancy may be low) and also a metabolism designed to sequester and store as many resources as possible whenever they are available. In past environments, this was an adaptive pattern. In the contemporary environment, where fats and sugars are widely available post-birth, it leads to obesity and health problems later in life (Gluckman *et al.* 2007).

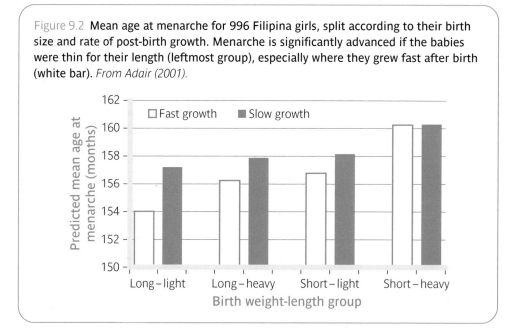

Figure 9.2 **Mean age at menarche for 996 Filipina girls, split according to their birth size and rate of post-birth growth. Menarche is significantly advanced if the babies were thin for their length (leftmost group), especially where they grew fast after birth (white bar).** *From Adair (2001).*

There is also evidence for major stresses early in life, such as difficult family environments or father absence, developmentally inducing earlier menarche and a faster life history strategy (Belsky 2007; see section 7.2). This would again make adaptive sense as long as family stress was a generally reliable cue that the environment was hostile and favoured earlier reproduction.

9.3 Imprinting

Our next category of plasticity, imprinting, is rather similar to developmental induction in that it involves key experience in a critical period early in life. However, it differs in that what the environment provides is a template of what some aspect of the world should look like.

9.3.1 Filial imprinting

The classic demonstrations of imprinting come from birds such as ducks and geese where the chicks do not stay in the nest but instead follow the mother about. These chicks face an adaptive problem, namely how to identify who their mother is. Evolution cannot provide a fixed template for this, since mothers look different from one another and moreover the appearance of the species will change over evolutionary time, so any fixed mechanism would become out of date. Instead, it solves the problem with a plastic mechanism: chicks imprint on whatever moving object is in their vicinity in the first 36 hours of life and follow this. Normally the only such object is the mother and the system works well, but it can be vulnerable to cue unreliability. For example, if the first moving thing goslings see is the boots of a famous Austrian biologist, then that is what they will imprint on and that is who they will follow around until they are adults (Figure 9.3).

9.3.2 Sexual imprinting

A different imprinting mechanism, sexual imprinting, solves the problem of how to identify what a potential mate looks like. Again, evolution cannot provide a fixed template since the phenotypic characteristics of the species will be subject to change over evolutionary time. The mechanism makes use of the fact that one's own mother is by definition a fertile member of one's species. Thus, learning what one's mother looks like, and then in adulthood seeking another individual who looks similar, should usually work as a strategy for identifying a mate.

Witte *et al.* (2000) studied a type of finch called the Javanese mannikin, *Lonchura leucogastroides*. They reared chicks either as normal or where one or both the parents had a red feather crest that was not normal for this species stuck onto its head. After the chicks matured, the researchers let them choose between a normal mate and a mate with a red feather crest. Individuals raised by parents with red feather crests preferred a mate with a red feather crest over one without, whereas individuals raised by unadorned individuals did not (Figure 9.4).

Sexual imprinting can last for an individual's whole life. Hansen *et al.* (2008) transposed chicks of the great tit, *Parus major*, into the nests of the blue tit, *Cyanistes caeruleus*, and vice versa. After such a switch, the adult birds seek mates and chase rivals not from their actual species, but from their foster species. The researchers showed that this behaviour does not reverse with age, so sexual imprinting is essentially fixed for life in these birds.

Figure 9.3 Austrian biologist Konrad Lorenz (1903 – 1989) famously showed that he could cause goslings to imprint on him and follow him around as if he were their mother, as long as he was close to them in a critical period in the first few days of life. Lorenz was one of the founders of the study of animal behaviour and received the Nobel prize in 1973. © *Science Photolibrary.*

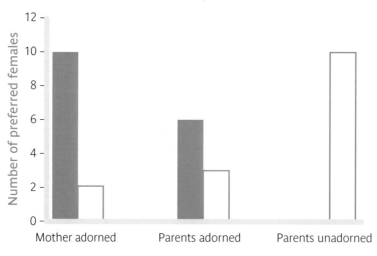

Figure 9.4 Adult male Javanese mannikins (*Lonchura leucogastroides*) choose a mate with an artificial crest (filled bars) more often than one without (open bars) if they were raised with either the mother or both parents adorned with such a crest, but not if their parents were unadorned. *From Witte et al. (2000).*

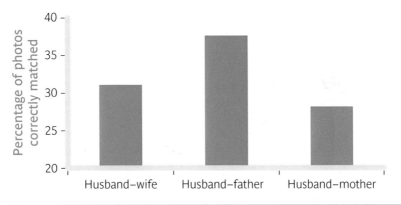

Figure 9.5 **Observers are more likely to correctly match a photograph of a woman's husband to one of her adoptive father than they are to match a photograph of the woman herself to her husband, or a photograph of the woman's adoptive mother to one of her husband. This suggests that women may be imprinting on their fathers and using this as a template for mate choice.** *From Bereczkei et al. (2004).*

There is some evidence for sexual imprinting in humans. Bereczkei *et al.* (2004) studied married Hungarian women who had been raised in an adoptive family. From photographs, observers blind to the hypothesis rated the similarity of the faces of (a) the woman and her husband, (b) the woman's adoptive father and her husband, and (c) the woman's adoptive mother and her husband. The judges were more likely to rate similarity between the adoptive father and the husband than between the woman and her husband or her adoptive mother and her husband, and the similarity was greater than expected by chance (Figure 9.5).

Imprinting need not be on other members of the species. Salmon imprint on the chemical composition of the water in their natal location, and Emlen (1970), in an experiment where he kept migratory birds in a planetarium and manipulated the apparent centre of rotation of the night sky, showed that the birds used celestial rotation to imprint which constellations indicate which compass directions.

9.4 Associative learning

Associative learning refers to any ability of an organism to make a new association between a stimulus and a behaviour as a result of experience. It comes in two main forms, classical conditioning and instrumental conditioning.

9.4.1 Classical conditioning

In classical conditioning, a response that is currently made to stimulus A comes to be made to stimulus B if A and B are experienced together often enough. The most famous example was studied by Russian psychologist Ivan Pavlov in dogs. Dogs spontaneously salivate when presented with food. Pavlov rang a bell whenever he was about to feed the dogs and, eventually, dogs would salivate to the sound of the bell alone, having formed an expectancy that bells mean food. In this example, presentation of food is called the unconditioned stimulus, the bell the conditioned stimulus, and salivation the response.

The adaptive value of classical conditioning is that it allows the animal to exploit regularities of its particular environment. If the approach of predators in some habitat is always preceded by the sound of cracking twigs, or by being in a particular place, or a particular time of day, then an animal can gain a competitive advantage by starting anti-predator vigilance whenever those cues are present and not waiting until the predator actually appears. Evolution cannot build those associations in, since they will be different for every local context, and so it instead provides a mechanism to make them in the light of experience.

A possible drawback of a system of classical conditioning is that conjunctions of unconditioned and conditioned stimulus could be spurious. For example, I might encounter a predator when it happens to be raining, even though there is no true correlation between rain and predator activity. I could end up with a costly learned association making me hide every time it rains. Classical conditioning systems have two defences against such spurious associations. First, they usually require *repeated* pairings of unconditioned and conditioned stimulus to form an association and, second, the association is reversible. If the conditioned stimulus occurs repeatedly without the unconditioned stimulus following, the association breaks down, in what learning theorists call extinction.

9.4.2 Instrumental conditioning

In Pavlov's classical conditioning experiment, the dogs were not rewarded for salivating. It was just something they did anyway and they learned to transfer it to a new context. This contrasts with instrumental conditioning, where the individual changes its behaviour as a result of the rewards or punishments that it has received. As any dog trainer knows, you can increase the frequency of a behaviour by always following that behaviour with food and decrease the frequency of a behaviour by always following it with punishment.

Instrumental conditioning allows the individual to do more of whatever happens to work in its local environment. If chasing a particular prey type often ends in food, then the animal will increase its frequency of chasing that prey, whereas if chasing a particular prey type often ends in pain, then the animal will decrease the frequency. In other words, instrumental conditioning allows trial-and-error learning about the best way to behave for where (and who) one is. Evolution cannot make this knowledge fixed since it will be different across time, space, and individuals, and instead plasticity is favoured.

Instrumental conditioning relies on there being a set of outcomes that are intrinsically rewarding or intrinsically unpleasant, to which behaviours can become associated. What these are varies from organism to organism as a function of that organism's evolutionary history, but they might include food, water, social contact, shelter, and, on the unpleasant side, lack of food or water, loud noise, lack of shelter, lack of social contact, pursuit, and so on.

9.4.3 Evolved constraints on associative learning

Associative learning is a very powerful set of processes that accounts for individuals' abilities to acquire quite different behaviours as a result of their individual history. However, it is important to stress that animals cannot learn just anything with equal facility. Rather, animals may be predisposed to form associations between particular types of situation and particular outcomes.

The classic example comes from learned taste aversions, which have been well studied in many species, especially rats. If rats taste a novel flavour and then become sick, they will learn a lasting aversion to food with that flavour, even on the basis of one exposure and even if several hours elapse between the food and the onset of the sickness. By contrast, rats have extreme difficulty learning to avoid food whilst a certain sound is playing or whilst light of a certain colour is being shone. It is as if they are prepared by evolution for a world in which the flavour of a food is a reliable indicator of the type of effect it will have, whereas the sound whilst eating is not such an indicator. This makes perfect adaptive sense since there is a reliable correlation in nature between the flavour of a food and its chemical properties, and not between the ambient sound and the food's chemical properties.

The evolutionary specificity of the system is even greater, in fact. If the individual has had even one prior experience of that taste without being sick, the strength of taste aversion formed is much reduced and some types of flavours are much more effective at producing aversions than others. Thus, evolution has given us the capacity to learn which foods are poisonous in our local environments, but given us prior expectations about the classes of associations and types of flavours that might be important (Rozin & Kalat 1971).

Another example of evolutionary preparedness to learn is found in the learning of fear reactions in rhesus monkeys. Captive-reared rhesus monkeys, *Macacca mulatta*, have no fear of snakes, although wild monkeys do. If a captive monkey sees film of a wild monkey making a fear reaction to a snake, it learns to avoid snakes (this is an example of social learning; see section 9.5). However, if it sees film of a wild monkey making a fear reaction to flowers, it does not learn to avoid flowers (Cook & Mineka 1989). There is an evolutionary preparedness for the idea that things like snakes could be dangerous, but no preparedness for predatory flowers.

What we take away from examples such as these is that the dynamics of learning (how many pairings are required for an association to form, which cues will be most easily paired with which outcomes, how soon the outcome has to follow the cue, how long the association will last) will differ according to the domain of learning, the relationship of the cues to important outcomes across the environment the organism has lived in, and the fitness payoffs of getting the behaviour right or wrong. As Paul Rozin and James Kalat put it in a seminal paper almost 40 years ago, 'animals may not only learn some things more easily than others, but they may also learn some things in a different way from others' (Rozin & Kalat 1971: 481). Which things they learn readily, and how, will depend on their evolutionary history. A beautiful example of this principle is that the vampire bat, *Desmondus rotundus*, is the only animal known that is unable to readily form learned taste aversions (Ratcliffe *et al.* 2003; Figure 9.6). It has no need to; in its natural environment, its food—the blood of live animals—is always fresh, and cannot be poisonous or the donor would not be alive!

9.4.4 Importance of associative learning in humans

Associative learning processes contribute importantly to the ability of humans and other animals to learn from experience. However, the movement in 20th-century psychology known as behaviourism attempted to elevate associative learning to the position of sole explanation of human behavioural flexibility and this is certainly wrong. For example, the behaviourists saw the child's acquisition of its native language as a kind of instrumental conditioning, with the child rewarded for making the right sounds and words.

Figure 9.6 **Learning abilities reflect evolutionary history: The fruit-eating bat** (*Brachyphylla cavernarum*) **forms learned taste aversions, but its blood-eating cousin the vampire bat** (*Desmodus rotundus*) **does not.** *Left: © Barry Mansell/Nature Picture Library; right: © Michael Lynch/istock.com.*

The influential linguist Noam Chomsky has instead argued that native language acquisition is a process much more akin to imprinting: the child has an evolutionarily prepared expectation that there will be a language spoken around it and during a critical period of its life it fills in the template of which language it is by extracting regularities from the stream of speech going on around it. This happens fast and essentially without reinforcement, and so is not usefully viewed as the outcome of instrumental conditioning.

Similarly, adherence to moral norms like the avoidance of sibling incest may not be the result of conditioning, but rather of a developmental induction process. Lieberman *et al.* (2003) found that the best predictor of how morally repugnant people find sibling incest is simply how long they lived with a member of the opposite sex of similar age during their childhood. Thus, there may be an inbreeding avoidance mechanism that is automatically activated in development by cues of opposite-sex sibling presence. This does not depend on rewards or punishments provided by family members. In fact, the mechanism can work in direct opposition to what the family values or rewards. In China and Taiwan, for example, there has sometimes been a social practice of adopting a female child to grow up alongside a son and become his wife when they mature. Wolf (1995) showed that the resulting marriages are generally unsuccessful, despite the wishes and desires of the families. This is consistent with the idea that having a co-resident member of the opposite sex automatically turns off sexual desire towards that person and turns on an inbreeding avoidance mechanism.

9.5 Social learning

Animals that have any kind of social relationships have a third option besides doing the same thing regardless of experience and altering their behaviour in the light of experience. They can copy the behaviour of other members of their species. Learning from what other individuals do is called social learning and there is increasing recognition of its importance in animal behaviour.

For example, young Norway rats, *Rattus norvegicus*, will only eat foods that they have either seen other rats eat, smelt on the breath of other rats, or which their mother has eaten and they have tasted in her milk (Galef & Clark 1971; Galef & Sherry 1973). Thus, they obtain information about what foods are edible in their local environment without having to learn by their own trial and error.

Many other types of behaviour, such as foraging techniques, can be socially learned. In Japanese quail, *Coturnix coturnix japonica*, social learning is involved in mate preferences. Galef & White (1998) put a female quail in a central cage with a (visible) caged male on either side. They measured how much time the female spent on the side with each male, as a measure of her preference for that male. They then introduced a second female into the cage of the less-preferred male, with whom that male mated. Having observed this, the original female subsequently spent more time with the male who had mated. The information that another female quail found the male attractive was information the original quail used in her own attractiveness decisions (Figure 9.7).

Figure 9.7 **Female Japanese quail (*Coturnix coturnix japonica*) spend more time with a previously unpreferred male after they have observed him mating with another female (the experimental condition), but not if they have observed him alone (the control condition). Error bars represent one standard error.** *From Galef & White (1998). Photo © Eric Isselée/istock.com.*

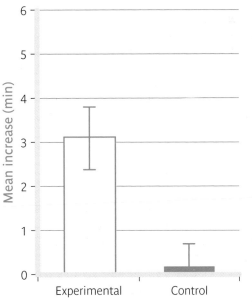

9.5.1 The evolution of social learning

When is social learning an adaptive strategy? There must be other individuals available to learn from, and so social learning is only available as an option to species which are social or where there are at least continuing parent–offspring interactions after birth. However, not all social species learn socially and even those that use social learning do not use it for everything. As ever, whether social learning evolves will depend upon the costs and benefits. Let us now consider what those costs and benefits are.

The great benefit of social learning is that it allows an individual to get information about how to behave adaptively in the local environment without going through the costly trial and error of learning it for him- or herself. For example, in the rat food-preference learning example, observing what older rats eat is a way of finding out what is safe to eat without going through the process of trying everything and risking poisoning oneself. The food that the other rats are eating cannot be poisonous or they would not be alive. Thus, social learning is a kind of short cut that reduces the cost of learning.

The potential cost of social learning is that what the other rats are doing might not be the most beneficial behaviour in the current environment. They might be wrong, the situation might have changed since they were young, or, at the very least, they might have different dietary tolerances from me and so what they do might not be a good guide to what I should do. How can we predict when this cost will be enough to outweigh the obvious benefits of social learning?

There is some good theoretical work in this area, which is mainly due to anthropologists Alan Rogers, Robert Boyd, and Pete Richerson (Rogers 1988; Boyd & Richerson 1995). The theory examines two alternative strategies: individual learning (finding out through one's own trial and error what behaviours work in the current environment) and social learning (copying others). The models assume that there is some cost c of individual learning. This reflects the time, energy, and danger of doing one's own trial and error. Social learning saves this cost. The models also assume that the environment also changes occasionally, so what was the best behaviour to perform in the past becomes obsolete and some other behaviour becomes locally correct. A potential problem with social learning is that the model from which one learns might have an obsolete behaviour.

Some of the key predictions from these models are sketched in Figure 9.8. The first is that the more costly it is to learn for oneself, the more advantageous social learning becomes, other things being equal. This just follows from the fact that the more costly individual learning is, the greater the saving one makes by getting the information for free from someone else.

The second prediction is more surprising (Figure 9.8b). When social learning is rare (i.e. most of the population is learning individually), it can be extremely advantageous, but as social learning becomes more common, its advantage reduces, and when everyone is learning socially, it is always better to learn individually. Why would this be?

Imagine I am a single social learner in a population of individual learners. Whoever I learn from has learned *his* behaviour by individual trial and error, and thus it is bound to be quite up to date. However, as the proportion of social learners in the population increases, so does the probability that anyone I learn from is also a social learner, and so the information is at two steps or more from the original trial and error learning that established it as correct. In general, as the proportion of social learners in the population increases, so too does the expected length of the chain of social learning that stands between me and the last individual that actually worked out the best behaviour by trial and error. This also means that the time elapsed between

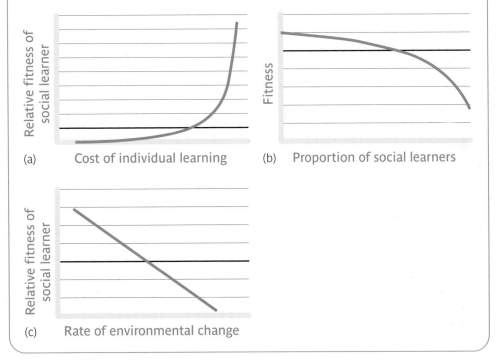

Figure 9.8 **Predictions from a theoretical model of the evolution of social learning.** (a) The fitness of a social learner (blue line) relative to an individual learner (black line) increases as the costs of individual learning increase. (b) The fitness of social learners is highest when they are rare, but declines as they increase in frequency and is less than that of individual learners when they are very common. (c) The fitness of social learners relative to individual learners is highest when environmental change is quite slow. *After Rogers (1988) and Boyd & Richerson (1995).*

when the information was gathered and the current moment becomes greater and greater, and so the more likely it is to be obsolete. In the extreme, if everyone is learning socially, then new information is never being gathered about the environment. Everyone is just copying one another, and the tradition is stuck and may well be obsolete. Thus, if everyone is learning socially all the time, I would always do better to learn individually.

The third prediction of the models is that social learning is most useful for aspects of the environment that change at an intermediate rate—slow relative to an individual's lifespan, but fast relative to evolutionary time. This is actually quite an intuitive result. If environmental change is very slow or non-existent, then there is no need for learning at all and evolution can produce a fixed behaviour. If environmental change is very fast, then I cannot rely on what my elders learned some time ago, and would always do better to find out for myself. Where the rate of environmental change is intermediate, that is the world changes over many generations, but nonetheless what was a good behaviour for the previous few generations is pretty likely also to be a good behaviour for me, then social learning is likely to be favoured.

What these models predict, then, is that if individual learning is sufficiently costly, and the environment is changing at an intermediate rate, then social learning can evolve, but organisms will never become exclusively reliant on indiscriminate social learning. There will always be a mix of social learning and individual learning in the population. The models do not specify whether this mix will take the form of some individuals doing just social learning and others doing just individual learning, or every individual doing a mixture of the two, but both individual learning and social learning will be present.

9.5.2 Cultural traditions

Social learning can lead to the formation of local cultural traditions. For example, song in both songbirds and whales is socially learned, and as a result different communities have different 'dialects' whose evolution over time can be studied just as human cultural traditions can be. We can demonstrate that these traditions are due to social learning rather than genetic differences between populations by fostering individuals into an adoptive nest. They acquire the tradition of their foster group, not their genetic relatives.

Cultural traditions are not restricted to song. Many animals use social learning to acquire locally appropriate ways of getting food. Black rats, *Rattus rattus*, living in pine forests in Israel have an efficient method of stripping the scales from pine cones (starting from the bottom) in order to access the nutritious seeds within (Figure 9.9). Rats captured outside pine forests and given pine cones seldom managed to learn this method, but all rats raised by a forest-raised mother acquired it (Terkel 1996; Galef & Laland 2005). Another famous example concerns macaque monkeys, *Macaca fascata*, living in a semi-wild colony on an island in Japan. They are provisioned with sweet potatoes, which are dumped on the sand, and prior to 1953 the monkeys brushed the sand off with their hands. One young female began instead to wash her sweet potatoes in the sea and this behaviour gradually spread until all troop members were doing it (Figure 9.9). The original individual is now long dead, but the tradition persists (see Galef 1996 for some sceptical discussion).

Figure 9.9 **Black rats (*Rattus rattus*) socially learn a technique for stripping pine cones, whilst Japanese macaques (*Macaca fascata*) socially learned to wash sweet potatoes before eating them.** *Photos: left: from Terkel (1996); right: from de Waal (1999).*

9.5.3 Human traditions and cumulative cultural evolution

Humans are clearly very dependent on social learning and, as a consequence, there are identifiable social traditions that persist over time in all human societies and in many domains of life. The discovery that cultural traditions are widespread elsewhere in the animal kingdom has provoked a debate about just what it is that makes human culture special.

A notable feature of most animal traditions is that they do not become any more complex over time. Potato washing in the Japanese macaques, for example, has persisted for many decades, but it is no more complex now than it was when one individual came up with it in 1953. Human traditions have a very different character. Consider agricultural methods or techniques for building boats. Over the generations, the traditions are added to and improved, and the designs (as measured by crop yields or speeds of boat) get better and better. After many generations, the designs are so sophisticated that no lone individual would be able to come up with them from scratch within a single lifetime. Instead, they represent an accumulation of many lifetimes of skill development. Thus, whereas the macaque tradition involved one individual learning something and subsequent generations exploiting that learning, human cultural traditions involve every generation adding to and improving the effectiveness of the behaviour. Thus, human traditions exhibit cumulative cultural evolution, which is thought to be very rare in nature (Boyd & Richerson 1996).

Many features of cumulative cultural evolution can be observed in the laboratory. In an ingenious study, Caldwell & Millen (2008) set participants the task of building, one person at a time, towers that were as tall as possible from uncooked spaghetti and modelling clay. Participants had 5 minutes to build their tower, but they could also observe the towers of the previous participants. The next participants could in turn observe their work. Thus, each run of the experiment constituted a miniature tower-building tradition.

The experiment produced definite traditions, as people built on and elaborated the types of towers that were established within their group. Figure 9.10 shows two sets of towers. The first group has developed a tower style based on a broad open skeleton of single spaghetti strands, whereas the second group has developed a tradition based around a thick central core, which is a bundle of strands.

There is cumulative evolution over time. As Figure 9.11 shows, towers tend to get taller down the 'generations' of builders. However, Figure 9.11 is based on the average of many runs of the experiment and masks interesting variation. In any particular run of the experiment, cultural progress is quite uneven, with the tower-building tradition sometimes stagnating or going backwards for a number of generations before discovering some improved approach.

This simple experiment has many similarities to spontaneously occurring human traditions. First, there is an effect of history. Two different populations might, through chance alone, set off in a different initial direction (even though the actual task is exactly the same for every group) and this difference then persists over many generations as a recognizable cultural or stylistic difference. However, within the parameters of a particular style, the tradition tends to improve and discover good design tricks over time, albeit at an uneven pace and with periods of stagnation. We might reasonably assume that if there were really only one good way of making spaghetti towers, then all groups would eventually find it, perhaps via different routes. However, if, as seems more likely, there are a number of possible tower architectures that are about equally good, then different groups might end up with different tower styles however long the experiment is continued.

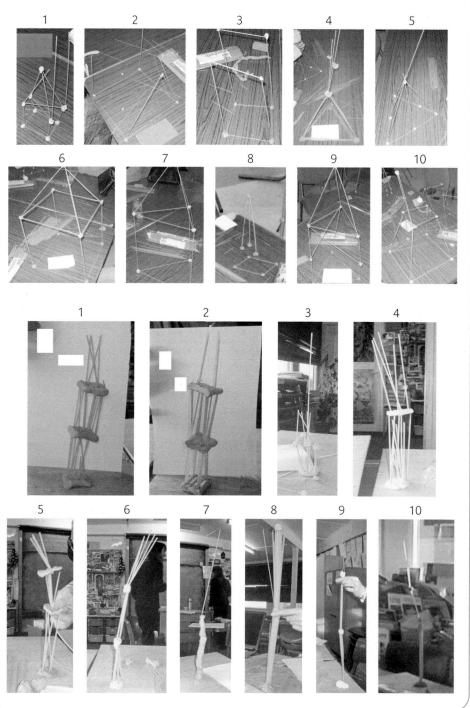

Figure 9.10 **Spaghetti towers from the experiment of Caldwell & Millen (2008). Each sequence represents one run of the experiment, from the first to the tenth builder. The top group explored designs based on a wide skeleton of single spaghetti strands, whereas the bottom group developed a tradition based on a core bundle of strands.** *From Caldwell & Millen (2008).*

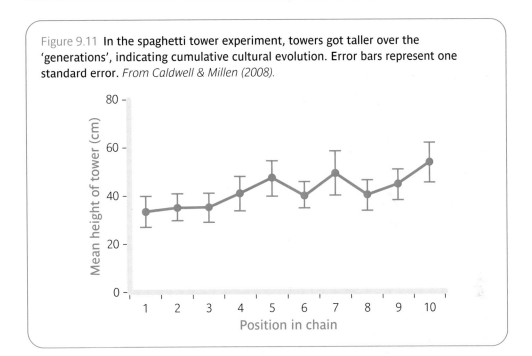

Figure 9.11 **In the spaghetti tower experiment, towers got taller over the 'generations', indicating cumulative cultural evolution. Error bars represent one standard error.** *From Caldwell & Millen (2008).*

Why do human traditions demonstrate cumulative cultural evolution whilst those of other animals seldom do? Part of the answer may simply be the extent of human investment in learning in general (see Chapter 10). However, there may also be specific evolved abilities that humans have, such as the ability to teach, to evaluate the effectiveness of different available alternatives, and to imagine improvements that are key. However it originated, cumulative cultural evolution is undoubtedly a large part of the reason that humans have become so ecologically dominant.

9.6 Learning and adaptation

We have already discussed the idea that learning is favoured by natural selection as a way of producing adaptive phenotypes under certain conditions, such as that the environment is variable. We now consider some more general issues regarding learned behaviours. Will they generally maximize the organism's fitness? How will they differ from behaviours that are not learned? What effect does the ability to learn have on the course of evolution?

9.6.1 Learning and natural selection are similar processes

Natural selection and associative learning are fundamentally rather similar processes. In both systems, some variation in behaviour is generated (by mutation in the former case, by trial and error in the latter). If the behavioural variant improves the organism's performance, it will increase in frequency (through selective advantage in the genetic case and through

Figure 9.12 In a scenario where there are 12 possible feeding behaviours, each with a different rate of return in terms of calories per hour, learning within the organism's lifetime and natural selection over the generations, or any combination of the two, will produce the same result, namely to push the behaviour towards the adaptive peak, which in this case is behaviour 7.

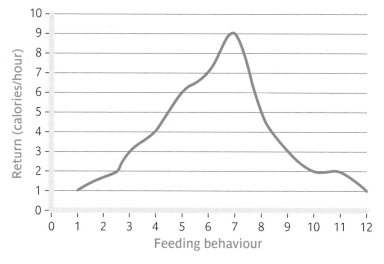

reinforcement by reward in the case of learning) and now new variants will be generated starting from an improved baseline.

We can specify this similarity rather more precisely by thinking of both learning and evolution as processes of hill climbing in a design space (see section 5.4.4). Imagine that there are 12 different ways of feeding in a particular environment and each is associated with a particular rate of return in terms of calories per hour (Figure 9.12). The most adaptive behaviour is the one with the highest return, in other words the one at the top of the hill.

If feeding behaviour is directly influenced by genetic variants, then individuals bearing alleles that cause a better feeding behaviour will have higher reproductive success and increase in frequency. Then more mutations may come along associated with better behaviours still, and these will in turn go to fixation. The net result will be that the phenotype of the animal will, over the generations, gradually move to the behaviour with the highest payoff.

Now consider the case where feeding behaviour is instead controlled by associative learning. Animals trying out better feeding behaviours will receive more reward and through this they will increase the frequency of that behaviour. They may then try out a variant which is better still, which will in turn be rewarded more and increase in frequency. Over many learning episodes, the overall result will be that the phenotype moves to the behaviour with the highest payoff. Exactly the same outcome would also result if the animal was using some social learning strategy, such as 'imitate any individual who is getting more food than me'. All three processes —genetic evolution, associative learning, and discriminate social learning—will lead to the selective retention of changes in behaviour that improve the rate of feeding return.

This means that learning and genetic evolution will often produce the same behavioural outcomes. If we examine an animal's behaviour and establish that it is using the best possible method of feeding, this confirms that the behaviour is adaptive, but says nothing about whether the behaviour is individually learned, socially learned, or developmentally robust, since all three processes would have led to the same consequence. The only real difference will be a vastly different speed of response if the environment changes (fastest for individual learning, intermediate for social learning, and much slower for genetic evolution).

This conclusion reminds us of the fundamental importance of distinguishing between the ultimate and proximate explanations for a behaviour (section 5.3.3). To say that a particular behaviour is adaptive is to claim that it increases reproductive success (e.g. that it is at the top of the hill on Figure 9.12). This leaves it completely open whether that behaviour has been arrived at by genetic evolution, individual learning, social learning, or more likely some combination of these. These are considerations of proximate mechanism, not ultimate function.

9.6.2 Learning does not always produce adaptive behaviour

The previous section stressed the way that learning leads the individual to behave in an adaptive way given the current environment. Although this is often true, there are conditions under which it is not the case and in this section we consider what they are.

Associative learning relies on there existing an intermediate outcome which natural selection has made rewarding because it has tended to be correlated with reproductive success over past environments. For example, people find sweet tastes rewarding because sweet tastes in the ancestral environment were exclusively associated with honey and ripe fruit, and these were always good things to have more of. Sweet foods are a powerful incentive and people will learn to do things to get access to them.

However, in the environment of the contemporary developed world, sweet tastes are usually produced by refined sugars and eating more refined sugars, far from promoting reproductive success, is often bad for health. Thus, learned behaviour in contemporary humans may often be maladaptive in terms of maximizing our long-term interests because the environment has changed in such a way that the rewarding outcome does not now correlate with reproductive success in the appropriate way. If anything, selection in humans in the West is now favouring individuals who do not find sweet things rewarding, but there is a long time lag whilst this change works through.

How about social learning? The consensus here is that, although the capacity to learn socially is adaptive in general, specific beliefs and behaviours learned in this way may not always promote the individual's reproductive success. For example, following cultural traditions about types of food to avoid might be a prudent policy overall, but there could be specific instances where the cultural tradition is in fact obsolete and one would do better to eat it. Theoretical work suggests that the frequency of this type of mismatch will increase as the costs of learning individually about the effectiveness of the behaviour increase. This makes sense, since the more difficult learning for oneself becomes, the more one will tend to rely on what others do, and those others will in turn have relied more on others who came before them.

9.6.3 Learning can guide evolution

As well as natural selection favouring the ability to learn, learning can also guide the process of natural selection. To consider why, consider the following scenario. An animal lives in an ecological context where it would be beneficial for individuals to be able to signal the presence of food to other individuals at a distance. Signalling in this way requires two separate adaptations: (a) the animal has to produce a call when it finds food and (b) other individuals have to understand what the call means and respond to it appropriately. If mutations providing adaptation (a) and adaptation (b) arise simultaneously, then they will increase reproductive success and spread. However, this is vanishingly unlikely. It is much more likely that either (a) or (b) will arise on its own. However, neither one provides any selective benefit whatever without the other being already present in the population, so either one arising alone would go extinct and selection would never be able to assemble the complete signalling system.

This is where learning comes in. If adaptation (a) arises by mutation, and individuals are able to understand the signal and respond to it by a process of learning, then (a) does have a selective advantage even without the mutations for (b) having arisen. Thus, (a) spreads. Once (a) is widespread, then if mutations for (b) subsequently arise, they are selectively favoured in their turn. You then end up with a signalling system that is not learnt, but learning has played a role in its evolution.

These kinds of interactions between learning and natural selection generally go under the name of the **Baldwin effect**. The Baldwin effect has two stages. In the first stage, a capacity to learn completes a gap in a genetically specified system, making that system beneficial and promoting the spread of the alleles underlying it. In the second stage, **genetic assimilation**, subsequent genetic mutations fill the gap in the system and remove the necessity for learning to be involved. The second stage may or may not follow the first; if the costs of learning are low enough, learning may remain involved.

The Baldwin effect may be rather powerful and important. Its significance is that it makes it easier for natural selection to form complex suites of novel adaptive behaviours where the parts all depend on each other (Hinton & Nowlan 1987).

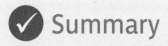 # Summary

1.   Behaviours cannot usefully be divided into those that are 'learned' and those that are 'innate'.

2.   'Learned' is not the opposite of 'evolved'. Instead, learning represents the operation of evolved mechanisms and is a way natural selection can produce locally adaptive behaviour.

3.   Learning and plasticity are favoured by selection where the environment is variable over the very long term, but consistent over the short term, and the costs of plasticity do not outweigh the benefits.

4.   Key forms of plasticity include developmental induction, imprinting, and associative learning.

5.   Animals do not possess one single learning mechanism, but instead different mechanisms for different domains, each with characteristic design features.

6.   Social learning can be favoured over individual learning where the environment changes relatively slowly and the costs of individual learning are substantial.

7.   Cultural traditions are found in many species, but cumulative cultural evolution is rare outside of humans.

8.   Learning may play a key role in the evolution of complex behavioural adaptations, through the Baldwin effect.

? Questions

1. The development of human language involves learning words and grammatical rules, rather than having them develop without the need for learning. Why do you think evolution has designed the language faculty this way?

2. How could the notion of evolutionary preparedness to learn, and the work on fear-learning in monkeys (section 9.4.3), help us to understand human phobias?

3. Female Japanese quail find a male more attractive if they have seen him mate with another female. The same does not seem to be true of human females. Why do you think this is?

4. Some of the longest-enduring cultural traditions in humans, lasting thousands of years, involve food taboos (e.g. restriction on eating pork) or notions about what happens after death. By contrast, techniques of food production change over much shorter timescales. Why do you think this might be?

→ Taking it further

Good review papers on developmental induction and its relevance to various aspects of human health and behaviour are Bateson *et al.* (2004), Belsky (2007), and Gluckman *et al.* (2007). Rozin & Kalat's (1971) classic paper on evolved constraints on learning is still worth reading today. On social learning in animals, Galef & Laland (2005) is a good review, and on human culture in particular see Boyd & Richerson (1996) and Henrich & McElreath (2003). Boyd & Richerson also have two important books on human culture in evolutionary perspective, one technical (Boyd & Richerson 1985) and one more accessible (Richerson & Boyd 2004). The Baldwin effect was first discussed as early as 1896 or possibly earlier, but a key paper in the contemporary revival of interest in it is Hinton & Nowlan (1987).

Our place in nature

In this chapter, we examine what is known of the recent evolutionary history of one particular species, namely the human being, *Homo sapiens*. We do this for several reasons. Understanding the historical context of human beings sheds considerable light on what it is that is so different about modern humans, and what it is that they share with their living relatives. It also helps us hazard some guesses about when, and why, certain important features of human life arose. Finally, examining human evolution gives us a chance to think more generally about how species are classified, how they relate to one another, and how they change.

In section 10.1, we review the general issues surrounding how living things are classified and how we use evidence to reconstruct their relationships to one another. Section 10.2 places humans into the context of the primates, the order of living things to which they belong, and gives a brief overview of what is known of the history of our lineage since our ancestors split with the ancestors of our closest living relatives, the chimpanzees. Section 10.3 discusses some of the main ways that humans differ from other living primates.

10.1 Reconstructing the tree of life

A central claim of evolutionary theory is that all living organisms are related to each other in a single, huge family tree or phylogeny, that is if we trace the family line of any human back through their parents, grandparents, great-grandparents, and so on, then sooner or later we will hit an ancestor that was also an ancestor of any other human that we care to choose. Somewhat further back, we will hit an ancestor who, as well as being an ancestor of all living humans, is also an ancestor of today's living chimpanzees. Further back still, we will meet an ancestor who is also the ancestor of all living mammals. Finally, vastly further back, we will reach an individual who is also an ancestor to all living humans and all living plants (and a great many other things besides).

These various ancestors will look rather different from one another. The ancestor that I share with you probably lived within the last few tens of thousands of years, and probably looked pretty much like us both. The ancestor that I share with the chimpanzee in my local zoo lived several millions of years ago, and was neither exactly a human nor exactly a chimpanzee. It probably looked like me in some ways, like the chimp in other ways, and like neither of us in still others. The ancestor that I share with all plants really did not look much like me at all. Nonetheless, it was my direct ancestor. Although my ancestors look less and less like me as we go further back in time, there is not really a clear point where we can say, that ancestor was a human, but its mother was not.

The point of this illustration is to remind ourselves that the history of life is continuous and it is hard to say where a particular kind of organism starts or stops. People often ask questions like 'When did humans first evolve?' or 'Which evolved more recently, chimpanzees or humans?' These questions do not really make sense. One cannot say that a creature evolved at a particular moment in time or that one is older than another. Instead, we can ask more specific questions, such as 'At what point in the tree of life did the ancestors of animal A diverge from those of animal B?' or 'At what point in evolutionary history did such-and-such a phenotypic characteristic appear in this lineage?' These are the kinds of questions that we will address in this chapter.

10.1.1 The species concept

The first way we classify living organisms is according to the species to which they belong. Species are uniquely identified by their biological name, following the system introduced by the great Swedish naturalist Carl Linnaeus (Figure 10.1). The name consists of two parts, the first, capitalized, representing the genus (plural genera, see next section) and the second, not capitalized, representing the species within the genus, as in *Canis lupus*, the wolf. Where relevant, a third term can be used to distinguish a particular subspecies (such as *Canis lupus signatus*, the Iberian wolf).

A species is defined by its reproductive isolation from other species. In other words, humans are a distinct species from gorillas because humans and gorillas do not interbreed. If there were even occasional interbreeding between humans and gorillas, then although humans look

Figure 10.1 **Carl Linnaeus (1707–1778), the father of biological taxonomy.**
With kind permission of The U.S. National Library of Medicine.

different from gorillas, they could be at most a subspecies of the same species. Although the criterion of reproductive isolation is clear in principle and usually in practice, at least for sexual species, some discussion is in order.

First, reproductive isolation can be partial. Where two populations are interfertile but seldom encounter each other, or tend not to choose each other as mates, then it is a matter of debate whether we have two species or one. Second, reproductive isolation leads to an accumulation of evolutionary differences, but does not do so instantaneously. Two populations which become separated because, for example, the forest is cut down in the middle of their range may ultimately speciate, but the year after the forest has been cut down, it would seem odd to describe them as separate species. Since they are now isolated, any evolutionary change occurring in one population will not spread to the other and so they have the potential to diverge. Exactly when we say that they are now species rather than populations or subspecies is somewhat arbitrary, and there is neither abrupt change nor a 'day zero'.

Grey areas: the genus Canis

We can illustrate these grey areas with the example of the genus *Canis* (dogs and wolves; Figure 10.2). By some authorities, there are at least six living species of wolf (the grey, red, Indian, Himalayan, Eastern, and Ethiopian). This diversity reflects the way wolves have become isolated into unconnected pockets over what was once a vast continuous range. There are also

Figure 10.2 **How many species do you see? From top left, coyote (*Canis latrans*); Indian wolf (*Canis indica* or *Canis lupus pallipes*); grey wolf (*Canis lupus*); domestic dog (*Canis lupus familiaris*, formerly *Canis familiaris*); dingo (*Canis lupus dingo*, formerly *Canis familiaris dingo* or even *Canis dingo*); and New Guinea singing dog (also *Canis lupus dingo*).** *Clockwise from top left: © Darren Bean/fotoLibra; Kenneth W. Fink/Ardea; Photodisc; Anup Shah/Nature Picture Library; Josh Bryen/fotoLibra; Monika Wisniewska/istock.com.*

dozens of subspecies of these species, some of which may be equally deserving of consideration as separate species as the Indian or Himalayan wolf is. On the other hand, it is plausible that there is actually only one wolf species, since complete reproductive isolation has not been demonstrated. Although there are differences between populations, for example in size and coloration, these could be handled at the subspecies level, and the specific name *Canis lupus* be applied to all.

The domestic dog was considered by Linnaeus to be a sister species to the wolf/wolves, called *Canis familiaris*. In 1993, in the face of evidence of fertile matings between dogs and wolves, the dog was reclassified as a subspecies of the grey wolf and renamed *Canis lupus familiaris*. The Australian dingo (and its subvariety, the New Guinea singing dog), which had formerly been considered a subspecies of domestic dog, now also became wolves, *Canis lupus dingo*, although there might be an argument for naming them as a sub-subspecies *Canis lupus familiaris dingo*. To make things worse, coyotes, *Canis latrans*, and golden jackals, *Canis aureus*, can sometimes interbreed with both dogs and wolves. Thus, although all canines are clearly related by descent and can be placed in a phylogeny, where to draw the species lines is far from straightforward.

Species in time

We have seen some of the problems with drawing species boundaries amongst living creatures, but the indeterminacy is much greater when it comes to extinct ones. For example, how do we decide whether a human-like fossil from 100,000 years ago belongs to the same species as living humans? The criterion of reproductive isolation cannot be employed, since two creatures living at different points in history could never meet and therefore never breed or fail to do so. The usual solution in such cases is to use different species names for two forms if they are as phenotypically different from one another as individuals from two different living species are. This criterion, as you can imagine, leaves particular decisions quite open to debate.

It follows from this that where a new species name begins in time is really a matter of decision rather than discovery. For example, it is generally accepted that our species name, *Homo sapiens*, is first applied to fossils from Africa which date from within the last 200,000 years. Earlier forms were quite like us but are given different species names. This does not mean that an abrupt event occurred in Africa 200,000 years ago so that a new species with a new essence suddenly appeared. Rather, there was a continuous sequence of evolving human-like creatures, and 200,000 years is the point at which we decide that the similarity has become sufficient to apply the same species name as living humans.

10.1.2 Higher taxonomic units

Taxonomic terminology is nested, with each taxonomic unit sitting inside a more inclusive grouping at the level above. Thus, our grey wolf, *C. lupus*, sits in the genus *Canis*, within the family Canidae (which also includes foxes), within the order Carnivora (along with cats, seals, bears, etc.), within the superorder Laurasiatheria (with such company as camels, bats, and moles), within the class Mammalia (the mammals), within the phylum Chordata (animals with back-bones and their close relatives), within the kingdom Animalia (the animals), within the domain Eukaryota (which includes everything except bacteria and their allies). This hierarchical system owes its origins to Linnaeus, just as the two-part species naming system does.

There are two important points to make about these taxonomic concepts. First, it is increasingly accepted that taxonomic groups should reflect not phenotypic resemblance, or ecological niche, but phylogeny, that is bats belong in Laurisiatheria with camels and not with the birds, even though they look like birds and they live rather like birds, because if you went back through the generations of their ancestors you would meet a common ancestor with camels much sooner than you would meet a common ancestor with any bird. A taxonomic unit should therefore be monophyletic. This means that it should contain an ancestor and all of its descendants (Figure 10.3). However, many of the groupings traditionally used in taxonomy do not turn out to be monophyletic. 'Reptile' is not a monophyletic group, since crocodiles are more closely related to birds than they are to turtles. A phylogenetically correct unit would thus be 'reptiles and birds'. Similarly, 'ungulate' is not a phylogenetically meaningful entity since camels are more closely related to wolves than they are to horses. Nonetheless, these terms have become entrenched in classificatory practice and you will often encounter them.

Second, groupings such as genus and family have an element of arbitrariness about them. They are not totally arbitrary, since they should be monophyletic. However, how deep a point on the tree to use to define the genus and how deep a point to use to define the family (and thus, the number of descendants to include within each label) is a matter of convention. Again, this applies even more strongly when we apply genus names to extinct creatures on the basis of fossil evidence.

10.1.3 Establishing phylogenies

In this section, we look briefly at how the phylogenies of different living organisms are established. All of the various methods depend on measuring some set of characters in each of

Figure 10.3 **Phylogeny of six vertebrate groups. The grouping 'reptiles and birds' is monophyletic, since it includes the ancestor at point A and all of its descendants. The grouping 'reptiles' is said to be paraphyletic, since it includes the ancestor at A and some but not all of its descendants (because it excludes the birds). The grouping 'warm-blooded animals', which would include mammals and birds, is said to be polyphyletic because the most recent common ancestor of warm-blooded animals was not thought to be warm-blooded. This means that warm-bloodedness has evolved independently at least twice.**

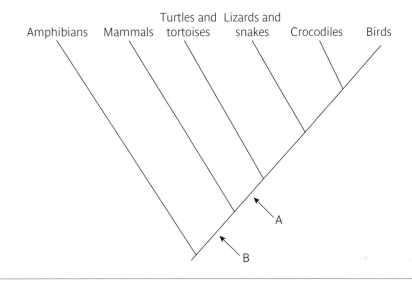

the organisms we wish to classify. These characters were traditionally morphological, like the presence of some skeletal or other anatomical feature. Increasingly these days, the characters are molecular, that is the sequence of nucleotides in a gene, the presence of a transposable genetic element in the genome, or the sequence of amino acids in a protein is used as a character for establishing phylogeny.

To be informative, characters have to exhibit some variation. If all of the species under comparison show the character identically, then that character carries no information for grouping them. On the other hand, if every species is different with respect to the character, then it is not informative either. Instead, we need to find characters which are shared by some but not all of the species under study.

In addition, the characters in question need to be similar due to retention from a common ancestor, rather than similar due to convergent evolution (the wings of birds and those of bats, for example, are not phylogenetically informative). For morphological characters, convergent evolution can be difficult to exclude. Molecular characters can be advantageous here. Since many changes at the molecular level are neutral with respect to function (see section 3.2.4), they are unlikely to evolve identically multiple times. Thus, we can take shared molecular features as strong evidence of common descent.

An example: the place of the whales

The key to phylogeny construction is the identification of shared derived characters unique to the descendants of each branch on the phylogeny. These characters stem from evolutionary events that occurred in the period immediately prior to each hypothesized branching. For example, Nikaido *et al.* (1999) studied seven species of mammal and took as their characters the presence or absence of a transposable genetic element (TE) at 20 different loci in the genome. Here, we will consider just a subsample of their data, pertaining to a cow, deer, hippopotamus, whale, and pig. They found TEs present at loci 4, 5, 6, and 7 in the whale and hippopotamus, but none of the other animals. This justifies joining the whale and hippopotamus in a branch. They also found TEs at four loci (8, 11, 14, and 15) in the cow and deer but no others, justifying joining the cow and deer. Finally, there were two loci (10 and 12) with TEs present in all the animals except the pig. Thus, the best phylogeny is the one shown in Figure 10.4, with the pig's ancestors diverging first, after which the TEs at loci 10 and 12 appeared, and then the whale/hippo and cow/deer lines dividing, followed by further changes within each of these lines before their divisions into the current species.

This methodology is fine, but there is always more than one logically possible phylogeny. The pig's ancestors could have diverged after the TE at locus 10 appeared, but the pig could have subsequently lost that. Perhaps the hippopotamus is not closely related to the whale, but the TEs at loci 4–7 have, by chance, independently arisen in each branch. To adjudicate between these different possibilities, we draw on the principle of parsimony. The most parsimonious phylogeny is the one that assumes the smallest number of evolutionary events. A phylogeny with whales and hippos in separate branches needs to assume at least eight separate mutations (the TEs at loci 4, 5, 6, and 7 each need to evolve twice, once in each branch), whereas the phylogeny where whales and hippos are in the same branch need assume only four mutations (occurring in the common ancestor of whales and hippos). Thus, the latter phylogeny is to be preferred.

In contemporary research, phylogenies are created by simultaneously considering many characters and many species. This is done using computer procedures that search through the vast set of possible trees for the most parsimonious or likely one. The output of these programs typically consists not just of a phylogeny, but also of a confidence statistic in the correctness of each branch. Figure 10.5 shows an example of a molecular phylogeny, for the Canidae. This phylogeny was produced by comparing about 15,000 base pairs of genetic sequence from the different animals. The results support the special affinity of the domestic dog, grey wolf, coyote, and golden jackal, discussed above.

The molecular clock revisited

Molecular data provide the additional possibility of estimating how long ago branching events occurred. This is based on the idea of the molecular clock, which we met briefly in section 3.2.4. The idea of the molecular clock is that, since much molecular change is neutral, and since the probability of mutation should be about the same in each generation, then molecular differences between two lineages will accumulate at a roughly constant rate over time. Thus, the more differences there are at the molecular level, the longer it is since two lineages diverged.

Molecular clock ideas are widely used. However, there is variation in the rate of change from molecule to molecule, and also from organism to organism. DNA sequences diverge more slowly in primates than rodents, and mitochondrial DNA (mtDNA) evolves much more slowly in turtles than in other organisms (Li *et al.* 1987; Avise *et al.* 1992). Aspects of the biology

Figure 10.4 **Using molecular data to infer the phylogeny of five mammals. The characters are presence or absence of an inserted transposable genetic element at a number of loci in the genome (? indicates that it was not possible to determine presence or absence). By seeing which elements are present in which subgroups of the species, researchers can infer both the shape of the phylogeny and the point in evolutionary history at which the insertions most likely occurred, shown by the horizontal bars across the branches. The numbers on the phylogeny refer to loci where insertions occurred at that point.** *From Nikaido et al. (1999) after Freeman & Herron (2004), p. 565.*

Locus	3	4	5	6	7	8	9	10	11	12	13	14	15
Pig	0	?	0	0	0	0	?	0	0	0	?	?	0
Hippo	0	1	1	1	1	0	1	1	0	1	1	0	0
Whale	1	1	1	1	1	0	?	1	0	1	1	0	0
Cow	0	0	0	0	0	1	1	1	1	1	1	1	1
Deer	0	0	0	0	0	1	?	1	1	1	1	1	1

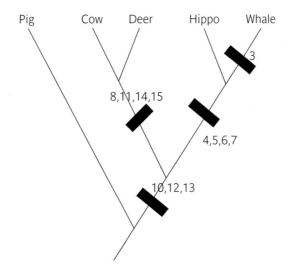

of the specific lineage, including generation time, metabolic rate, and the effects of natural selection, are implicated in these differences. Thus, for any particular application, the molecular clock has to be calibrated for the lineages and molecules under study. This is usually done by taking a branching event whose date is known from the fossil record to establish an evolutionary rate, which then allows inferences about the date of other events in the history of the same lineage.

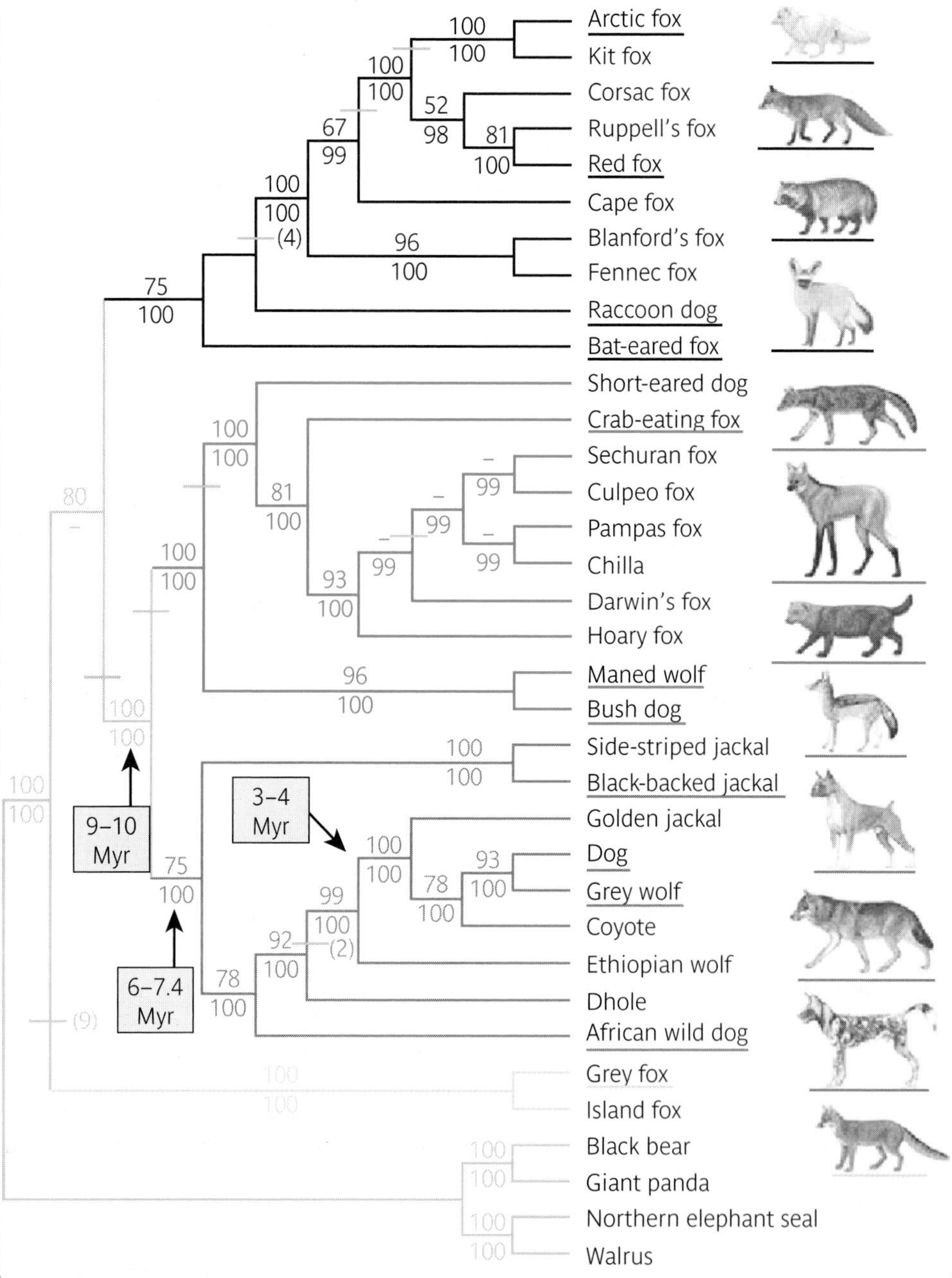

Figure 10.5 **Estimated phylogeny of canid species based on 15,000 base pairs of genetic sequence. The numbers on each branch represent two types of confidence statistic in the correctness of that grouping (100 = total confidence). Myr, million years.** *From Lindblad-Toh et al. (2005).*

10.1.4 Fossils

So far, we have discussed how evolutionary history is inferred from living organisms. We now consider how fossil evidence fits into this. Fossils are remains or traces of once-living things, usually preserved in rocks. Fossil formation is extremely rare, depending as it does on exactly the right geological conditions. It tends to be the hard parts of organisms, such as bones and teeth, which fossilize.

Fossils allow ancient creatures to be placed into phylogenies in much the same way as is done for living ones. The characters used are usually morphological, for obvious reasons. However, with relatively recent remains, DNA can sometimes be extracted. Thus, there is sometimes molecular evidence on the place of ancient organisms in the phylogeny. This has proved useful for the study of human evolution, particularly in the case of the Neanderthals (see section 10.2.3).

Fossil evidence, where it is available, has a number of key uses. Fossils can often be dated reasonably accurately by using the concentration of radioactive isotopes in the surrounding rock to estimate their age. This allows time brackets to be attached to points on the phylogeny and also allows calibration of molecular clocks.

Fossil whales

Fossils can also provide confirmatory evidence of ancestral states predicted by the phylogeny of living creatures. For example, according to the molecular phylogeny described in Figure 10.4, whales belong to the artiodactyls (the even-toed ungulates). Living artiodactyls have a shared derived character in the form of a pulley-like shape to a leg bone called the astragalus. The phylogeny thus predicts that ancestral whales also had this feature before they took to the water and lost their legs. In 2001, two fossils dating from around 50 million years ago, *Pakicetus attocki* and *Ichthyolestes pinfoldi*, were described (Thewissen *et al.* 2001). Both specimens had legs and the pulley-like astragalus identifying them as artiodactyls, but both also had features of the ear bones that are found today only in whales. Since these two specimens can be related to later fossils in which the hind legs become smaller and smaller, the fossil evidence confirms that whales descended from a legged ancestor by a shift to water living and loss of limbs, and the shape of the astragalus before those limbs were lost confirms that the ancestor was indeed an artiodactyl.

Fossil flatfish

Fossils can also shed light on how a phenotypic characteristic got from A to B by furnishing examples of intermediate states. A wonderful example of this comes from flatfish such as plaice. These fish have both eyes on the same side of their body and lie horizontally in the water. We know that they are phylogenetically related to fish that lie vertically in the water and have one eye on each side of the head. Indeed, larval flatfish are like this, with the eye migrating across the midline during development. However, it is hard to see how evolution moved from the ancestral state to the current flatfish in gradual stages. Could there be a form that has changed its body axis, but whose eye has not yet migrated? And if there were, what use would this eye be?

Friedman (2008) examined fossils of flatfish ancestors from an Italian site dating from around 45 million years ago. Their body form is indicative of a horizontal orientation, but they have one eye on each side of the head. However, the eye on the lower side is partially migrated upwards. This would allow the fish to lift its head and use the lower eye to look along the sea bed, so the

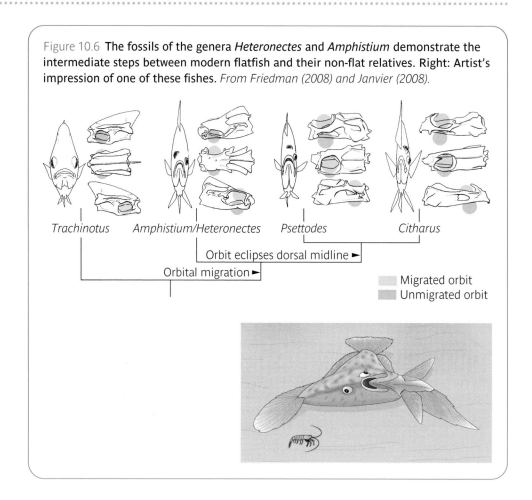

Figure 10.6 **The fossils of the genera** *Heteronectes* **and** *Amphistium* **demonstrate the intermediate steps between modern flatfish and their non-flat relatives. Right: Artist's impression of one of these fishes.** *From Friedman (2008) and Janvier (2008).*

eye was not completely useless. However, the further the eye migrated, the less head lifting would be required to do so, and this created a selection pressure for greater and greater eye asymmetry until the lower eye crossed the midline (Figure 10.6).

10.2 Humans as primates

Armed with an appreciation of the general issues and methods involved in reconstructing the tree of life, we now turn to humans. Humans belong to the order of the primates. The primates are a group of mammals consisting of as many as 400 species, mainly found in the tropics, whose ancestors are thought to have diverged from those of their closest non-primate relatives about 80 million years ago.

Primates are distinguished from other mammals by a number of shared derived features. One of the five digits can be opposed to the other four, and so primates are adept at grasping using their hands and often their feet. (Humans have lost the opposability of the big toe on their feet, but retain the opposability of the thumb on the hand.) Primates are relatively highly dependent on vision (rather than olfaction) and tend to have forward-facing eyes, binocular vision, and some degree of colour vision ability. Primates also have relatively large brains for their body size. The ancestral habitat, and habitat of most living primates, is arboreal, or tree dwelling.

10.2.1 The primate phylogeny

The primate phylogeny first branches into the strepsirrhines and haplorrhines, a branching dated to soon after the ancestral primates themselves appeared. There are rather more than 100 species of strepsirrhine. Small strepsirrhines called lorises, pottos, and galagos occupy nocturnal forest niches in Africa and Southeast Asia. Most strepsirrhine species, however, belong to the lemurs, a group restricted to the island of Madagascar. Madagascar has no monkeys or apes, and here the lemurs have diversified to produce a range of large and diurnal species as well as small and nocturnal ones (Figure 10.7).

Within the haplorrhines, the tarsiers are the most divergent group. These are a single genus of small, nocturnal animals found in Southeast Asia. The tarsiers resemble lorises and indeed are sometimes classified along with the strepsirrhines in a (paraphyletic) grouping called prosimians.

Figure 10.7 **Major branches on the primate phylogeny. First, the strepsirrhines (shown, bamboo lemur *Hapalemur griseus*) branch from the haplorrhines. The haplorrhines then branch into the tarsiers (shown, *Tarsius tarsier*) and the rest, which split into the platyrrhines or New World monkeys (shown, spider monkey *Ateles geoffroyi*), and the catarrhines, or Old World monkeys and apes (shown, olive baboon, *Papio anubis*).** *Left to right: © Pete Oxford/Nature Picture Library; Dariusz Lewandowski/ istock.com; James Margolis/istock.com; istock.com.*

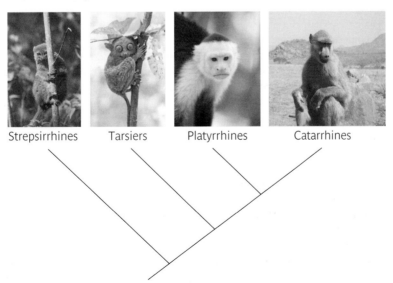

Strepsirrhines   Tarsiers   Platyrrhines   Catarrhines

Figure 10.8 **The diversity of apes. From the left, gibbons and orang-utans are found in tropical Asia, chimpanzees and gorillas in tropical Africa, and humans on every continent.** *Left to right: © Pailoolom/Fotolia.com; Photodisc; Corel; Photodisc; Digital Vision.*

The monkeys and apes form the remainder of the haplorrhines. They first branch into the platyrrhines and catarrhines (a branching estimated at 40 million years ago). The former are four families of monkeys found in the Americas, whilst the latter are the monkeys and apes of the Old World. Note that this means that 'monkey' is not a monophyletic group, since the last common ancestor of all monkeys was also the ancestor of the apes. Molecular phylogeny of the catarrhines shows clearly that the deepest divergence in this group is between the Old World monkeys on the one hand and the apes on the other. There are around 80 species of Old World monkey, which include many familiar animals such as baboons, macaques, and colobus. They are found across large areas of Africa and Asia. The apes consist of around 20 species of tailless primates, with rather more found in Asia than Africa, which include humans (Figure 10.8).

10.2.2 Humans are apes

In this section, we first describe the living non-human apes, then discuss the shape of the ape phylogeny.

The gibbons, or lesser apes

The apes are sometimes divided into the great apes, and the gibbons, or lesser apes. There are 13 species of gibbon (in four genera) all found in Southeast Asian forests. Gibbons form long-term pair bonds and their territorial social groups typically consist of a monogamous pair plus one or more dependent offspring. Gibbons are specialized in their hand, arm, and shoulder anatomy for brachiation, a mode of locomotion involving swinging from branch to branch using only the arms (Figure 10.8). This makes them the fastest moving of all tree-dwelling mammals.

The orang-utans

The first great ape we will meet is the orang-utan, which lives only in Indonesia. There are two forms, the Bornean and Sumatran (considered either as separate species, *Pongo pygmaeus* and *Pongo abelii*, or as subspecies, *Pongo pygmaeus pygmaeus* and *Pongo pygmaeus abelii*), although there is evidence that the genus was once more widespread and perhaps contained more species. Orang-utans are primarily fruit eaters and highly arboreal. They are the only apes to live in a solitary manner much of the time. Males defend a group of females, but they do so by having a large range that encompasses the smaller individual female ranges, rather than by the harem living together. The consequence is a kind of asocial polygyny, with males much larger than females.

The gorillas

The gorilla is the first of the African great apes. There are two species, *Gorilla beringei* in Congo, Rwanda, and Uganda, and the critically endangered *Gorilla gorilla* further west. Gorillas are herbivores and live on the ground rather than in trees. Their form of locomotion is known as knuckle walking and involves support from the upper surface of the curled digits of the hand, rather than the palm being on the ground as is the case for monkeys. Gorillas are highly sexually dimorphic and live in a classic polygynous harem group structure, with a single dominant male and multiple females.

The chimpanzees

The chimpanzees are two species of African great ape, the common chimpanzee, *Pan troglodytes*, found north of the Congo River, and the bonobo, *Pan paniscus*, found to the south. The bonobo is also known as the pygmy chimpanzee, although it is not in fact much smaller than the common chimpanzee. Like gorillas, chimpanzees knuckle walk, but they are smaller and more arboreal. Common chimpanzees are omnivorous and occasionally hunt, whilst bonobos are more frugivorous. Both types of chimpanzee live in fluid, multiple male, multiple female social groups with a promiscuous mating structure.

The ape phylogeny

It has long been recognized, on the grounds of multiple shared characters, that humans belong to the apes and the great apes in particular. Thus, the term 'ape' is often used incorrectly, as in 'A comparative study of learning in humans and great apes'. Since humans are apes, this title should really read 'A comparative study of learning in great apes'.

The view of ape relatedness, common until the last few decades, had the gibbons diverging first, then the great apes dividing into humans on the one hand, and the other great apes on the other (within those, with the gorilla closest to the chimpanzee). This view is influenced by phenotypic similarity, since the other great apes are all hairy quadrupeds and humans are not. However, outward resemblance is not the basis for classification, phylogeny is. When molecular data began to be available, it provided quite surprising results. Humans were not equally related to all the great apes. Instead, they were strikingly closely related to chimpanzees. The level of similarity—up to 99% resemblance in aligned DNA sequences—was as high as often found in animals from the same genus. Moreover, chimpanzees of both kinds were more closely related to humans than they were to orang-utans or even to gorillas.

The contemporary view of ape relationships is shown in Figure 10.9, which comes from a recent study using molecular genetic data. As you can see, the human–chimpanzee divergence is quite late in the phylogeny, leaving humans and chimpanzees as a small family more closely related to each other than to other apes. Indeed, the depth of separation of humans and chimpanzees—taken to be around 5–7 million years ago—is only about twice that of the two chimpanzee species from each other. Humans are not just apes, and not just African great apes, but on molecular grounds could easily be placed in the genus of the chimpanzees.

10.2.3 Hominins: human ancestors after the human–chimpanzee divergence

Although there are no living animals more closely related to us than the chimpanzees, we do have a considerable number of fossils attesting to intermediate forms, which are designated with a variety of genus and species names, but collectively known as hominins. A full review of

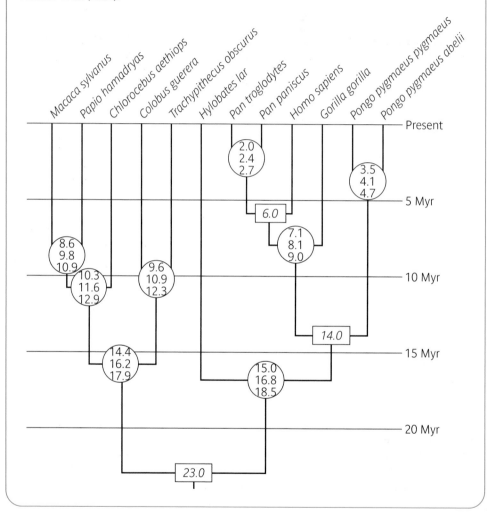

Figure 10.9 **Molecular phylogeny of 12 catarrhine primates. The first five animals are Old World monkeys and *Hylobates lar* is a gibbon. The numbers on the branch points are molecular clock estimates of divergence times (the central number is the estimate, and those above and below give the 95% confidence interval; units are millions of years). The three dates with no confidence intervals (6.0, 14.0, and 23.0) are those which have been taken from the fossil record to calibrate the clock.** *From Raaum et al. (2005).*

the hominins, and of the considerable debate that surrounds their phylogenetic relationships, is beyond the scope of this book, but we will briefly examine some key points. The hominins are not a single evolving lineage; rather there is a branching of multiple forms, many of which go extinct and only some of which are on the line leading to living humans.

Figure 10.10 **Two fossil skulls crucial to our understanding of our history: Left:** *Sahelanthropus tchadensis,* **close to the branch point of ancestral hominins and chimpanzees.** *From Wood (2002).* **Right: One of the Herto crania from Ethiopia, 160,000 years old and on the verge of anatomical modernity.** *From White et al. (2003).*

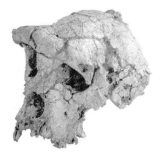 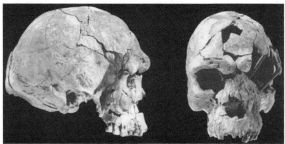

Early hominins, like the living gorillas and chimpanzees, are restricted to Africa. There are a number of fossils from Chad, Kenya, and Ethiopia dating from close to the time of the human–chimpanzee divergence, which have been given the genus names *Sahelanthropus, Orrorin,* and *Ardipithecus* (Figure 10.10). These are generally chimpanzee-like in the size of their brains and facial structure, but there are some hints of bipedal locomotion. (Chimpanzees can walk upright for short distances, but to do so more habitually requires a suite of anatomical changes throughout the body.) Evidence on these first hominins is still extremely fragmentary, however; *Sahelanthropus,* the most ancient, was only described in 2002.

The australopithecines

A group of hominins dating from around 4 million to 1 million years ago are given the general name of australopithecines, and at this point the fossil evidence becomes fuller. The australopithecines were around the size of living chimpanzees. They show unambiguous evidence of bipedal adaptations in the feet, pelvis, legs, and in the position of the aperture through which the spinal cord enters the skull, which is to the rear of the skull in quadrupedal apes and underneath the skull in humans. However, they have short legs and long arms, and other features suggesting that they may have spent some of their time in the trees. If the bodies of the australopithecines suggest humanity, their skulls are more chimpanzee-like, with a protruding face and a brain little larger than that of a living chimpanzee. However, the canine teeth are reduced compared with other apes, an evolutionary direction continued in humans. Wear on fossil teeth suggests an omnivorous diet with a large fruit component.

The context of early hominin evolution is a period in which the climate of Africa was cooling and drying, and previously continuous tropical forest was becoming more patchy and giving way to open savannahs. Apes in general did not do well in this period, with many forms disappearing, presumably because they were highly adapted to dense forest. This provides a scenario for the evolution of bipedal locomotion. The adaptations for bipedalism would have been more costly the more time was spent in trees. On the other hand, they would have become more beneficial the more time was spent moving on the ground between patches of forest. This is for two reasons: bipedal locomotion across open ground is more energy efficient (at least at slow speeds) and an upright hominin absorbs less heat, since it presents a smaller surface to the overhead sun (Wheeler 1991). The drying of the climate and the fragmenting of forests could thus have made the benefits of increased bipedalism come to outweigh the costs.

Later in their history, the australopithecines diverged into the so-called robust (e.g. *Australopithecus boisei*) and gracile (e.g. *Australopithecus africanus*) forms. The former are more robust only in their teeth and jaws, with enlarged molars and anchors for powerful chewing muscles, suggesting perhaps a diet of coarse plant matter. It is the gracile branch that is thought to be ancestral to modern humans, with the robusts disappearing around 1 million years ago and leaving no descendants.

Origins of the genus Homo

The first fossils to which researchers agree to apply the same genus name—*Homo*—as ourselves date from around 2.5 million years ago, attracting such names as *Homo habilis*, *Homo rudolfensis*, and later *Homo ergaster* and *Homo erectus*. These early *Homo* are significant for a number of reasons. First, they are the animals in which brain size (relative to body size) begins to move clearly away from that of chimpanzees and towards the human pattern. Second, around the time of their emergence, stone tools start to appear, often in association with animal bones. This suggests that, whereas the robust australopithecines were specialized in processing coarse plant foods, early *Homo* had begun to depend on meat to a greater extent. Third, early *Homo* are the first hominins to migrate outside of Africa and are found in Southeast Asia from at least 1.5 million years ago. Full modern bipedalism is in place by this time and so the degree of long-range mobility was presumably high.

The archaics

Beginning around 0.8 million years ago, new forms of hominin begin to appear, in Africa first. They are larger overall and their brains are particularly expanded so that the sizes of both body and brain are within the range of living humans. They are associated with more complex stone tools than *H. erectus*. These animals are often referred to as 'archaic *H. sapiens*'. This is a confusing relic of history, since although several species names are used for the archaics nowadays—*Homo heidelbergensis* and *Homo neanderthalensis*, for example—*H. sapiens* is not usually one of them!

A branch of the archaics that lived until relatively recently is the Neanderthal, *H. neanderthalensis*. This is a robust, cold-adapted form that had appeared in Europe by around 300,000 years ago, and did not go extinct until perhaps 30,000 years ago. We will return to the Neanderthals in the next section. Note that whilst the transition to the archaics had occurred in Africa, *H. erectus* continued largely unchanged in Asia for a remarkably long period, only disappearing within the last 50,000 years.

To add a further twist to the Asian story, in 2004, some remarkable fossils from the Indonesian island of Flores were described. They date from within the last 20,000 years, which is well after modern humans arrived in the region. They were claimed as a new species, *Homo floresiensis* (Brown *et al.* 2004). *H. floresiensis* was much smaller than a modern human, or even than *H. erectus*, but it had a number of *erectus*-like features such as a relatively small brain. It has been suggested that *floresiensis* is a dwarfed descendant of *H. erectus*. Dwarf species are common on islands, where limited resources may select for curtailment of growth and reproduction at smaller sizes. The interpretation of the Flores finds remain controversial, but it is clear that *H. erectus* and its line persisted in Asia until relatively recently.

10.2.4 Origins of *Homo sapiens*

Fossils that are almost indistinguishable from the bones of modern humans begin to be found within the last 200,000 years, with the oldest dates in Africa (Ethiopia, 160,000 and 130,000; South Africa 120,000–70,000; Israel 100,000; Figure 10.10). Researchers accord these the name *H. sapiens*, and refer to them as anatomically modern humans (AMH). As you may have noted, the dates of extinction of the Neanderthals in Europe (30,000 years ago) and of *H. erectus* in Asia (less than 50,000) are well after the appearance of AMH in Africa. AMH were clearly expanding, as we see from their presence in the Middle East by 100,000 years ago, in Asia and Australasia by 60,000 years ago, and in Europe by 30,000 years ago.

This poses a very interesting question, which has preoccupied palaeoanthropologists greatly in the last few decades. Were the expanding AMH from Africa a distinct species, which replaced the Neanderthals in Europe and *erectus* in Asia without interbreeding with them, or were they a subspecies that may have interbred, or alternatively, was the whole complex of AMH–Neanderthal–*erectus* just one big evolving species with regional variation?

The first possibility, that AMH were a new species that replaced the other living hominins essentially without interbreeding, is known popularly as the 'out of Africa' model and is the one that has most evidence in its favour. The key lines of evidence are the following:

1. Morphology: If there was some evolutionary continuity between ancient *H. erectus* populations in Asia and contemporary Asians, or between Neanderthals and contemporary Europeans, you might expect to find morphological features in Europeans or Asians that were retained from those ancestors. This has not been convincingly demonstrated. Instead, AMH from all over the world look relatively like each other and most resemble the early African AMH fossils described above.

2. Genetics of living humans: Molecular phylogenies have been constructed using a number of sequences from the human genome, such as mtDNA. mtDNA is particularly useful, since it is passed on only through mothers and is thus not recombined. It also has a high mutation rate, meaning that it is informative even about the short time span of the history of *H. sapiens*. One of the largest studies of human mtDNA was by Ingman *et al.* (2000), who sequenced the complete mitochondrial genome of 53 individuals from around the world. The resulting molecular phylogeny is shown in Figure 10.11. As you can see, the deepest branches in the phylogeny separate some Africans from other Africans. Europeans, Asians, and Australian aborigines are all rather closely related to each other, and all come off one sub-branch of the African tree. Native Americans (the Warao and Guaraní) sit within the group of East Asians.

   What this suggests is that all current humans can be traced back to a small set of ancestors at the root of the tree. The small size of the population at this point reduced the diversity of mtDNA to a very few lineages. After this bottleneck event, the population began to expand and their mitochondria to diverge, and this expansion and divergence began in Africa, hence the deep branches between contemporary Africans. Much later, small groups representing a subset of Africans migrated into the other continents and their relatives began to expand and diverge, giving us the more recent branchings amongst living non-Africans. Molecular clock calculations from these data

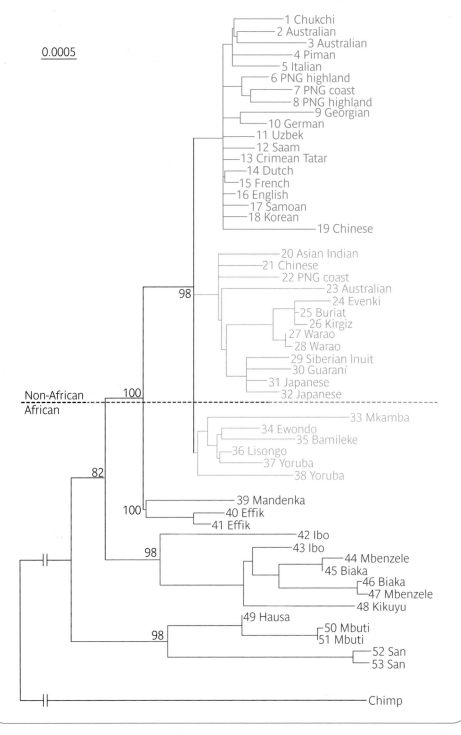

Figure 10.11 Molecular phylogeny of the mitochondrial genome of 53 living humans. As you can see, the deepest divergences are between Africans, with all non-Africans coming off a younger node in the phylogeny, supporting the idea that all contemporary humans expanded from a small population within Africa in the last 200,000 years. The numbers on the nodes are bootstrap values, which represent the degree of statistical confidence in that branching. PNG, Papua New Guinea. *From Ingman et al. (2000).*

0.0005

1 Chukchi
2 Australian
3 Australian
4 Piman
5 Italian
6 PNG highland
7 PNG coast
8 PNG highland
9 Georgian
10 German
11 Uzbek
12 Saam
13 Crimean Tatar
14 Dutch
15 French
16 English
17 Samoan
18 Korean
19 Chinese
20 Asian Indian
21 Chinese
22 PNG coast
23 Australian
24 Evenki
25 Buriat
26 Kirgiz
27 Warao
28 Warao
29 Siberian Inuit
30 Guaraní
31 Japanese
32 Japanese

98

100

Non-African
African

33 Mkamba
34 Ewondo
35 Bamileke
36 Lisongo
37 Yoruba
38 Yoruba

82

39 Mandenka
40 Effik
41 Effik
100

42 Ibo
43 Ibo
44 Mbenzele
45 Biaka
46 Biaka
47 Mbenzele
48 Kikuyu
98

49 Hausa
50 Mbuti
51 Mbuti
52 San
53 San
98

Chimp

estimated the African bottleneck at 172,000 years ago and the expansions out of Africa at 40,000–60,000 years ago.

Inference from the rest of the genome is made more difficult by recombination, but ingenious reconstructions using data from multiple genetic loci also support the model of a small ancestral African population around 150,000 years ago for all living humans, with expansions into Eurasia and Australasia later, and the Americas most recently of all (Fagundes *et al.* 2007; Hellenthal *et al.* 2008).

3. Neanderthal DNA: The phylogenetic relationship between AMH and Neanderthals could of course be easily settled if molecular evidence could be obtained from Neanderthals, but this seemed an unlikely prospect until a few years ago. In 1996, a team succeeded in extracting a short sequence of mtDNA from a fossil Neanderthal bone (Krings *et al.* 1997). Analysis of this sequence showed the Neanderthal to be much more like a living human than a chimpanzee, yet outside the range of variation of living humans. DNA has now been successfully extracted from the bones of a dozen Neanderthals and their genomes are all more similar to each other than they are to any living human. The most extensive study obtained a sequence 1 million base pairs long from a 38,000-year-old fossil found in Croatia (Green *et al.* 2006). Molecular clock calculations place the divergence of the ancestors of living humans and of the Neanderthals at around 500,000 years ago (compared with within the last 200,000 years for all living humans).

The best inference from these remarkable data is that the AMH entering Europe replaced Neanderthals with little or no interbreeding. If there had been even a modest amount, we should expect to find some living humans with some Neanderthal mutations in their mtDNA and no such people have been found.

History of modern humans since their origins

The data thus support the view that all of us alive descend most immediately from a small population which lived in Africa within the last 200,000 years (for a summary of hominin evolution, see Figure 10.12). This population gave rise to great colonizers, peopling all the other continents of the Old World by around 40,000 years ago. Stone tool technologies also became strikingly more complex and varied by around 40,000 years ago. During this great colonization, AMH would no doubt have met hominins of the other species, but since they do not seem to have interbred we can have no idea what their interactions might have been (Neanderthals, remember, were as big as AMH in both body and brain). Whatever happened, only AMH survive.

Most of this long, colonizing journey took place in the later portion of an epoch called the Pleistocene. Pleistocene climate was characterized by repeated and relatively fast oscillations between warm, wet climates (interglacials) and glacial periods in which the temperate latitudes had ice ages and the tropics became relatively arid. The most recent glacial maximum was at around 19,000 years ago. It is only in the melting and warming after this maximum that humans entered the Americas, coming across the Bering Strait from Northeast Asia. The current relatively warm period, starting 10,000 years ago, is called the Holocene.

Agriculture, metalworking, towns, and cities have all only appeared within the Holocene and even then unevenly, with some populations only making these transitions in recent history. Thus, for about 95% of the period since people had anatomies indistinguishable from yours, they were living as hunter-gatherers, in various different environments, using only stone, bone, or wooden tools. Even more strikingly, the period of humans living with agriculture, settled towns, and metalworking constitutes about one-thousandth of the time since the ancestors of humans split from those of the living chimpanzees.

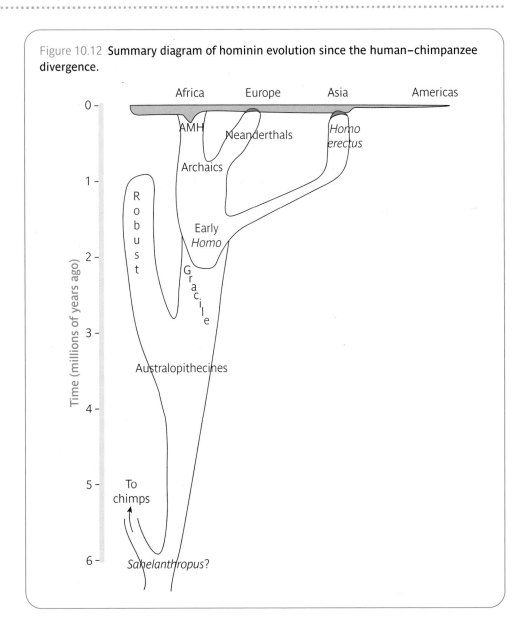

Figure 10.12 **Summary diagram of hominin evolution since the human–chimpanzee divergence.**

10.3 What makes humans different?

In the remainder of this chapter, we examine some of the most striking biological differences between humans and their nearest relatives. Bipedalism has already been touched upon and will not be discussed further. Instead, we focus on six key and interconnected areas: meat eating, tool use, brain size, life history, learning, and language.

10.3.1 Meat eating

Humans are omnivores and contemporary human populations vary in the extent to which they depend upon animal products. Nonetheless, the use of animal foods is an almost universal human feature and one that is particularly important in the absence of agriculture. Chimpanzees do hunt and consume meat, but it constitutes a small proportion of their diets and hominin evolution evidently involved a shift to greater dependency on animals—whether hunted or scavenged—as food. Not only is the archaeological evidence of tools and cut bones clear, but humans have reduced the size of their guts relative to other apes. Guts are generally smaller in carnivores than herbivores due to the relative ease of extracting nutrients from animal as compared with plant foods.

Meat provides a high-quality (but difficult to obtain, see below) resource, which could be used to fund the metabolic cost of humans' large brains. Indeed, carnivory is doubly advantageous in this regard, since guts and brains are amongst the most metabolically expensive tissues. Reducing expenditure of energy on the former would free up resources for increasing the size of the latter (Aiello & Wheeler 1995).

10.3.2 Tool use

All human societies use an array of tools to solve the problems set them by their environment. Humans are not unique in using tools; chimpanzees, for example, adapt sticks to fish for termites and use stones to crack nuts. However, the extent of dependency on tools, their variety, and their sophistication are human hallmarks. Tool use is related to meat eating, since the use of many tools is for butchery or hunting.

Tools are an ancient aspect of hominin life, going back at least 2.5 million years. However, for long periods of that time the types of tools appear to have remained relatively unchanged. Moreover, the appearance of new tool types is not well correlated with anatomical changes. When AMH appear at least 150,000–200,000 years ago, for example, the tool kit does not show much increased complexity for tens of thousands of years. In a sense, this should be no surprise. Evolution is a continuous process and may be affecting some traits (e.g. morphology) at different times or rates from others (e.g. tool use). However, the lack of coupling between brain size change and the tool record is interesting.

10.3.3 Brain size

One of the most striking features of humans is the large size of their brains relative to those of chimpanzees, a difference that is not accounted for by their larger body sizes. Figure 10.13 shows estimates of cranial capacity in hominins, taken from fossils. The capacity of the cranium gives us some idea of the size of the brain. As you can see, with the origins of the genus *Homo*, there is an increase above australopithecine levels and there is a further sharp increase through the archaics to modern humans.

There are many theories concerning the functional significance of large brains in humans, including the social brain hypothesis (see section 8.3.2), which associates brain enlargement with increasingly large and complex social groups. Brain enlargement must also relate to humans' increasing occupation of the learning niche (see section 10.3.5). Whatever the cause, the existence of large brains has made complex demands on the rest of our biology, as we will see in the next section.

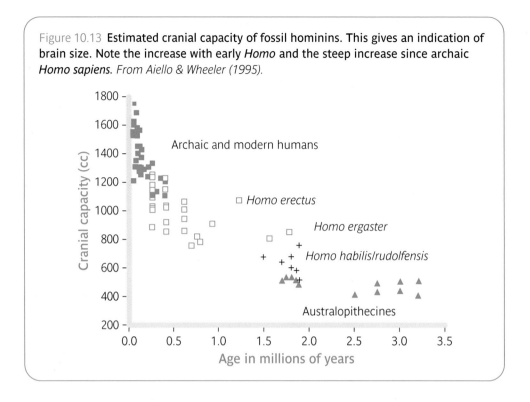

Figure 10.13 **Estimated cranial capacity of fossil hominins. This gives an indication of brain size. Note the increase with early *Homo* and the steep increase since archaic *Homo sapiens*.** *From Aiello & Wheeler (1995).*

10.3.4 Life history

Humans have made some striking changes in their life history compared with other apes. Lifespan has generally increased, and even relative to this the period of development and dependency known as childhood is greatly elongated in humans.

Increasing brain size means increasing head size, but increasing head size is ultimately constrained by limits on the size of the birth canal. Thus, human babies are born with relatively underdeveloped brains and do a period of rapid brain growth post-birth. There then follows a long period of slow growth, which is only ended by the adolescent growth spurt immediately prior to puberty.

Both increased longevity and elongated childhood may be associated with increased brain size. Animals with large brains tend to be long lived because the brain is an energetically costly investment early in development. Having made this investment, there is selection to slow down life history, increase somatic maintenance, and stay alive long enough to recoup the investment. To look at it the other way around, if there is high mortality and selection for fast life history, costly investment in building a brain is likely to be diverted to reaching reproductive state sooner and so there are no short-lived, large-brained animals. Increasing brain size in humans was thus associated with living longer.

As for childhood, having created such a large and complex brain, it takes time for it to mature and make the connections it needs to function adaptively; an advantage is created for a learning period, during which the animal should remain relatively small and absorb information from the environment around it. The slow physical growth and fast skill learning of childhood can be interpreted this way. It is only ended by a rapid spurt to adult size and the development of reproductive abilities.

Having children dependent for such a long time makes the costs of offspring much higher for humans than for other apes. We solve this collectively; fathers, grandparents, and other relatives provision infants and mothers in humans to a far greater extent than is true for other primates. Thus, our family structure follows directly from the extended developmental period. Note that for human women as compared with chimpanzee females, most of the additional life-span is after menopause, that is in the post-reproductive period. Thus, women may have been selected to use their additional years to invest mainly in their offspring's reproduction rather than their own (see section 7.4). Because of human cooperative provisioning of infants, humans actually maintain a higher reproductive rate than other great apes, despite their children being a burden for much longer.

10.3.5 The learning niche

The most striking thing about humans is the diversity of ways in which they make their livings and the dazzling array of skills involved. For example, reviewing their experience of hunter-gatherer populations, Kaplan *et al.* (2000: 171) itemize some of the skills they have witnessed:

> Arboreal animals have been shot with arrows from the ground or a tree, driven by climbing, shaken down from branches, frightened into jumping to the ground, brought down by felling a tree with an axe, lured by imitated calls, lured by making captured infants emit distress calls, captured by the spreading of sticky resin on branches to trap them, and captured by scaffolding constructed from tree branches and vines. Ground-dwelling prey are shot with arrows, driven to other hunters or capture devices, run down upon encounter, slammed to death against the ground, strangled at the neck, or suffocated by stepping on them while they are trapped in a tight spot. Burrowing prey are dug out, chopped out of tree trunks, stabbed through the ground with spears, frightened to the point at which they bolt from the burrow, smoked out, and captured by introducing a lasso through a small hole. Aquatic prey are shot on the surface or below it, driven into traps, poisoned, discovered on muddy bottoms by systematically poking the bottom of a pond, and speared underwater by random thrusts in drying lakes. The widely varied kill techniques are tailored to a wide variety of prey under a wide variety of conditions.

What is striking is that each of these strategies would require a lot of experimentation, learning, and local experience to get right. In contrast, gorillas simply browse on plants, which does not require much skill acquisition. Chimpanzee foraging is more complex, involving nuts, insects, and hunting as well as ripe fruit, but its complexity is still not in the league of human foraging. The consequence of human specialization on extracting high-quality resources using skill-based learned techniques is that the apprenticeship is long. Chimpanzees are capturing enough food to support themselves by the age of about 10 years, and have reached their maximum productivity by about 15 years. Human hunter-gathers do not produce as much as they consume until around the age of 20 years, and they do not reach their maximum effectiveness at foraging until the age of 30 years or so (Figure 10.14). This is mainly because they are still learning the skills involved. Indeed, early in life, people tend to concentrate on relatively easy targets such as ripe fruits.

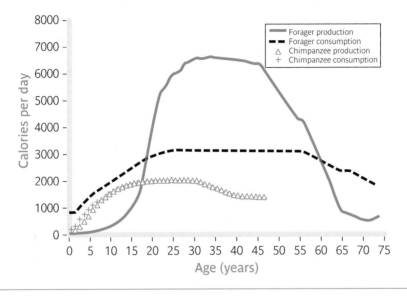

Figure 10.14 Production and consumption of calories per day by age for male hunter-gatherers ('foragers') and chimpanzees. Male chimpanzees are supporting themselves by the age of 10 and never produce more than they consume. Men's production only equals their consumption at around age 20. Between 20 and 60, production greatly exceeds consumption, and they are often provisioning kin and mates. Note that this age pattern corresponds surprisingly closely to the pattern of wage earnings of men in industrial societies. *From Kaplan et al. (2000).*

High-skill foraging makes humans able to occupy an incredible breadth of different environments in many ingenious ways, but the necessary cost is the long life, long juvenile period, and need for collective provisioning of young that we see in all human societies. There can be no doubt that it is our occupation of this learning-based niche that set the stage for our managing to invent agriculture, domesticate animals, and create technologies undreamed of by our forebears. This is particularly true given that much of our learning is social learning, which when coupled with sufficiently sophisticated learning strategies gave rise to the cumulative cultural evolution (see section 9.5.3) that characterizes human societies.

10.3.6 Language

Another striking feature of humans is their use of language. Other apes have complex systems of communication, but human language has some special features. It is productive, meaning that there is not a fixed repertoire of meanings, but rather a stock of meaningful units (words) that can be combined according to rules to yield an essentially unlimited set of novel meanings (sentences). It also allows reference to things that are not present. Some animals have alarm calls for specific types of predator, for example, but in these systems to make the 'eagle' call is to assert that an eagle is coming. In human language, by contrast, one can communicate information about eagles and how best to deal with them by using words, whether there is one around or not. To use the word 'eagle' is in no way to assert that an eagle is imminent.

These features make human language a powerful medium for social learning. I can tell you to avoid a particular food, without you having to risk trying it or even without you having to see me try it, and thus the efficiency of social learning is greatly increased. Note that the socially transmitted information includes information about other group members; see section 8.5 for a discussion of the special role in language of gossip and the maintenance of cooperation by indirect reciprocity.

The evolution of the capacity for language doubtless involved multiple adaptive changes to both the brain and the vocal tract. One change that seems to have been important is a mutation to a gene called *FOXP2*, which is fixed in humans but absent from other apes (although Neanderthals had it; Krause *et al.* 2007). Very rare humans who have a mutation to this gene seem to have largely normal cognitive abilities, but specific motor deficits and problems with both the production and the comprehension of language. Intriguingly, this very gene is implicated in the development of song in songbirds as well, suggesting a special relationship with the learning of vocal sequences (White *et al.* 2006).

 Summary

1. Living things are classified hierarchically according to their phylogenetic relationships. Key lower-level groupings are the species, which is based on reproductive isolation, and the genus.

2. Phylogenetic relationships are established using shared derived morphological or molecular characteristics.

3. Fossil evidence can confirm phylogenetic hypotheses and be informative about intermediate but now extinct states.

4. Humans belong to the primates, and more specifically the haplorrhines, the apes, and the great apes. Their closest living relatives are the two species of chimpanzee.

5. There is fossil evidence for a series of intermediate forms between ourselves and the chimpanzees, which tend to have more human-like characteristics over time and all of which are now extinct.

6. All living humans seem to have descended from a small population living in Africa within the last 200,000 years.

7. Compared with our closest relatives, humans show marked changes in the domains of locomotion, diet, brain size, life history, skill learning, and language.

Questions to consider

1. Look at Figure 10.5. Are our traditional classificatory terms 'fox' and 'dog' phylogenetically meaningful?

2. In what sense might it be correct to describe the tarsiers and lorises as 'primitive primates' even though they have been evolving for exactly the same amount of time as any other primate?

3. Although inference from scattered fossils is difficult, it seems that sexual dimorphism in size may be less in archaic *H. sapiens* as compared with early *Homo* and the australopithecines. What might this be telling us?

4. Why are human bodies relatively hairless?

→ Taking it further

For a wonderfully readable voyage through the phylogenetic history of life, see Dawkins (2004). For a more detailed academic treatment of hominin and human evolution than has been possible here, see Lewin & Foley (2003). Fossil discoveries continue apace; for the description of the *Sahelanthropus* finds, see Brunet *et al*. (2002) and Guy *et al*. (2005); for the Herto crania, which are the earliest currently known *H. sapiens*, see White *et al*. (2003), and for the debate sparked by the discovery of *H. floresiensis*, see Argue *et al*. (2006). The ability of population genetics to tell us about the history of the human population improves all the time; see, for example, Hellenthal *et al*. (2008). For a remarkable account of the uneven transition from hunter-gatherer to agricultural life during the Holocene, and its consequences for subsequent human history, see Diamond (1997).

Evolution and contemporary life

We have now reviewed all of the areas of evolutionary biology covered by this book. This final chapter asks how what we have learned applies to people living in the contemporary environment. This is an important issue because many researchers in the human sciences accept that human evolution is something that happened, but feel it was all in the past and thus that evolutionary theory offers little insight into why people do what they do now. Needless to say, I see this as a misapprehension and in this chapter I will argue for the value of understanding evolution for understanding the present.

Section 11.1 takes issue with the view that human evolution is something that happened back in the past, by providing evidence that natural selection is still going on. Section 11.2 argues, moreover, that evolution leaves a legacy and contemporary behaviour can never be understood without appreciating this. Section 11.3 considers how evolutionary ideas and understandings fit in with the accounts of human behaviour provided by other psychological, sociological, and anthropological theories that do not make reference to evolution. Finally, in sections 11.4 and 11.5 we conclude by considering two key issues that any adequate human science must provide a good account of: how do we explain cross-cultural variation and how much of our behaviour is adaptive. In doing this, we encounter three of the main contemporary evolutionary approaches to human behaviour: evolutionary psychology, human behavioural ecology, and gene-culture co-evolution theory.

11.1 Human evolution is still going on

It is a strange conceit of human beings to imagine that, having reached their current state, their evolution has stopped. However, we know that this is not so. Not all individuals conceived are equally likely to survive and reproduce, and so natural selection in humans is still going on. Indeed, it may be going faster than in previous ages. There are two main lines of evidence for current selection: evidence from the genome (section 11.1.1) and evidence based on measuring reproductive success at the phenotypic level (section 11.1.2). Finally, one recent study combines evidence from both levels (section 11.1.3).

11.1.1 Ongoing selection in the human genome

At the genetic level, there is very strong purifying selection in force against most new mutations. We know this because a large proportion of all conceptions are spontaneously aborted and this is often because the foetus is carrying some major genetic mutation. In addition, individuals with major mutations often have Mendelian diseases, which impair survival or reproduction. However, purifying selection only maintains the status quo and there is strong evidence that directional selection is also at work in humans.

Most genes in the human genome show some allelic variation, although the extent to which the alleles differ, and their relative frequencies, varies from gene to gene. Much of this allelic variation will be neutral and some will be deleterious mutations that have not yet been removed from the gene pool, but some of it may represent adaptive mutations that have not yet reached fixation. Geneticists can detect alleles that are under positive selection since such alleles increase in frequency more quickly than recombination can randomize their associations with adjacent genetic sequences. This generates a characteristic pattern of linkage disequilibrium between selected alleles and surrounding DNA (see section 2.3.1).

Researchers have found the hallmarks of current positive selection for around 7% of human genes (Voight et al. 2006; Wang et al. 2006). This means that for each of these genes (as many as 2,000 in number) there is an allele which is not yet at fixation but which is rapidly increasing in frequency in at least part of the human population. Hawks et al. (2007) tested the population frequency distribution of these alleles against the assumption that adaptive evolution had gone on at a constant rate throughout human history. This assumption was rejected. Instead, it looks like the rate of selection has greatly accelerated in the last 40,000 years.

Why would this be? One reason is that the human population has grown much bigger in the last few thousand years and larger populations generate more variation to be selected. The second reason is that, in the last 40,000 years, and particularly the last 10,000 years, humans have been constantly colonizing new niches; first, the areas outside the tropics, then the agricultural and herding niches rather than those of hunter-gatherers. These transitions would have changed the adaptive landscape and provided new problems to solve, so we might expect them to lead to a flurry of evolutionary change. For example, many of our currently most deadly infectious diseases only became a problem for humans once we began to live in denser groups and in close proximity to domesticated animals, events that have only occurred in the last few thousand years. This creates a whole new set of selection pressures, and genes involved in immunity are particularly likely to show evidence of current selection (Wang *et al.* 2006).

11.1.2 Ongoing selection at the phenotypic level

Biologists often do not know which, if any, genes contribute to variation in a trait, but they can nonetheless look for evidence of selection at the phenotypic level by testing whether there are non-random associations between a phenotypic characteristic and some component of reproductive success. When this is done for humans, it is easy to find evidence of stabilizing selection. For example, babies of around average birthweight have fewer health problems than those that are extremely large or extremely small. However, again, stabilizing selection only maintains the status quo. Is there any evidence of directional selection at the phenotypic level?

Nettle & Pollet (2008) reviewed the evidence for positive selection on male wealth across eight contemporary and recent human societies. It seems to be something of a cross-cultural universal that in societies with accumulatable wealth, richer men have higher reproductive success, and indeed we found positive relationships in rural African, historical European agrarian, and contemporary industrial societies. The strengths of the relationships were different in the different types of society (Figure 11.1) and the relationships also came about in rather different ways. In the rural African societies, rich men had more wives, and hence many more children, and the slope of the relationship was very steep. In contemporary industrial societies, there was no polygyny and family sizes were about the same for all social classes, but rich men were much less likely to remain childless than poorer men were. Even the weaker relationship found in the contemporary industrial societies was as strong as many selective gradients reported by field biologists working on other species.

Thus, natural selection on male wealth is still going on, even in contemporary Britain and the USA. We do not know if this selection is leading to a change in gene frequencies, because we do not know if the attributes that make men likely to become wealthy exhibit heritable variation, but if there are alleles in any way, however indirectly, associated with greater likelihood of accruing wealth, then they will be increasing in frequency.

11.1.3 From genotype to phenotypic consequence: *DRD4* in the Ariaal of Kenya

A rare study of ongoing selection that included both genotypic and phenotypic information is Eisenberg *et al.* (2008). The study concerns a polymorphism of a gene called *DRD4*, which codes for the receptor of a brain neurotransmitter called dopamine. Dopamine is involved in the control of reward-driven behaviours such as eating and the approach to novel situations. This gene is highly polymorphic in the human population. There are two main groups of alleles, those with seven repeats of a particular 48 base pair sequence, and those with a different number of repeats, most commonly four. An individual with at least one copy of the seven-repeat allele is said to have a 7R+ genotype, and an individual with no such copy, a 7R– genotype.

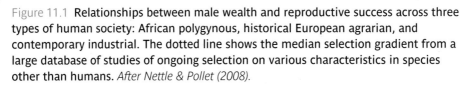

Figure 11.1 **Relationships between male wealth and reproductive success across three types of human society: African polygynous, historical European agrarian, and contemporary industrial. The dotted line shows the median selection gradient from a large database of studies of ongoing selection on various characteristics in species other than humans.** *After Nettle & Pollet (2008).*

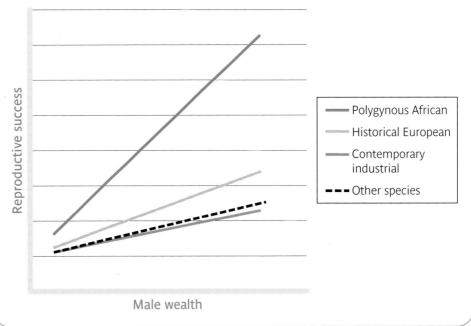

It had previously been noted that 7R+ genotypes are more common in populations that are nomadic or have a recent history of migration. Thus, it is possible that something about the reward-seeking behaviour of 7R+ individuals increases fitness when the population have to move around to find resources, but not when it is settled. A unique opportunity to test this hypothesis came from the Ariaal of Kenya. This is an ethnic group that traditionally practised nomadic livestock herding. Part of the group has been settled for about 35 years and has begun some agriculture, whilst the remainder continues to be nomadic. This is a very impoverished part of the world and both parts of the population are under nutritional stress.

The frequencies of the 7R+ genotypes were the same in the settled and nomadic populations, which is not surprising since they only have a generation or two's separation. However, in the settled population, individuals with the 7R+ genotype were considerably more underweight than those with 7R−, whilst in the nomadic population exactly the reverse was true (Figure 11.2). Given that body mass in this population is going to be a major determinant of survival and reproductive success, this suggests that 7R− is under positive selection in the settled group and 7R+ is under positive selection in the nomads. This concurs with the frequency distribution of the two alleles across populations.

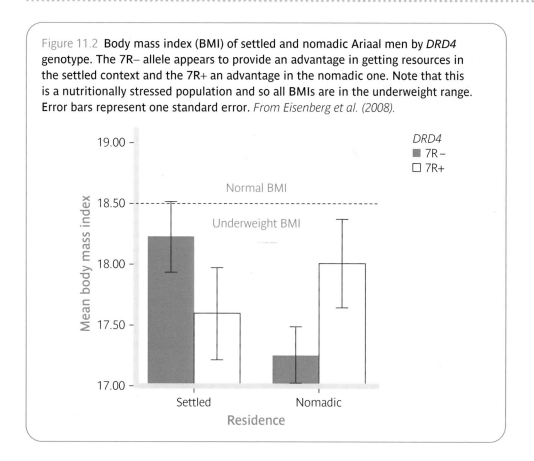

Figure 11.2 **Body mass index (BMI) of settled and nomadic Ariaal men by** *DRD4* **genotype. The 7R– allele appears to provide an advantage in getting resources in the settled context and the 7R+ an advantage in the nomadic one. Note that this is a nutritionally stressed population and so all BMIs are in the underweight range. Error bars represent one standard error.** *From Eisenberg et al. (2008).*

11.2 Evolution leaves a legacy

The previous section has shown unambiguously that evolution is still going on in human populations. However, natural selection on a trait does not have to be still operative for evolutionary explanation to be relevant to that trait. Consider animals in zoos. The selective pressures on these animals have been to a considerable extent relaxed by the provision of abundant food, the removal of predation, and the treatment of disease. However, zoo animals still behave in the ways that they are adapted to behave. The nocturnal animals are still active in the dark, even though there is now no selective advantage for them in doing this. The large carnivores are still motivated to cover a large range, even though they have no actual need to hunt. The herbivores are still vigilant when out of cover, even though there is now no predation. Thus, to a considerable extent, behaviour reflects the environment of evolutionary adaptedness and not necessarily the current one.

This principle has a number of applications to contemporary human behaviour. For example, New *et al.* (2007) showed that when viewing complex scenes, young Americans spontaneously allocate more attention to animals than to other types of object. This is true even when the other objects are matched for visual properties and mobility. What is interesting about this is that these participants would not encounter animals particularly often, and animals would not much affect their reproductive success either way. They would do better to allocate attention to automobiles, which are the major threat to their survival in the environments that they actually live in, but they allocated rather little attention to automobiles compared with elephants.

This makes sense if we consider that the psychological mechanisms that control attention bear the legacy of thousands of years of evolution in which keeping track of animals—either to eat them or to avoid being eaten by them—was essential to reproductive success and automobiles played no part. These people now live in an environment where fitness depends on the ability to deal with automobiles, but the evolved psychological mechanisms may not be particularly well prepared for the task. Similarly, people tend to fear things like snakes, spiders, and open spaces with no cover. This may have made some sense in ancestral environments, but makes much less in an environment where snakes and spiders pose no danger at all and the real dangers are from such things as electric shocks and road accidents.

Arguments based on the idea that we are adapted to past environments but not the current one are called mismatch hypotheses. We have already met some others earlier in this book. For example, the ability of women to alter the sex ratio in favour of girls at times when they were short on calories may have been adaptive across a range of ancestral environments. In contemporary Britain, however, the main consequence of the existence of such a mechanism is to make dieting women have more daughters, which provides no actual selective advantage (section 7.2.2). Mismatch arguments are appealing and often plausible, but we return to the topic with some cautionary words about them in section 11.5. Nonetheless, the general point that contemporary behaviour cannot be understood without some appreciation of evolutionary history is a valid one.

11.3 The place of evolutionary theory in the explanation of current behaviour

In this section, we consider how evolutionary theory contributes to providing explanations for current human behaviour. Throughout this book, we have stressed how behaviour can be usefully understood using the adaptationist framework, that is using the idea that behaviours tend to persist because they are adaptive to generate hypotheses about why organisms behave the way they do. However, many of the explanatory frameworks you will have encountered in the human sciences make no overt reference to adaptive fitness or evolution at all. In psychology, behaviour is explained with reference to particular cognitive or brain mechanisms, in anthropology, with reference to particular cultural traditions, and in sociology, to social and political context. None of these explanations usually mentions evolution. What is the relationship between these very different kinds of explanation and the evolutionary ones we have considered in this book?

The answer is definitely *not* that evolutionary explanations are mutually exclusive alternatives to these other more traditional approaches. Unfortunately, you will read many people holding up false choices like 'Is men's greater aggression due to evolution or due to social factors?' The answer is not one or the other, or even a bit of both. Rather, 'evolution' and 'social factors' are not alternatives to each other any more than 'learned' is the opposite of 'evolved' (section 9.1). Evolution is a general framework for understanding living things and 'social factors' is a class of proximate mechanism. To avoid these kinds of conceptual confusions, we need a framework for categorizing different types of explanation and appreciating how to fit them together.

11.3.1 Tinbergen's four questions

Conceptual confusion of the kind exemplified by the question 'Is aggression the result of society, of the brain, or of evolution?' arises all too frequently in behavioural science. The reason for this was understood nearly 50 years ago by Niko Tinbergen (1963). Tinbergen pointed out that when we ask *why* an animal performs a particular behaviour, we can actually mean a number of different things. In fact, there are four major types of question we could be asking, which are the following:

1. *Proximate causation* or *proximate mechanism*: What are the events preceding the behaviour that contribute to its occurrence? These events could be external causes, such as a particular state of the environment which triggers the behaviour, or internal causes, such as particular hormones or parts of the brain that are involved.

2. *Ultimate causation* or *function*: What are the effects of performing the behaviour on reproductive success and, thus, why has natural selection retained the ability to perform that behaviour?

3. *Ontogeny* or *developmental course*: How does this behaviour develop over the course of the individual's life?

4. *Phylogeny* or *evolutionary history*: When in the history of that species did the capacity to produce this behaviour evolve?

Tinbergen's point is that each of these is a valid and important question, and each has its own answer, which can be worked on somewhat independently of the answers to the other three. More specifically, you can never correctly answer question (2) with an answer to question (1), or vice versa. Nor is it a question of having to choose one of the four answers over the others. We ultimately need answers to all four questions and also to appreciate how the four fit together. However, in the immediate term, we just need to be clear which one it is we are answering and not get into fruitless debates where answers to different questions are set against each other as if they were alternatives.

11.3.2 An example: human infant crying

Zeifman (2001) provided a recent example of Tinbergen's four questions in action by considering the phenomenon of the crying of human infants. There are many reasons that infants cry: because of neural activity in a part of their brain called the limbic system, because they are alone, in order to signal to their mothers they need care, because the young of other mammalian species cry, and so on. These explanations are all so different from each other that it is hard to assess their relative merit. Zeifman shows that the four-question framework can be used to organize these different parts of the story into a coherent whole. A summary of some key points is shown in Table 11.1.

Table 11.1 **Why do human infants cry?** *Adapted from Zeifman (2001).*

Question	Answer	Evidence
Proximate mechanism	*External*: Physical separation from caregiver; cold; lack of food *Internal*: Limbic system; endogenous opioids involved in crying cessation	Observational and experimental evidence Experimental evidence from animal models Effects of milk and sweet-tasting substances Behaviour of babies born to methadone-addicted mothers
Ultimate function	To elicit care and defence from mothers	Triggers of crying Less crying in cultures where infants are carried more Women of reproductive age respond to infant cries Crying found in infant-carrying species when infant is put down and in infant caching species only when mother is at nest
Ontogeny	Peaks at 6 weeks, declines to 4 months, then stable until 12 months After 12 months may be more strategically directed to extort care and not so associated with separation from caregiver	Observational and experimental evidence
Phylogeny	Crying in response to being put down found in all non-human primates (for whom the ancestral state is for the infant to be carried)	Comparative evidence

The answers to the four questions do interweave in various ways. The fact that the main proximate cause of crying in young infants is physical separation from the caregiver and that non-human primates generally carry rather than cache their young are relevant to understanding crying's ultimate function, and successfully predicts that there will be less crying in cultures where babies are carried more often, a prediction that turns out to be true. The fact that there is a shift during ontogeny from crying mainly when *not* with the caregiver to crying in the company of the caregiver to obtain a resource suggests that crying has different functions at different ages. Thus, investigating one of the four questions helps suggest possible answers to the others, but all four need to be investigated, often using different types of evidence.

To take an evolutionary approach to behaviour is to include question number (2), the question of ultimate function, in one's enquiry. This does not imply that question (2) is the only important one. The others are important too. Nor is it to claim that the behaviour is either 'innate' or 'genetically determined'. Learned behaviours are just as subject to evolutionary analysis as unlearned ones (see Chapter 9). Crying in humans is clearly subject to learning and variation from culture to culture, but as Zeifman shows, this does not prevent a broad evolutionary framework being applied.

11.3.3 Evolution in relation to the human sciences

The crying example allows us to be more precise about how evolutionary explanations fit in with the explanations offered by the traditional human sciences. Research in these disciplines has generally been concerned with different aspects of proximate mechanism (or more rarely, ontogeny). For example, a sociologist of crying might research why babies from families in some social classes cry more than those from others (external proximate mechanism). An anthropologist might research why babies cry more in some cultures than others (again, external proximate mechanism). A psychologist might study how neural activity in a particular part of the brain initiates crying (internal proximate mechanism), whereas a developmental psychologist might look at the time course of crying's development (ontogeny). All of these research activities are valid and none of them precludes *also* asking the questions of ultimate function and of phylogeny.

Indeed, it is not just that evolutionary enquiries are compatible with investigations of proximate mechanism. They may actually help because if we understand *why* natural selection has produced a particular behaviour or capacity, it might help us search in the right place for the internal and external proximate mechanisms. This dialogue goes the other way, too; understanding the proximate mechanism and ontogeny of a behaviour will often provide key evidence for developing functional hypotheses.

Thus, researchers in all of the human sciences can often benefit from an appreciation of the evolutionary function and history of the behaviours or capacities that they are investigating. This is not an alternative to the mechanistic understandings that they have already developed, but a way of making sense of why those mechanisms are as they are and hopefully of making novel predictions.

In the final sections of this book we will briefly consider two of the most challenging and open evolutionary questions concerning contemporary humans: how is cross-cultural variation best to be explained and how much of human behaviour is adaptive?

11.4 How should cross-cultural variation be explained?

A striking feature of contemporary humans is their diversity. Some humans are monogamous, some polygynous; some cultures punish women severely for premarital sex, and others regard it as entirely normal. The amount people allocate to strangers in cooperation experiments varies from place to place.

The psychological differences between populations may go deeper than we once thought. For example, Uskul *et al.* (2008) studied groups of fishermen, farmers, and livestock herders from Turkey's Black Sea coast. Previous cross-cultural research has suggested that farmers tend to collaborate and consult with each other, and have a socially interdependent outlook, whilst herders rely more on autonomous individual decision making. This makes sense, given that agricultural activities involve working together with permanent neighbours on joint projects such as irrigation and land management, whereas a herder has his own mobile capital, his herd, which he can decide to take anywhere.

Uskul *et al.* gave their participants a number of cognitive tasks. One example is shown in Figure 11.3. The participant has to choose whether the bottom object goes with A or with B.

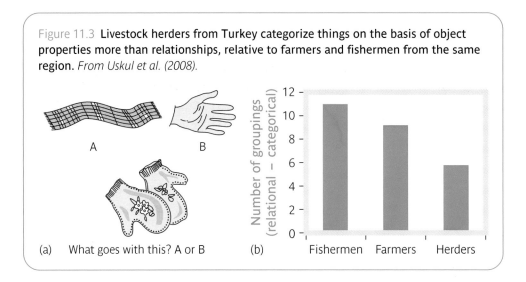

Figure 11.3 **Livestock herders from Turkey categorize things on the basis of object properties more than relationships, relative to farmers and fishermen from the same region.** *From Uskul et al. (2008).*

Either response is possible. To link the gloves with the hand is to emphasize a relationship: the glove and hand work together. To link the gloves with the scarf is to emphasize a type of object: both are winter wear. The fishermen and farmers were particularly likely to categorize on the basis of relationships, whilst the herders were relatively more likely to categorize on the basis of object type (Figure 11.3).

What this suggests is the development of a mindset based on individual objects and their properties in the herders, and based on relationships in the fishermen and farmers. This could well be adaptive given the different life problems that these three groups need to solve, and so the results are not difficult to interpret from the point of view of ultimate function. However, what the proximate mechanism is for the difference is less clear.

We can eliminate genetic differences between the groups immediately, as the participants all came from the same ethnic and linguistic groups, differing only in their subsistence activity. Thus, some kind of phenotypic plasticity is involved, but how exactly does it work? Evolutionists have considered two main classes of phenotypic plasticity that may underlie cross-cultural differences, known as evoked culture and transmitted culture. We now review them both.

11.4.1 Evoked culture

The idea of evoked culture is that natural selection has produced animals able to exhibit the best phenotype for the environment they find themselves in, across several different environments. They do this by having a range of phenotypes they can produce and a set of cues, either early in development, or in their current context, to which they respond by changing their state. For such a system to evolve, all of the different environmental conditions must have been experienced recurrently over evolutionary time and a number of other conditions must also be met (see section 9.1).

Evoked culture has been compared to a jukebox, where a particular environmental cue presses a button and the appropriate song comes out. There are some good examples of evoked culture in operation in humans. For example, it is well known that men in rural subsistence cultures are attracted to female bodies with more body fat than men from developed urban

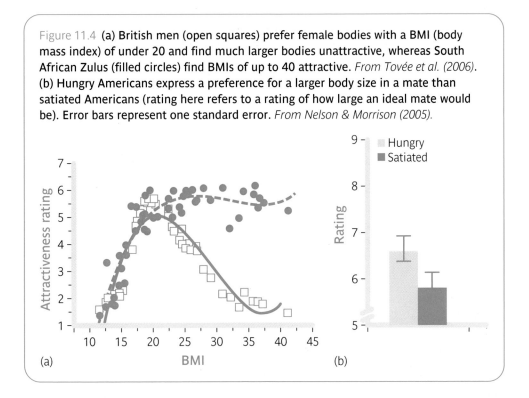

Figure 11.4 **(a) British men (open squares) prefer female bodies with a BMI (body mass index) of under 20 and find much larger bodies unattractive, whereas South African Zulus (filled circles) find BMIs of up to 40 attractive.** *From Tovée et al. (2006).* **(b) Hungry Americans express a preference for a larger body size in a mate than satiated Americans (rating here refers to a rating of how large an ideal mate would be). Error bars represent one standard error.** *From Nelson & Morrison (2005).*

settings (Tovée *et al.* 2006; Figure 11.4a). This makes adaptive sense; in the rural subsistence context, undernutrition is a real constraint on fertility and an optimal mate would be one displaying sufficient reserves. In the developed setting by contrast, obesity is more likely to be an issue, and a mate signalling dietary restraint is likely to be a healthier one.

The cross-cultural difference in preference could well be evoked culture, but what are the relevant cues? Nelson & Morrison (2005) performed the remarkably simple experiment of assessing American men's preferences for female body size either when they were hungry or when they had just eaten. The hungry men's preferences were shifted a small but significant amount towards larger body sizes (Figure 11.4b).

Thus, it looks as if hunger serves as a cue to evoke preference for fatter mates. Of course, the function of this mechanism is not to make individuals change their preferences between meals. That is just a small side-effect. Rather, where hunger is widespread or chronic, mate preferences will be shifted enduringly towards larger bodies.

Evoked culture differences have some particular properties, to be contrasted with transmitted culture, which we will meet in the next section. They will appear very fast when the triggering cues appear and will only endure as long as the triggering cues persist, that is you could make Western men like rural African men simply by making them hungry and make rural African men more like Western men just by making food more abundant. The exception is where the system relies on developmental induction by early life cues, in which case there will be a time lag of a generation or so for behaviour to respond if the environment changes radically. Evoked culture, however, does not rely on generations of accumulated cultural history or on social learning. It is a more immediate individual response to the current situation. When South African men move to the UK, for example, their body preferences soon converge on the UK norm. For this reason, some people prefer not to refer to evoked culture as 'culture' at all, since although it leads to cross-cultural differences in behaviour, it does not involve persistent socially learned traditions.

11.4.2 Transmitted culture

Transmitted culture refers to inter-population differences that are the result of generations of social learning, with each generation learning from and modifying the norms of the previous one, who in turn learned from the previous one, and so on. Transmitted culture is a key part of the human ability to adapt to multiple different environments. As we saw in section 9.5.3, social learning in humans leads to cumulative cultural evolution, which produces better and better solutions to problems in the local environment.

Transmitted culture has rather different properties to evoked culture. It can lead to the preservation of apparently arbitrary norms. For example, the fact that Muslims do not eat pork, whereas Hindus do not eat beef, cannot be evoked culture, since there are places in India where there are both Hindus and Muslims experiencing the same environment and yet maintaining different phenotypes. Instead, history matters for transmitted culture. Muslims do not eat pork because they learned this norm from earlier generations of Muslims, who in turn learned it from still earlier ones. This does not mean that it is not adaptive, but it means that it might not be; it could be something that got fixed by benefit or by chance many centuries ago and now persists neutrally or even maladaptively (see section 11.5).

Where cultural differences are transmitted rather than evoked, we expect to see a rather different pattern. There may often be a long lag when the environment changes but the culture stays the same, since people are depending on what previous generations did rather than on input within their own lifetime. Also, the environment that a population came from will often predict their behaviour better than the one they are currently in. Finally, cultural traditions that are transmitted will themselves exhibit a kind of evolution. They will develop, mutate, split, and change in complexity over time, as each generation learns something that is slightly different from what the previous generation did.

It is hard to say whether the cognitive differences between farmers and herders described by Uskul *et al.* (2008) are best viewed as evoked or transmitted culture. It is plausible that performing cooperative tasks would evoke an appropriate mode of cognition. On the other hand, there is some evidence for cultural lag, with regions of the world such as the southern states of the USA, which were once peopled by herders, retaining individualistic social norms for many generations after the herding itself has ceased to be the mainstay of the economy (Cohen & Nisbett 1996). This suggests a role for social learning and transmitted culture.

11.5 How much of our behaviour is adaptive?

In this section, we consider the issue of how much of our contemporary behaviour is adaptive, that is to what extent do the decisions that contemporary humans make about what work to do, when to marry, how many children to have, and who to name in their wills actually maximize their reproductive success?

The answer we give to this question will depend on what kinds of proximate mechanisms for decision making we believe that humans have evolved. There is a diversity of opinion amongst evolutionary scientists on this question. In fact, there are three major current styles of evolutionary approach and the central differences between them concern the extent to which current behaviour is adaptive. They are generally known as evolutionary psychology, human behavioural ecology, and gene-culture coevolution theory. In the next three sections, we review each one briefly and illustrate the differences between them by considering how each one might approach the excessive consumption of fats and sugars that is common in the contemporary developed world.

11.5.1 Evolutionary psychology: behaviour adaptive for the ancestral environment

The evolutionary psychology approach has been developed largely by psychologists concerned with understanding why particular brain and cognitive mechanisms work in the way that they do. The hypotheses it creates are adaptationist ones, but the adaptation in question is not so much to the current environment as to the ancestral environments in which the human brain evolved. The ancestral environment is not, of course, any particular time or place. It is a statistical composite of all the contexts in which human ancestors have lived, weighted towards the most recent. However, since the period humans spent as hunter-gatherers so dwarfs the period since the transition to agriculture (see section 10.2.4), the environment of evolutionary adaptedness is often conceptualized as a kind of archetypal hunter-gatherer society.

Evolutionary psychology does not, contrary to accusations sometimes levelled at it, assume that human behaviour will be the same in all contexts. It allows for phenotypic plasticity. However, the plasticity it allows for is mainly of the evoked culture kind (see 11.4.1) and adaptive plasticity of this kind can only evolve if all of the different environmental states have been experienced recurrently over evolutionary time. Thus, humans can be expected to behave in ways that would have been adaptive in ancestral environments and will only behave adaptively for the contemporary environment to the extent that this resembles an ancestral one. In other words, evolutionary psychology makes heavy use of mismatch arguments.

Let us see how an evolutionary psychology approach might apply to the contemporary problem of obesity in the developed world. Much of this obesity is proximately caused by people eating more sugar and fat than is good for them. In ancestral environments, sugars (in the form of fruit and honey) and fats (animal products) were rather scarce, transient resources, and so it made sense to be motivated to fill up on them whenever they were available. This caused no health problems as the supply was never sufficient for people to overeat on them. It causes health problems now because the unprecedented abundance of synthetic fats and sugars is not something our evolved appetites have equipped us to handle. As Nesse & Williams (1995: 148) put it:

> Our dietary problems arise from a mismatch between the tastes evolved for Stone Age conditions and their likely effects today . . . Fat, sugar and salt were in short supply through nearly all of our evolutionary history. Almost everyone, most of the time, would have been better off with more of these substances, and it was consistently adaptive to want more and to try to get it.

This argument certainly has some face validity, given the high intrinsic palatability of sweet and fatty foods to humans.

Issues with evolutionary psychology

Evolutionary psychology is a thriving discipline. Its central premises are valid and it is true that there can be mismatches between current and ancestral environments. However, some caution and sophistication are required concerning how evolutionary psychological theories, and in particular mismatch arguments, are deployed. The reasons for this are the following:

1. Time: Evolution can happen a lot more quickly than was once thought. Even a modest selection gradient (like the dotted line in Figure 11.1), when coupled with reasonable heritability, can generate one standard deviation's change in a phenotypic characteristic in around 25 generations. This is still 500 years, but it is a lot less time than the time lag since the Pleistocene. Thus, it is perfectly plausible that there has been significant adaptation to agrarian and even city life.

2. Variety of past environments: The term 'hunter-gatherers' actually lumps together some very different ways of making a living. Human hunter-gatherers varied enormously in the types of ecological problems that they confronted. They ranged from almost complete to rather little dependence on large game animals, from large groups to small, from highly mobile to more concentrated around aquatic resources, from deserts to lakes. For at least 40,000 years, the hunter-gatherers who formed our ancestors were to be found anywhere from the equator to the temperate latitudes, experiencing enormous changes in ecology as the glacial periods gave way to interglacials. Thus, at the very least, we would expect evolution to have a high degree of evolved plasticity in humans for coping with different types of environment. This is not a problem for all mismatch arguments. Certain features (such as scarcity of sugar) may have been recurrent across all these ancestral environments, but different in the contemporary one, and these would be the features most likely to lead to maladaptive behaviour in the present. However, it is a problem for any attempt to explain current human tendencies as evolved for any very specific ancestral ecology.

3. Diversity of human social organization: The diversity of human social organization, as documented by anthropologists, is impressive. There are matrilineal and patrilineal societies, societies with marriage for life and societies where relationships come and go, polygynous and polyandrous societies, and so on. Again, this is not contrary to evolutionary psychological explanations per se, since they allow for extensive phenotypic plasticity via such mechanisms as evoked culture. However, some early evolutionary psychology underemphasized cross-cultural variation and perhaps the most exciting current areas of research concern how evolved plastic mechanisms lead to the observed cross-cultural variation in behaviour (e.g. Gangestad *et al.* 2006).

4. Nature of human adaptations: As discussed in section 10.3.5, humans display a suite of characteristics suggesting that they are highly adapted for the learning niche: large brains, extended juvenile period, slow life history, and long life. This adaptive package allowed humans to colonize a variety of different habitats by acquiring locally relevant adaptive knowledge during their lifetimes. This suggests we should expect a sophisticated ability to find new ways of behaving adaptively when the environment changes.

5. Nature of the current environment: Finally, it is easy to assert that the contemporary urban environment is so different from the environments of our ancestors that we are not adapted to making good decisions in it, but less easy to show in exactly which respects it lies outwith the range of ancestral environments. In all environments, humans have

to figure out a good way of getting resources, learn to eat the right stuff, master the local technology, retain social allies, avoid disease, attract a mate, trade-off energy spent in reproduction against the accumulation of resources and the maintenance of alliances, and trade-off investment in a current relationship against a future one, investment in children against investment in grandchildren, and so on. Modern urban environments embody these timeless dilemmas just as earlier ones did, and although the solutions might be different, the problems are the same to a surprising extent.

These problems all counsel against over-using (or using too simplistically) the idea that we are 'Stone Age' minds living in modern societies, although they are certainly not arguments against invoking ultimate function and evolutionary history in trying to understand human cognitive and neural mechanisms.

11.5.2 Human behavioural ecology: behaviour adaptive for the current environment

The second major evolutionary approach to modern humans is known as human behavioural ecology. Human behavioural ecology comes out of anthropology, and uses many tools developed for the study of animal behaviour. Its main focus has been on how the behaviour seen in various non-Western, usually subsistence, societies can be understood as an adaptive response to the local ecological conditions. Thus, human behavioural ecology has been quite successful at explaining why some societies are matrilineal and some patrilineal, or why some are monogamous and some polygynous, and many other traits.

Human behavioural ecology is also an adaptationist approach, but it is the current environment, not the ancestral one, for which behaviour is seen as adaptive. Thus, a large part of the research effort of human behavioural ecology is directed at the careful measurement of the current environment, and the assessment of what behavioural options are available to people in it and what their payoffs might be. The behaviours that maximize reproductive success are then predicted using some kind of optimality model. The attention to current context is a feature that human behavioural ecology shares with ecology more generally, but also with the more traditional social sciences. Central to anthropology and sociology, for example, is the idea that current context shapes behaviour, and thus that current context must be deeply understood. Human behavioural ecology makes little use of the ideas of the environment of evolutionary adaptedness or of mismatch.

Human behavioural ecology assumes that humans have considerable phenotypic flexibility, presumably through having evolved a general motivation to maximize reproductive success, and extensive abilities to learn and to experiment in pursuit of this goal. However, human behavioural ecologists have not really been concerned with what the psychological processes leading to behavioural flexibility are. They are really interested in answering the question of ultimate function of behaviour (see section 11.3.1) and are agnostic about the proximate mechanisms.

How might a human behavioural ecologist approach the issue of modern overconsumption of sugars and fats? The first thing that careful research would reveal is that it is not *everyone* in the contemporary USA who shows this behaviour. It is socially patterned, with poor diet and obesity being much more common amongst those on low incomes (Figure 11.5a). The reason this might be the case is further elucidated by considering the relative costs of different foods. Figure 11.5b plots the energy density of different food types (the calories per kilo, as it were)

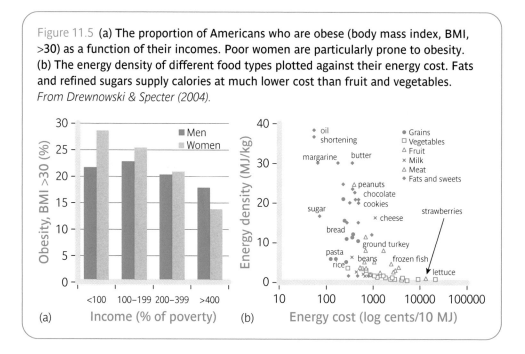

Figure 11.5 (a) The proportion of Americans who are obese (body mass index, BMI, >30) as a function of their incomes. Poor women are particularly prone to obesity. (b) The energy density of different food types plotted against their energy cost. Fats and refined sugars supply calories at much lower cost than fruit and vegetables. *From Drewnowski & Specter (2004).*

against their current cost in the USA (the dollars per calorie). You can see that fats, sugar, cookies, chocolate, and so on are dense in energy and provide each calorie much more cheaply than do lean meat, fruit, or vegetables (Drewnowski & Specter 2004). Thus, it could be that people with limited dollars to spend actually have no better strategy available to them than getting their calories from the cheapest source. In support of this view, UK women with healthy diets spend around £600 more on food per year than those women with what would be considered poor diets (Cade *et al.* 1999).

Darmon *et al.* (2002) produced a mathematical model, incorporating the actual costs of different foods, in which they predicted diets using the assumptions that (a) the people need to take in a certain number of calories per day, (b) they want as balanced a diet as possible, but (c) they have a fixed amount of money that they are able to spend per day. The predictions of the model (Figure 11.6) are that as the budget gets smaller, greater and greater proportions of the calories have to come from cheap fats and sweets, and less and less from animal products and particularly from expensive fruits and vegetables. This is exactly the behaviour that poor people in Western countries exhibit.

This puts quite a different slant on the obesity problem. Rather than everyone doing something that would have made sense in the ancestral environment but makes no sense now, people may be choosing the best diet that is available given the resources that they have to spend. This means that the high sugar and fat consumption of the poor is, in fact, adaptive. By adaptive here, we do not mean that this diet promotes the best possible health. In the long term, it causes a lot of health problems. Rather, we mean that given a limited budget, this is the only way available to obtain one's daily ration of calories. When the budget available is increased, people obtain the calories in a way that is more healthy in the long run.

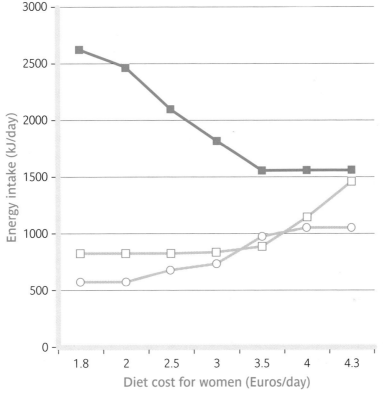

Figure 11.6 **Predicted diet composition for women needing to obtain 1748 kcal of energy per day but with different budgets to spend, from the mathematical model of Darmon *et al.* (2002). Filled squares, fats and sweets; open squares, meat, fish, and eggs; open circles, fruit and vegetables.** *Reproduced from Drewnowski & Specter (2004).*

11.5.3 Gene-culture co-evolution: adaptive social learning

The third evolutionary approach to contemporary human behaviour comes from the literature on social learning and cultural evolution. This area is sometimes called gene-culture coevolution theory, dual inheritance theory, or just cultural evolution. The idea here is that humans have evolved capacities to use social learning to obtain locally appropriate behaviour. This produces cumulative local cultural traditions which proximately cause the different behaviour of individuals in different populations.

This is also an adaptationist approach, in that it is based on the idea that the ability to learn socially increases fitness. However, although the ability to learn socially might be an adaptation in general, not every behaviour so learned maximizes reproductive success. For example, in the process of acculturation, I might obtain the adaptive belief that a certain plant is poisonous and also the maladaptive or arbitrary belief that this plant has to be danced around whenever it is encountered. Social learning would persist as long as the information so learned was on balance much more useful than harmful, but I could still acquire a lot of useless beliefs in the process. When the environment changes, there can also be a cultural lag of a number of generations before behaviour adaptive for the new situation is learned and spreads.

Gene-culture co-evolution theory makes quite specific predictions about when behaviour will be adaptive for the current environment and when it will not (see section 9.6.2). At equilibrium, there is always a mixture of reliance on social learning and individual checking that what is so learned is actually useful, and the ratio of social to individual learning depends on the cost of individual learning. Where it is relatively easy to learn for oneself what the fitness consequences of alternative behaviours are, culture will change quickly and usually produce the optimal behaviour for the current context. Where learning for oneself is hard, there can be long cultural lags where an outmoded behaviour persists, or even the spread of a behaviour that is not actually good for reproductive success.

How might gene-culture co-evolution theory approach the problem of overconsumption of sugars and fats? One way might be to look at how dietary preferences are actually formed. It is quite plausible that to a considerable extent people learn to eat what their parents ate. This will lead to a local cultural tradition of diet, which may well be optimized for the local environment (in which case there is no tension between a gene-culture co-evolution account and a human behavioural ecological one; the former is merely providing a proximate mechanism for the behaviour predicted by the latter to be generated). However, if, for example, people become more sedentary, the diet may now be too energy rich and obesity may ensue. This is a kind of mismatch argument too, but the time lag is not the time required for genetic evolution to happen, but the time required for cultural change to happen. The maladaptive cultural state will not endure forever. Instead, some innovators will originate a different dietary pattern, obtain a better payoff, and then perhaps be emulated, leading to a spread of their dietary norms through the population. The exact dynamics of the spread may be uneven, and can be modelled if enough is known about who learns from whom and what the social structure of the population is.

11.5.4 Synthesis

The different emphases of the three main evolutionary approaches are summed up in Table 11.2. All are united by an interest in Tinbergen's question of ultimate function, and thus they are all evolutionary approaches. However, they are concerned with proximate mechanisms to different extents and differ very considerably in their predictions about issues such as how quickly humans can respond with adaptive behaviour if their environment changes in some unprecedented way.

Although I have stressed the differences between the approaches in the foregoing discussion, they also have considerable scope for complementarity. Human behavioural ecology concentrates on ultimate function but tends to be agnostic about proximate mechanisms, relying on some general notion of flexibility or learning. However, merely to evoke learning is a non-explanation; there are many types of learning, each produced by different types of evolved decision-making mechanisms with different design features (see Chapter 9). Thus, evolutionary psychology, with its focus on cognitive mechanisms, can enrich human behavioural ecology's explanations with an account of how behavioural flexibility actually works. Gene-culture co-evolution theory also provides one set of mechanisms for locally adaptive behaviours to become established, namely through social learning, and thus can also enrich the ultimate

Table 11.2 **Key features of the different evolutionary approaches to human behaviour**

Approach	Focus	For which environment is behaviour adaptive?	If environment changes in an unprecedented way, how quickly can adaptive behaviour respond?
Evolutionary psychology	Internal proximate mechanisms (e.g. brain circuits) and ultimate function	Environment of evolutionary adaptedness (EEA); contemporary environment only to the extent it resembles the EEA	At the speed of genetic change
Human behavioural ecology	Ultimate function	Current environment	At the speed of individual learning
Gene-culture coevolution	External proximate mechanisms (e.g. social learning dynamics) and ultimate function	Environment in which that cultural tradition developed	At the speed of cultural change

explanations of human behavioural ecology. Social learning is only possible because there are evolved cognitive mechanisms to make us learn from each other, and thus there is room for collaboration between evolutionary psychologists and gene-culture co-evolutionists on what the nature of these mechanisms is.

On the central issue of whether current behaviour is best seen as optimized for the genetically ancestral environment, the current environment, or the culturally ancestral environment, the answer seems to be that it depends. Different problems need careful study on a case-by-case basis. Some really complex questions, such as why the fertility rate has become so extremely low in developed countries, can probably only be answered by invoking a combination of flexible responses to changing conditions, cultural dynamics, and mismatch with the ancestral environment (Borgerhoff Mulder 1998). Thus, all three approaches have a useful contribution to make.

Human evolutionary behavioural science

Many evolutionary researchers, such as myself, try to incorporate elements of all three traditions into our work to give the best explanations possible for particular cases. There is no clear consensus on what this broad evolutionary approach should be called. Evolutionary psychology *sensu lato* (which means 'in the broad sense') has been suggested, as has human evolutionary behavioural science (Sear *et al.* 2007). Perhaps our best hope is that it would not need a separate name or separate subdiscipline at all. Instead, researchers in all the existing human sciences will simply incorporate evolutionary perspectives, and the framework of Tinbergen's four questions, into what they do. Thus, the Darwinian worldview, with all the beauty and explanatory power that it provides, will become as automatic for those studying humans as it already is for those studying all the other organisms on this earth.

✔ Summary

1. Natural selection in the contemporary human population is still going on and has been observed at both the genetic and the phenotypic levels.

2. There are at least four different types of explanation we can give for a behaviour in terms of proximate mechanism, ultimate function, ontogeny, and phylogeny. The existing human sciences have been mainly concerned with proximate mechanism, and evolutionary approaches can enrich these by adding the angle of ultimate function.

3. Cross-cultural differences in humans are abundant and can be explained by evolved mechanisms of evoked and transmitted culture.

4. There are three major styles of evolutionary approach to modern human behaviour: evolutionary psychology, human behavioural ecology, and gene-culture co-evolution theory. They often make different predictions about the degree to which current behaviour is adaptive, but they are also complementary to a considerable extent.

❓ Questions to consider

1. What will happen to the frequencies of the 7R– and 7R+ *DRD4* alleles in the Ariaal population over the coming generations? What about in the human population more generally?

2. People in developed Western societies have very small families (averaging fewer than two children per woman). Speculate on why this might be, from an evolutionary psychology, a human behavioural ecology perspective, and a gene-culture co-evolution theory perspective.

3. Think about some aspect of people's behaviour that you have experienced in your everyday life and that has puzzled you. How does taking an ultimate, evolutionary perspective help you to understand what is happening?

➡ Taking it further

For core texts in evolutionary psychology, see Tooby & Cosmides (1992) and Buss (1995, 2005); in human behavioural ecology, Winterhalder & Smith (2000); in gene-culture co-evolution theory, Henrich & McElreath (2003) and Richerson & Boyd (2004). Discussions of the relative merits of the three approaches to human evolutionary behavioural sciences are to be found in Smith (2000), Barrett *et al*. (2002), and Laland & Brown (2002). An attempt to integrate across them is made by Sear *et al*. (2007). Two scientific societies that serve the human evolutionary behavioural science community are the Human Behavior and Evolution Society (HBES; http://www.hbes.com) and the European Human Behaviour and Evolution Association (EHBEA; http://www.ehbes.com). Both of these hold annual conferences. Journals with a main focus on carrying human evolutionary behavioural science research are *Evolution and Human Behavior*, *Human Nature*, *Evolutionary Anthropology*, *Evolutionary Psychology*, and the *Journal of Evolutionary Psychology*.

Glossary

adaptation As a noun, any characteristic of an organism that improves its survival and reproduction in its local environment; as a verb, the process by which adaptations are produced.

adaptationist stance The strategy of assuming that the characteristics or behaviours displayed by an organism enhanced ancestral reproductive success, and forming hypotheses about how they did so.

allele An alternative form of a *gene* or other DNA sequence.

alloparenting Parental care provided to offspring that are not the individual's own.

altruism Behaviour that has a positive effect on another individual's lifetime reproductive success and a negative effect on the actor's own lifetime reproductive success.

amino acid Molecular building block of proteins. There are 20 main types of amino acid making up the proteins in living organisms.

analogy A characteristic present in two or more different species that does not derive from their common ancestor, but is instead due to *convergent evolution*.

anisogamy Sexual reproduction involving gametes of different sizes, a small one (the male) and a large one (the female).

asexual reproduction Reproduction that does not involve the fusion of gametes from different individuals. Reproduction is asexual in prokaryotes such as bacteria, and asexual reproduction can be found in many single-celled eukaryotes, fungi, and plants, as well as a few animals.

association study A study in which the frequency of alleles in a sample of individuals with a particular phenotypic characteristic is compared with the frequency in a sample who lack the characteristic, in the hope of localizing a genomic region involved.

australopithecines Extinct hominins found in Africa between 4 and 1 million years ago. They were more bipedal than chimpanzees, but their brains were no larger. Probably multiple species or genera.

Baldwin effect An interaction between learning and genetic evolution, in which the capacity of the animal to learn allows genetic adaptations to be selected for, which would not otherwise be able to evolve. May involve *genetic assimilation*.

base The nucleotide units in the DNA chain. There are four bases: adenine (A), guanine (G), cytosine (C), and thymine (T).

base pairing The principles governing how bases on one strand of DNA bind to bases on the other. A always and only pairs with T, and C always and only with G.

Bateman's principle The principle that males gain more reproductive success from each additional mating partner than females do and, relatedly, that the variance in male reproductive success is larger than that in female reproductive success. Bateman's principle only holds where the total investment in each reproductive episode is lower for males than females.

behaviour genetics The quantitative genetic study of behavioural and psychological traits.

brachiation Locomotion through trees by swinging from the arms, like a child on monkey bars.

broad-sense heritability A heritability estimate that includes all sources of genetic influence, that is additive, dominance, and epistatic.

by-product benefits Benefits to other individuals that are produced as an unselected consequence of individuals following their immediate self-interest.

catarrhines The monkeys and apes of the Old World.

central dogma of genetics The idea that changes in DNA sequence can lead to changes in proteins, but changes in proteins cannot change the sequence

of DNA. Another way of stating this is that alterations in the genotype lead to alterations in the
phenotype, but alterations in the phenotype do
not generally lead to alterations in the genotype.

cheating   Taking the benefits of joint ventures
without paying the cost.

chromosome   DNA-bearing structure in the nucleus
of eukaryotic cells. Chromosomes consist of DNA
wound around proteins called histones.

classical conditioning   A form of associative learning where a response that was initially made to
one stimulus (the unconditioned stimulus) comes
to be made to a second stimulus (the conditioned
stimulus) that is repeatedly paired with the first.

classical genetics   The knowledge about the
functioning of genes that was worked out before it
was established that genes were DNA. Compare
molecular genetics.

co-dominant   Of two alleles, when one copy of each
is present, both are expressed in the phenotype.

codon   A sequence of three adjacent bases in
DNA or RNA that encodes a specific amino acid.

coefficient of additive genetic variance   A statistic
estimating the amount of additive genetic variation
in a population that affects a particular phenotypic
trait. Differs from a heritability, from which it is
derived, by being independent of the amount of
environmental variation affecting the trait.

coefficient of relatedness   The probability that
the alleles which two individuals have at a locus
are identical by descent. It can also be thought
of as the size of the expected increment in genetic
similarity between two relatives above and beyond
that expected for two unrelated individuals.

conditioned stimulus   See *classical conditioning*.

convergent evolution   The process whereby a
similar characteristic evolves independently in
different species due to the same selection pressures being at work in each case.

cooperation   Behaviour that provides a benefit to
another individual and has been selected because
it does so.

cue   An environmental event or feature that
triggers an evolved response of some kind.

cultural evolution   See *gene-culture co-evolution theory*.

cumulative cultural evolution   The existence of
socially learned traditions which become more
complex and better adapted over time.

deoxyribonucleic acid (DNA)   The large molecule
that encodes genetic information. DNA consists
of two sugar–phosphate strands wound around
each other in a double helix, bound together by a
sequence of pairs of *bases*.

developmental induction   A type of phenotypic
plasticity where environmental cues early in life
switch the organism onto a permanently different
developmental track.

diploid   Of a cell or organism, containing two sets of
genes, one from each parent. Compare with *haploid*.

direct reciprocity   A form of cooperation in which
individual A provides a benefit to individual B, who
at some point provides a benefit back, leaving both
individuals better off.

directional selection   The mode of selection in
which the highest fitness is found in individuals
either above (positive directional selection) or
below (negative directional selection) the average
population value of the characteristic.

dizygotic   Of a pair of twins, developing from two
separate sperm and eggs.

DNA fingerprinting   The forensic technique for
identifying individuals from biological samples,
which exploits the fact that simple sequence repeats
are highly variable from individual to individual.

dominance effects   Increases in phenotypic similarity when both alleles at a locus are shared by
two individuals above and beyond the sum of the
increases in phenotypic similarity when either one
or the other is shared.

dominance hierarchy   An ordering of the members of a social group according to their ability to
displace each other from a contested resource.

dominant   Of an allele, producing the same phenotypic effect when one copy is present as when two
copies are present. Of an individual in a social
group, higher ranked in a dominance hierarchy.

dual inheritance theory   A near synonym for *gene-
culture co-evolution theory*. The eponymous inherit-
ances are genetic and cultural.

environment of evolutionary adaptedness The range of historically recurrent circumstances that shaped the current adaptations of an organism.

epistatic effects Increases in phenotypic similarity when a combination of alleles at different loci is shared by two individuals, above and beyond the sum of the increases in phenotypic similarity when the individual alleles are shared.

eukaryote An organism whose cells have a distinct nucleus. All animals and plants are eukaryotes. Compare with *prokaryote*.

eusociality A social system where non-reproductive individuals work to promote the reproduction of a reproductive caste, such as a queen.

evoked culture Cross-cultural differences caused by environmental cues activating evolved plasticity in psychological or physiological mechanisms.

evolutionarily stable strategy (ESS) A behavioural policy that, when common in a population, cannot be invaded by any alternative strategy.

evolutionary psychology In a broad sense, any adaptationist perspective on modern human behaviour. In a narrower sense, specific hypotheses about the nature of human cognitive mechanisms based heavily on mismatch arguments and the idea of the environment of evolutionary adaptedness.

evolvability Of a genome, its propensity to generate mutations with varying phenotypic effects that natural selection can exploit to create novel phenotypes. Of a phenotypic characteristic, its ability to be reached by natural selection operating on variants that resemble it to a greater or lesser extent.

exon The parts of a gene that code for protein.

exponential growth Growth in which the rate of increase gets faster as the population grows larger.

extra-pair mating In animals that form durable pair bonds, a mating that occurs with an individual other than the social partner.

extrinsic mortality Mortality caused by background environmental factors, as opposed to intrinsic mortality, which is responsive to changes in behaviour.

Falconer's estimate of heritability A simple heritability estimate defined as twice the difference in correlation between monozygotic and dizygotic twin pairs.

female philopatry A tendency for females to remain in their natal location. Associated with male dispersal.

fitness The population frequency of an allele in one generation relative to its frequency in the previous one. Fitness is used in a related sense to mean the absolute or relative reproductive success of classes of individuals. More loosely, it is sometimes used to designate an individual's score on some phenotypic parameter likely to correlate with genetic fitness.

fixation An allele has reached fixation when it is the only allele present in the population at that locus, that is when its population frequency is 1.

fossil Preserved remains or traces of long-dead animals, usually found in rocks.

frequency-dependent selection Selection where the relative fitness of two types changes as their proportions in the population change. In negative frequency-dependent selection, one type has an advantage when rare, which declines as it becomes more common. The outcome is a stable mixture of the two types. In positive frequency-dependent selection, the advantage of one type becomes greater as its frequency increases; the outcome here is that the type increases with accelerating speed until it reaches fixation.

gamete Haploid cells such as sperm and eggs, which combine in fertilization to produce a new diploid individual.

gene DNA sequence coding for a protein.

gene-culture coevolution theory An evolutionary approach to modern human behaviour which stresses that human reliance on social learning leads to dynamics of behaviour that cannot be predicted by considering genetic evolution alone.

gene family A set of genes in a genome that are recognizably related, even though they may have come to differ in both structure and function to some extent. Gene families arise from a single ancestral gene by duplication and differentiate by the accumulation of mutations.

genetic assimilation   A process where a behaviour that previously required learning in order to develop comes, through fixation of genetic variants, to develop without the need for learning.

genetic code   The set of mappings between specific sequences of three RNA or DNA bases and specific amino acids that is used during the synthesis of proteins.

genetic correlation   A situation where directional selection for one characteristic changes the population value of another characteristic as well because of pleiotropy or linkage between the genes affecting them.

genetic drift   Random changes in allele frequencies in finite populations.

genome   The complete set of DNA (coding and non-coding) in an organism or cell.

genomics   That part of molecular genetics concerned with sequencing and analysing genomes. Genomics became possible in the 1980s with technologies that allowed sizeable DNA sequences to be 'read'.

genotype   The combination of alleles possessed by an individual or cell.

genus (pl. genera)   A group of closely related species.

germ line   The lineage of cells that produces new individuals, that is the *gametes*.

group selection   Selection brought about by different survival or reproduction of groups rather than of individuals, and thus tending to produce adaptations that enhance group rather than individual interests.

Hamilton's rule   The inequality $c < rb$, which describes the conditions under which an allele could spread by kin selection.

haploid   Of a cell or organism, containing one set of genes. In humans, only sperm and egg cells are haploid. Compare with *diploid*.

haplorrhines (also spelled haplorhines)   The group of primates containing the tarsiers, monkeys, and apes.

Hardy–Weinberg equilibrium   The principle that allele frequencies remain constant over the genera-tions, with heterozygotes and homozygotes present in predictable proportions. The Hardy–Weinberg equilibrium is the background expectation in an infinite population with no natural selection and random mating.

heritability   The proportion of variation in a phenotypic characteristic that is associated with genetic variation. See also *narrow-sense heritability* and *broad-sense heritability*.

heritable   Of a characteristic, affected by variation in genes and therefore transmissible genetically from parent to offspring.

hermaphrodite   An individual with both male and female reproductive functions. In simultaneous hermaphrodites such as snails, an individual is both male and female at the same time, whereas in sequential hermaphrodites such as clown anemone fish, the individual is male and then female (or vice versa) during the course of its life.

heterozygote advantage   A situation where a heterozygote genotype (*ab*) has higher fitness than either of the two homozygotes (*aa* or *bb*).

heterozygous   Having two different alleles at a particular locus.

Holocene   The current geological epoch, beginning 10,000 years ago.

hominin   A member of the group consisting of humans and their extinct relatives. Sometimes used to include the chimpanzees, genus *Pan*, but used in the narrower sense in this book.

homology   A characteristic present in two or more different species, which is inherited from their common ancestor.

homozygous   Having two copies of the same allele at a particular locus.

human behavioural ecology   An adaptationist approach to contemporary human behaviour, which stresses the ability of humans to use phenotypic flexibility to produce behaviour adapted to their current ecological context.

identical by descent   Of two alleles, descended from the same ancestral copy.

imprinting   A type of phenotypic plasticity in which early experience provides a template of what some aspect of the environment should look like. In filial imprinting, it is a template of the parent, in sexual imprinting, a template of a mate, and there are other types such as habitat imprinting.

inbreeding　Mating with biological relatives.

inclusive fitness　An individual's direct fitness plus any additional fitness of its relatives which is a consequence of its actions, discounted by the coefficient of relatedness.

incomplete dominance　A situation where a heterozygous individual has an appearance that is intermediate between the appearance of the two homozygotes.

independent segregation　The principle whereby which copy of one chromosome a gamete contains is independent of which copy of another chromosome it contains. Independent segregation means that alleles of genes on different chromosomes travel independently down the generations. Compare with *linkage*.

indirect reciprocity　A form of cooperation in which individual A provides a benefit to individual B as long as B has provided benefits to others in the past.

instrumental conditioning　A form of associative learning in which behaviours that lead to rewards increase in frequency and behaviours that lead to punishments decrease in frequency.

intragenomic conflict　Any situation where there are fitness differences between genes within the same organism, for example due to distortion of segregation, or differences in the way the genes are transmitted to offspring.

intrasexual competition　Competition between members of the same sex for reproductive opportunities.

intron　A DNA sequence that is not translated into protein found in the midst of a gene. Compare with *exon*.

isogamy　Sexual reproduction involving gametes of the same size.

kin selection　Natural selection on the basis of benefits provided to copies of the underlying allele that reside in different bodies.

learned taste aversion　A form of associative learning where an animal that has once been made sick following eating a particular food thereafter avoids all foods with the same taste.

lek　A display site where females gather to choose a mate and males compete to be chosen.

levels of selection debate　The theoretical debate within biology concerning which level of biological organization (genes, individuals, populations, or species) natural selection is best seen as acting at.

life history theory　The branch of evolutionary theory dealing with how selection acts on the allocation of energy to reproduction over time.

linkage or linkage disequilibrium　When two alleles have a tendency to be inherited together, they are said to be linked. Linkage arises because the two genetic loci are on the same chromosome and, in general, the closer their physical positions, the tighter the linkage.

linkage study　A study that localizes the genes for a particular phenotypic characteristic by examining the tendency of the characteristic to be co-inherited with genetic markers of known position in an extended family.

locus　A particular physical position within the genome.

male philopatry　A tendency for males to remain in their natal location. Associated with female dispersal.

matriline　The mother and her kin, such as maternal grandparents and maternal uncles.

meiosis　Special cell division process that produces a haploid gamete from a diploid cell. *Recombination* occurs during meiosis.

Mendelian disease　A disease state associated with allelic variation in a single gene. Because only one gene is involved, these diseases follow the Mendelian laws of inheritance fairly neatly.

messenger RNA (mRNA)　Type of RNA molecule produced from DNA by transcription, which acts as an intermediary in order for the DNA sequence to be turned into *protein*.

microsatellite　A simple sequence repeat with a short repeated motif, generally of one to six base pairs.

minisatellite   A simple sequence repeat with a repeated motif of a dozen to a few dozen base pairs.

mismatch hypothesis   A hypothesis that a behaviour maladaptive in the current environment would have been adaptive in the ancestral one and that evolution has not yet had time to change it.

mitosis   The normal cell division process whereby a cell produces a daughter that is genetically almost identical to itself. Compare with *meiosis*.

modern synthesis   The synthesis of Darwinian evolutionary theory with Mendelian genetics. Also called *neo-Darwinism*.

molecular clock   The hypothesis that neutral genetic changes accumulate at an approximately constant rate through time and that molecular divergence can therefore be used to estimate date of divergence of two lineages.

molecular genetics   That part of genetics concerned with the structure and functioning of genes at the molecular (DNA) level. Molecular genetics began with the ascertainment of the structure of DNA by Watson and Crick in 1953.

monogamy   A mating system in which one male mates with one female.

monophyletic   Of a group of species, containing an ancestral form and all of its descendants.

monozygotic   Of a pair of twins, developing from a single fertilized egg.

morphological   Concerning the form of phenotypic characteristics such as bones.

mutation   Any change in a DNA sequence.

mutation–selection balance   The equilibrium amount of genetic diversity resulting from the injection of new variants by mutation and their elimination by selection.

mutual-benefit behaviours   Behaviours with more than one individual involved that increase the lifetime reproductive success of all parties.

narrow-sense heritability   A heritability estimate that includes only the additive effects of similarity at different loci.

natural selection   The population process by which characteristics that make individuals well adapted to the local environment increase in frequency relative to less well-adapted alternatives.

Neanderthals   Extinct hominins found in Europe from around 300,000 to 30,000 years ago. They were similar in body and brain size to modern humans, but were not the ancestors of modern Europeans.

neo-Darwinism   The synthesis of Darwinian evolutionary theory with Mendelian genetics. Also called the *modern synthesis*.

neolocality   The social practice, common in contemporary developed economies, of married couples establishing households away from both sets of parents.

neutral theory of molecular evolution   The branch of genetic theory that models molecular changes that have no effect on fitness.

non-coding   Any DNA that is not translated into protein.

non-shared environmental influences   Aspects of the environment that are no more similar for family members than for randomly selected individuals.

optimality modelling   The construction of formal models that compute expectations about how a characteristic should look if selection has optimized it to perform some specific function. Empirical data can then be compared to the model's predictions to understand better how the characteristic has come to be as it is.

parent–offspring conflict   The conflict of interests that exists between parents and their offspring about the amount of energy allocated to parental care, with the optimal amount for offspring being higher than the optimal amount from the point of view of the parent.

parsimony   The principle that the phylogeny that implies the smallest number of distinct evolutionary events is likely to be the correct one.

paternity uncertainty   The probability that a female's young are not the genetic descendants of her male social partner.

patriline   The father and his kin, such as paternal grandparents and paternal uncles.

phenotype   The observable characteristics of an organism.

phenotypic gambit   The strategy of forming adaptationist hypotheses directly about the phenotype without needing to know what the genetic or developmental mechanisms that produce the pattern in the phenotype are.

phenotypic plasticity   The capacity of an organism with a given genotype to alter aspects of its phenotype in response to input from the environment.

phylogeny   A diagram showing the family tree or pattern of ancestral relationships between several different species.

plastic   Able to change as a result of environmental input.

platyrrhines   The monkeys of the Americas.

pleiotropy   The production of several distinct phenotypic effects by an allele.

Pleistocene   The geological epoch spanning 1.8 million to 10,000 years ago. Characterized by repeated cycles of glaciation and warming.

polyandry   A mating system in which one female mates with several males, but each male mates with just one female.

polygenic trait   A trait in which variation is influenced by allelic variation in many different genes.

polygyny   A mating system in which one male mates with several females, but each female mates with just one male.

polymorphism   Condition in which several different alleles at a particular genetic locus exist at appreciable frequency in the population.

polyploid   A cell or organism containing more than two sets of genes. Polyploidy is found in many plants, in amphibians, and elsewhere.

Price equation   A mathematical framework for expressing expected evolutionary change. The Price equation states that the expected evolutionary change in a characteristic is given by the covariance of that characteristic with fitness, plus any distortion of the characteristic in the process of reproduction.

prokaryote   An organism whose cell(s) lack a distinct nucleus. Bacteria are prokaryotes. Compare with *eukaryote*.

promiscuity   A mating system where any male can mate with any female.

prosimians   The primates that are not monkeys or apes.

protein   Large organic molecule made up of a chain of *amino acids*. Proteins serve many vital functions in living organisms.

proximate explanation   The account of how, within an individual, a particular characteristic is produced, for example which genes, environmental triggers, hormones, or parts of the brain are involved.

pseudogene   A non-coding sequence of DNA that shows a high degree of similarity to a sequence elsewhere in the genome or in a related species that does code for a protein.

Punnett square   Diagram used to work out the combination of offspring genotypes from parental genotypes. The possible types of gamete from each parent constitute the rows and columns, and each square within the diagram represents a possible offspring genotype.

purifying selection   Selection acting against mutations at a locus.

quantitative genetics   The branch of genetic theory that investigates the inheritance of polygenic traits, using the principle that the net effect of genotypic similarity across multiple loci is to increase phenotypic similarity.

recessive   Of an allele, producing the same phenotypic effect when one copy is present as when no copy is present, but producing a different phenotypic effect when two copies are present.

recombination   The event during meiosis that results in DNA being exchanged between the two copies of a chromosome.

reproductive isolation   Not interbreeding with other populations. Reproductive isolation is the criterion for a population being a distinct species.

reproductive success   The number of viable descendants produced by an individual.

ribonucleic acid (RNA)   A single-stranded molecule consisting of a sugar–phosphate backbone and a sequence of bases. RNA is the basis of the genetic material in some viruses, and in other organisms is involved in the *transcription* and *translation* of DNA into protein.

ribosome   Cellular structure outside the nucleus where proteins are synthesized.

robust   Of a phenotypic characteristic, developing in much the same way across a wide variety of environmental context. Of an individual, having bones which are thick for their length.

senescence   The process of intrinsic somatic deterioration that occurs when organisms approach their maximum age.

sequence   Of DNA or RNA, the order in which the bases occur along the sugar–phosphate backbone.

sexual dimorphism   Phenotypic difference between males and females.

sexual selection   Natural selection on the ability to gain mates. Sometimes divided into intrasexual and intersexual selection, where the former is selection on the ability to compete with rivals of the same sex and the latter is selection on the ability to attract members of the opposite sex.

sexually antagonistic selection   A selective regime where the optimal phenotype for a male is not the same as the optimum for a female. To the extent that the male and female phenotype are genetically correlated, a trade-off between male and female optimality will result.

shared derived characters   Morphological or molecular innovations found in a group of organisms which identify that group as phylogenetically related to each other.

shared environmental influences   Aspects of the environment that are more similar for family members than for randomly selected individuals.

simple sequence repeat   A sequence of DNA where the same motif is repeated a number of times in a row. Minisatellites and microsatellites are simple sequence repeats.

single-gene characteristic   Any characteristic where allelic variation in a single gene accounts for differences in the phenotype.

single-nucleotide polymorphism (SNP)   A polymorphism in the population that is due to a single base being different in some individuals.

social brain hypothesis   The hypothesis that social bonds are cognitively demanding and cause the evolution of increased brain size.

social learning   Learning from other members of the population.

somatic line   The cells of the body other than the gametes.

species   A group of organisms capable of reproducing and producing fertile offspring together.

stabilizing selection   A mode of selection where the highest fitness is at the average of the population distribution of the characteristic, with reduced fitness at both extremes.

strepsirrhines (also spelled strepsirhines)   The group of primates containing the lemurs, lorisis, pottos, and galagos, but excluding tarsiers, monkeys, and apes.

synonymous substitution   Any mutation of a single base that does not alter the amino acid produced.

trade-off   Any situation where increasing one trait associated with fitness necessitates decreasing another trait that is also associated with fitness.

transcription   Process whereby an mRNA negative is read off the DNA sequence.

transitions   A single-base genetic mutation where the change is between an A and a G, or between a C and a T.

translation   Process whereby an mRNA sequence is turned into a sequence of amino acids constituting a protein.

transmitted culture   Cross-cultural differences that are due to social learning creating local traditions.

transposable element   A DNA sequence able to copy itself into new locations in the genome.

transversions   A single-base genetic mutation where the change is between dissimilar bases such as A and C, or G and T.

ultimate explanation   The explanation of why the alleles that create a particular characteristic have displaced their competitors; that is, of why the particular characteristic gave its bearers a selective advantage over those that lacked it.

unconditioned stimulus   See classical conditioning.

uxorilocality   The social practice of married couples living at the wife's natal location.

virilocality   The social practice of married couples living at the husband's natal location.

References

Adair, L.S. (2001). Size at birth predicts age at menarche. *Pediatrics 107*: e59–66.

Aiello, L.C. & P. Wheeler (1995). The expensive-tissue hypothesis: The brain and the digestive system in human and primate evolution. *Current Anthropology 36*: 199–221.

Alexander, R.D. (1987). *The Biology of Moral Systems*. New York: de Gruyter.

Alexander, R.D., J.L. Hoogland, R.D. Howard, K.M. Noonan & P.W. Sherman (1979) Sexual dimorphism and breeding systems in pinnipeds, ungulates, primates and humans. In N.A. Chagnon & W. Irons (eds), *Evolutionary Biology and Human Social Behavior: An Anthropological Perspective*, pp. 402–35. Scituate, M.A.: Duxbury Press.

Almond, D. & L. Edlund (2007). Trivers–Willard at birth and one year: Evidence from US natality data 1983–2001. *Proceedings of the Royal Society of London, B: 274*: 2491–6.

Anderson, K.G., H. Kaplan & J.B. Lancaster (1999). Paternal care by genetic fathers and stepfathers I: Reports from Albuquerque men. *Evolution and Human Behavior 20*: 405–31.

Anderson, K.G., H. Kaplan & J.B. Lancaster (2007). Confidence of paternity, divorce, and investment in children by Albuquerque men. *Evolution and Human Behavior 28*: 1–10.

Argue, D., D. Donlon, C. Groves & R. Wright (2006). *Homo floresiensis*: Microcephalic, pygmoid, *Australopithecus* or *Homo*? *Journal of Human Evolution 51*: 360–74.

Armitage, K.B. & O.A. Schwartz (2000). Social enhancement of fitness in yellow-bellied marmots. *Proceedings of the National Academy of Sciences of the USA 97*: 12149–52.

Avise, J.C., B.W. Bowen, T. Lamb, A.B. Meylan & E. Bermingham (1992). Mitochondrial DNA evolution at a turtle's pace: Evidence for low genetic variability and reduced microevolutionary rate in the Testudines. *Molecular Biology and Evolution 9*: 457–73.

Bakker, T.C.M. (1993). Positive genetic correlation between female preference and preferred male ornament in sticklebacks. *Nature 363*: 255–7.

Barclay, P. (2006). Reputational benefits for altruistic punishment. *Evolution and Human Behavior 27*: 325–44.

Barclay, P. & R. Willer (2007). Partner choice creates competitive altruism in humans. *Proceedings of the Royal Society, B 274*: 749–53.

Barrett, L., R.I.M. Dunbar & J. Lycett (2002). *Human Evolutionary Psychology*. Basingstoke: Palgrave Macmillan.

Barton, N.H. & P.D. Keightley (2002). Understanding quantitative genetic variation. *Nature Reviews Genetics 3*: 11–21.

Bateman, A.J. (1948). Intra-sexual selection in *Drosophila*. *Heredity 2*: 349–68.

Bateson, M., D. Nettle & G. Roberts (2006). Cues of being watched enhance cooperation in a real-world setting. *Biology Letters 3*: 412–4.

Bateson, P. & M. Mameli (2007). The innate and the acquired: Useful clusters or a residual distinction from folk biology? *Developmental Psychobiology 49*: 818–31.

Bateson, P., D. Barker, T. Clutton-Brock, *et al.* (2004). Developmental plasticity and human health. *Nature 430*: 419–21.

Bellis, M.A. & R.R. Baker (1990). Do females promote sperm competition? Data for humans. *Animal Behaviour 40*: 997–9.

Belsky, J. (2007). Childhood experiences and reproductive strategies. In R.I.M. Dunbar & L. Barrett (eds), *Oxford Handbook of Evolutionary Psychology*, pp. 237–71. Oxford: Oxford University Press.

Bereczkei, T., P. Gyuris & G.E. Weisfeld (2004). Sexual imprinting in human mate choice. *Proceedings of the Royal Society, B: Biological Sciences 271*: 1129–34.

Bielby, J., G.M. Mace, O.R.P. Bininda-Emonds *et al.* (2007). The fast–slow continuum in mammalian life history: An empirical re-evaluation. *American Naturalist 169*: 748–57.

Birkhead, T.R. (2000). *Promiscuity: An Evolutionary History of Sperm Competition*. London: Faber.

Boag, P.T. & P.R. Grant (1981). Intense natural selection in a population of Darwin's finches (*Geospizinae*) in the Gálapagos. *Science 214*: 82–5.

Bodnar, L.M., H.N. Simhan, R.W. Powers, *et al.* (2007). High prevalence of vitamin D insufficiency in black and white pregnant women residing in the Northern United States and their neonates. *Journal of Nutrition 137*: 447–52.

Borgerhoff Mulder, M. (1988). Reproductive success in three Kipsigis cohorts. In T. H. Clutton-Brock (ed.), *Reproductive Success: Studies of Individual Variation in Contrasting Breeding Systems*, pp. 419–35. Chicago: University of Chicago Press.

Borgerhoff Mulder, M. (1998). The demographic transition: Are we any closer to an evolutionary explanation? *Trends in Ecology and Evolution 13*: 266–70.

Borgerhoff Mulder, M. (2000). Optimizing offspring: The quality–quantity trade-off in agropastoral Kipsigis. *Evolution and Human Behavior 21*: 390–410.

Bouchard, T.J. & M. McGue (2003). Genetic and environmental influences on human psychological differences. *Journal of Neurobiology 54*: 4–45.

Bouchard, T.J., M. McGue, Y.-M. Hur & J.M. Horn (1998). A genetic and environmental analysis of the California Psychological Inventory using adult twins reared apart and together. *European Journal of Personality 12*: 307–20.

Boyd, R. (2006). Reciprocity: You have to think different. *Journal of Evolutionary Biology 19*: 1380–2.

Boyd, R. & Richerson, P.J. (1985). *Culture and the Evolutionary Process*. Chicago: University of Chicago Press.

Boyd, R. & Richerson, P.J. (1995). Why does culture increase human adaptability? *Ethology and Sociobiology 16*: 125–43.

Boyd, R. & Richerson, P.J. (1996). Why culture is common but cultural evolution is rare. *Proceedings of the British Academy 88*: 73–93.

Brooks, R. (2000). Negative genetic correlation between male sexual attractiveness and survival. *Nature 46*: 67–70.

Brown P., T. Sutikna, M. Morwood, R.P. Soejono, Jatmiko, E.W., Saptomo, E.W. & R.A. Due (2004): A new small-bodied hominin from the late Pleistocene of Flores, Indonesia. *Nature 431*:1055–61.

Brunet, M., F. Guy, D. Pilbeam, *et al.* (2002). A new hominid from the Upper Miocene of Chad, Central Africa. *Nature 418*: 145–51.

Brunner, H.G., M. Nelen, X.O. Breakefield, *et al.* (1993). Abnormal behaviour associated with a point mutation in the structural gene for monoamine oxidase A. *Science 262*: 578–80.

Buchan, J.C., S.C. Alberts, J.B. Silk & J. Altmann (2003). True paternal care in a multi-male primate society. *Nature 425*: 179–81.

Bulmer, M.G. & G.A. Parker (2002). The evolution of anisogamy: a game-theoretic approach. *Proceedings of the Royal Society, B: Biological Sciences 269*: 2381–8.

Burt, A. & R. Trivers (2006). *Genes in Conflict: The Biology of Selfish Genetic Elements*. Harvard: Harvard University Press.

Burtt, E. (1951). The ability of adult grasshoppers to change colour on burnt ground. *Proceedings of the Royal Entomological Society of London 26*: 45–8.

Buss, D.M. (1989). Sex differences in human mate preferences: Evolutionary hypotheses tested in 37 cultures. *Behavioral and Brain Sciences 12*: 39–49.

Buss, D.M. (1995). Evolutionary psychology: A new paradigm for psychological science. *Psychological Inquiry 6*: 1–30.

Buss, D.M. (2005). (ed.) *Handbook of Evolutionary Psychology*. New York: Wiley.

Buss, D.M. (2008). *Evolutionary Psychology: The New Science of the Mind*. 3rd edn. Boston: Allyn & Bacon.

Cade, J., H. Upmeier, C. Calvert & D. Greenwood (1999). Costs of a healthy diet: Analysis from the UK women's cohort study. *Public Health Nutrition 2*: 505–12.

Caldwell, C.A. & A.E. Millen (2008). Experimental models for testing hypotheses about cumulative cultural evolution. *Evolution and Human Behavior 29*: 165–71.

Catchpole, C.K. (1980). Sexual selection and the evolution of complex songs among European warblers of the genus *Acrocephalus*. *Behaviour 74*: 149–66.

Charpentier, M.J.E, R.C. Van Horn, J. Altmann & S.C. Alberts (2008). Paternal effects on offspring fitness in a multimale primate society. *Proceedings of the National Academy of Sciences of the USA 105*: 1988–92.

Choi, J.-K. & S. Bowles (2007) The coevolution of parochial altruism and war. *Science 318*: 636–40.

Cicirello, D.M. & J.O. Wolff (1990). The effects of mating on infanticide and pup discrimination in white-footed mice. *Behavioral Ecology and Sociobiology 26*: 275–9.

Clark, R.D. & E. Hatfield (1989). Gender differs in receptivity to sexual offers. *Journal of Psychology and Human Sexuality 2*: 39–55.

Clutton-Brock, T.H., S.D. Albon, & F.E. Guinness (1986). Great expectations: Dominance, breeding success and offspring sex ratios in red deer. *Animal Behaviour 34*: 460–71.

Clutton-Brock, T.H., S.D. Albon & F.E. Guinness (1988). Reproductive success in male and female red deer. In T.H. Clutton-Brock (ed.), *Reproductive Success: Studies Of Individual Variation In Contrasting Breeding Systems*, pp. 325–43. Chicago: University of Chicago Press.

Coghlan. A., E.E. Eichler, S.G. Oliver, *et al.* (2005). Chromosome evolution in eukaryotes: A multi-kingdom perspective. *Trends in Genetics 21*: 673–92.

Cohen, D. & R.E. Nisbett (1996). *Culture of Honor: The Psychology of Violence in the South*. Boulder: Westview Press.

Consuegra, S. & C. Garcia de Leaniz (2008). MHC-mediated mate choice increases parasite resistance in salmon. *Proceedings of the Royal Society, B: Biological Sciences 275*: 1397–403.

Cook, M. & S. Mineka (1989). Observational conditioning of fear to fear-relevant versus fear-irrelevant stimuli in rhesus monkeys. *Journal of Abnormal Psychology 98*: 448–59.

Craig, I.W. (2007). The importance of stress and genetic variation in human aggression. *BioEssays 29*: 227–36.

Crow, J.F. (2001). The origins, patterns and implications of human spontaneous mutations. *Nature Reviews Genetics 1*: 40–7.

Daan, S., C. Dijkstra & J.M. Tinbergen (1990). Family planning in the kestrel (*Falco tinninculus*): The ultimate control of covariation of laying date and clutch size. *Behaviour 114*: 83–116.

Daly, M. & M. Wilson (1982). Whom are newborn babies said to resemble? *Ethology and Sociobiology 3*: 69–78.

Daly, M. & M. Wilson (1983). *Sex, Evolution and Behavior*. 2nd edn. Belmont, CA: Wadsworth.

Daly, M. & M. Wilson (1988). Evolutionary social psychology and family homicide. *Science 242*: 519–24.

Darmon, N., E.L. Ferguson & A. Briend (2002). A cost-constraint alone has adverse effects on food selection and nutrient density: An analysis of human diets by linear programming. *Journal of Nutrition 132*: 3764–71.

Dawkins, R. (1986). *The Blind Watchmaker*. London: Longman.

Dawkins, R. (1996). *Climbing Mount Improbable*. London: Viking.

Dawkins, R. (2004). *The Ancestor's Tale*. London: Weidenfeld & Nicolson.

Dawkins, R. (2006a). *The God Delusion*. London: Bantam Books.

Dawkins, R. (2006b). *The Selfish Gene*. 3rd edn. Oxford: Oxford University Press.

Dennett, D. (1995). *Darwin's Dangerous Idea*. New York: Simon & Schuster.

De Waal, F. (1999). Cultural primatology comes of age. *Nature 399*, 635–6.

DeWitt, T.J. (1998). Costs and limits of phenotypic plasticity: Tests with predator-induced morphology and life history in a freshwater snail. *Journal of Evolutionary Biology 11*: 465–80.

Diamond, J. (1997). *Guns, Germs and Steel: The Fates of Human Societies*. New York: Norton.

Dingemanse, N.J., C. Both, P.J. Drent & J.M. Tinbergen (2004). Fitness consequences of avian personalities in a fluctuating environment. *Proceedings of the Royal Society, B: Biological Sciences 274*: 847–52.

Dobzhansky, T. (1973). Nothing in biology makes sense except in the light of evolution. *American Biology Teacher 35*: 125–9.

Drewnowski, A. & S.E. Specter (2004). Poverty and obesity: The role of energy density and energy costs. *American Journal of Clinical Nutrition 79*: 6–16.

Dunbar, R.I.M. (1993). Coevolution of neocortical size, group size and language in humans. *Behavioral and Brain Sciences 16*: 681–735.

Dunbar, R.I.M. & S. Shultz (2007a). Evolution in the social brain. *Science 317*: 1344–7.

Dunbar, R.I.M. & S. Shultz (2007b). Understanding primate brain evolution. *Philosophical Transactions of the Royal Society, B 362*: 649–58.

Eisenberg, D.T.A., B. Campbell, P.B. Gray & M.D. Sorenson (2008). Dopamine receptor genetic polymorphisms and body composition in undernourished pastoralists: An exploration of nutrition indices among nomadic and recently settled Ariaal men of northern Kenya. *BMC Evolutionary Biology 8*: 173.

Emlen, S.T. (1970). Celestial rotation: Its importance in the development of migratory orientation. *Science 170*: 1198–201.

Encode Project Consortium (2007). Identification and analysis of functional elements in 1% of the human genome by the ENCODE project. *Nature 447*: 799–816.

Endler, J.A. (1986). *Natural Selection in the Wild*. Princeton: Princeton University Press.

Fagundes N.J.R., N. Ray, M. Beaumont, *et al.* (2007). Statistical evaluation of alternative models of human evolution. *Proceedings of the National Academy of Sciences of the USA 104*: 17614–19.

Fehr, E. & S. Gächter (2002). Altruistic punishment in humans. *Nature 415*: 137–40.

Fessler, D.M.T. (2002). Reproductive immunosuppression and diet: An evolutionary perspective of pregnancy sickness and meat consumption. *Current Anthropology 43*: 19–61.

Fisher, R.A. (1930). *The Genetical Theory of Natural Selection*. Oxford: Clarendon Press.

Flinn, M.V. & B.G. England (1995). Childhood stress and family environment. *Current Anthropology 36*: 854–66.

Foerster, K., T. Coulson, B.C. Sheldon, *et al.* (2007). Sexually antagonistic genetic variation for fitness in red deer. *Nature 447*: 1107–10.

Frank, S.A. (2003). Repression of competition and the evolution of cooperation. *Evolution 57*: 693–704.

Freeland, S.J. & L.J. Hurst (1998). The genetic code is one in a million. *Journal of Molecular Evolution 47*: 238–48.

Freeman, S. & J.C. Herron (2004). *Evolutionary Analysis*. 3rd edn. London: Prentice Hall.

Friedman, M. (2008). The evolutionary origin of flatfish asymmetry. *Nature 454*: 209–12.

Galef, B.G. (1996). Tradition in animals: Field observations and laboratory analyses. In M. Bekoff & D. Jamieson (eds), *Readings in Cognitive Ethology*, pp. 91–106. Cambridge, MA: MIT Press.

Galef, B.G. & M.M. Clark (1971). Parent–offspring interactions determine time and place of first ingestion of solid food by wild rat pups. *Psychonomic Science 25*: 15–6.

Galef, B.G. & K.N. Laland (2005). Social learning in animals: Empirical studies and theoretical models. *BioScience 55*: 489–99.

Galef, B.G. & D.F. Sherry (1973). Mother's milk: A medium for the transmission of cues reflecting the flavor of the mother's diet. *Journal of Comparative and Physiological Psychology 83*: 374–8.

Galef, B.G. & D.J. White (1998). Mate choice copying in Japanese quail, *Coturnix coturnix japonica*. *Animal Behaviour 55*: 545–52.

Gangestad, S.W. & J.A. Simpson (2000). The evolution of human mating: Trade-offs and strategic pluralism. *Behavioral and Brain Sciences 23*: 573–87.

Gangestad, S.W. & R. Thornhill (1997). The evolutionary psychology of extrapair sex: The role of fluctuating asymmetry. *Evolution and Human Behavior 18*: 69–88.

Gangestad, S.W., R. Thornhill & C.E. Garver (2002). Changes in women's sexual interests and their partners' mate-retention tactics across the menstrual cycle: Evidence for shifting conflicts of interest. *Proceedings of the Royal Society, B: Biological Sciences 269*: 975–82.

Gangestad, S.G., M.G. Haselton & D.M. Buss (2006). Evolutionary foundations of cultural variation: Evoked culture and mate preferences. *Psychological Inquiry 17*: 75–95.

Geschwind, D.H., B.L. Miller, C. DeCarli & D. Carmeli (2003). Heritability of lobar brain volumes in twins supports genetic models of cerebral laterality and handedness. *Proceedings of the National Academy of Sciences of the USA 99*: 3176–81.

Gibson, M. & R. Mace (2003). Strong mothers bear more sons in rural Ethiopia. *Proceedings of the Royal Society of London (Biology Letters) 270*: S108–9.

Gibson, M. & R. Mace (2007). Polygyny, reproductive success and child health in rural Ethiopia: Why marry a married man? *Journal of Biosocial Science 39*: 287–300.

Gintis, H. (2009). *Game Theory Evolving*. 2nd edn. Princeton: Princeton University Press.

Gluckman, P.D., M.A. Hanson & A.S. Beedle (2007). Early life events and their consequences for later disease: A life history and evolutionary perspective. *American Journal of Human Biology 19*: 1–19.

Goddard, M.R., H. Charles, J. Godfray & A. Burt (2005). Sex increases the efficacy of natural selection in experimental yeast populations. *Nature 434*: 636–40.

Gottschall, J. & Wilson, D.S. (2005). *The Literary Animal: Evolution and the Nature of Narrative*. Chicago: Northwestern University Press.

Gould, S.J. & R.C. Lewontin (1979). The spandrels of San Marco and the Panglossian paradigm: A critique of the adaptationist programme. *Proceedings of the Royal Society, B: Biological Sciences 205*: 581–98.

Grace, H.J. (1981). Prenatal screening for neural tube defects in South Africa: an assessment. *South African Medical Journal 60*: 324–9.

Grafen, A. (2002). A first formal link between the Price equation and an optimization program. *Journal of Theoretical Biology 217*: 75–91.

Grafen, A. (2006). Optimization of inclusive fitness. *Journal of Theoretical Biology 238*: 541–63.

Grant, P.R. (1986). *Ecology and Evolution of Darwin's Finches*. Princeton: Princeton University Press.

Green, R.E., J. Krause, S.E. Ptak, *et al.* (2006). Analysis of one million base pairs of Neanderthal DNA. *Nature 444*: 330–6.

Griffin, A.S. & S.A. West (2003). Kin discrimination and the benefit of helping in cooperatively breeding vertebrates. *Science 302*: 634–6.

Gross, M.T. (1991). Evolution of alternative reproductive strategies: frequency-dependent sexual selection in male bluegill sunfish. *Philosophical Transactions of the Royal Society, B 332*: 59–66.

Guy, F., L.E. Lieberman, D. Pilbeam, *et al.* (2005). Morphological affinities of the *Sahelanthropus tchadensis* (Late Miocene hominid from Chad) cranium. *Proceedings of the National Academy of Sciences of the USA 102*: 18836–18841.

Hansen, B.T., L.E. Johannessen & T. Slagsvold (2008). Imprinted species recognition lasts for life in free-living great tits and blue tits. *Animal Behaviour 75*: 921–7.

Hatcher, M. (2000). Persistence of selfish genetic elements: Population structure and conflict. *Trends in Ecology and Evolution 15*: 271–7.

Hawkes, K., J.F. O'Connell & N.G. Blurton Jones (1997). Hadza women's time allocation, offspring provisioning and the evolution of long post-menopausal lifespans. *Current Anthropology 38*: 551–77.

Hawks, J., E.T. Wang, G.M. Cochran, H.C. Harpending & R.K. Moyzis (2007). Recent acceleration of human adaptive evolution. *Proceedings of the National Academy of Sciences of the USA 104*: 20753–8.

Helle, S. (2008). A tradeoff between reproduction and growth in contemporary Finnish women. *Evolution and Human Behavior 29*: 189–95.

Helle, S., V. Lummaa & J. Jokela (2004). Accelerated immunosenescence in pre-industrial twin mothers. *Proceedings of the National Academy of Sciences of the USA 101*: 12391–6.

Hellenthal, G., A. Auton & D. Falush (2008). Inferring human colonization history using a copying model. *Plos Genetics 4*: e1000078.

Henrich, J. & McElreath, R. (2003). The evolution of cultural evolution. *Evolutionary Anthropology 12*: 123–35.

Heston, L.L. (1966). Psychiatric disorders in foster home reared children of schizophrenic mothers. *British Journal of Psychiatry 112*: 819–25.

Hinton, G.E. & S.J. Nowlan (1987). How learning can guide evolution. *Complex Systems 1*: 495–502.

Hoffman, M.B. (2008). Law and biology. *Journal of Philosophy, Social Science & Law 8* (http://www6.miami.edu/ethics/jpsl/, accessed 3rd December 2008).

Holden, C.J. & R. Mace (2003). Spread of cattle led to the loss of matriliny in Africa: A co-evolutionary analysis. *Proceedings of the Royal Society of London, B 270*: 2425–33.

Holden, C.J., R. Sear & R. Mace (2003). Matriliny as daughter-biased investment. *Evolution and Human Behavior 24*: 99–112.

Hoogland, J.L. (1983). Nepotism and alarm calling in the black-tailed prairie dog (*Cynomys ludovicianus*). *Animal Behaviour 31*: 472–9.

Hoogland, J.L. (1985). Infanticide in prairie dogs —lactating females kill the offspring of close kin. *Science 230*: 1037–40.

Hopcroft, R.L. (2005). Parental status and differential investment in sons and daughters: Trivers–Willard revisited. *Social Forces 83*: 169–93.

Houle, D. (1992). Comparing evolvability and variability of quantitative traits. *Genetics 130*: 195–204.

Hurford, J. (2007). *The Origins of Meaning* (Volume 1 of *Language in the Light of Evolution*). Oxford: Oxford University Press.

Hurst, L.D., A. Atlan & B.O. Bengtsson (1996). Genetic conflicts. *Quarterly Review of Biology 71*: 315–59.

Ingman, M., H. Kaessmann, S. Paabo & U. Gyllensten (2000). Mitochondrial genome variation and the origin of modern humans. *Nature 408*: 708–12.

Isvaran, K. (2007). Intraspecific variation in group size in the blackbuck antelope: The roles of habitat structure and forage at different spatial scales. *Oecologia 154*: 435–44.

Jablonski, N.G. & G. Chaplin (2000). The evolution of human skin coloration. *Journal of Human Evolution 39*: 57–106.

Janvier, P. (2008). Squint of the fossil flatfish. *Nature 454*: 169–70.

Jeffares, D.C., T. Mourier & D. Penny (2006). The biology of intron gain and loss. *Trends in Genetics 22*: 17–22.

Jobling, M.A., M.E. Hurles & C. Tyler-Smith (2004). *Human Evolutionary Genetics: Origins, Peoples & Disease.* New York: Garland.

Johnson, K., E. DuVal, M. Kielt & C. Hughes (2000). Male mating strategies and the mating system of great-tailed grackles. *Behavioural Ecology 111*: 132–41.

Jones, A.G., D. Walker & J.C. Avise (2001). Genetic evidence for extreme polyandry and extraordinary sex-role reversal in a pipefish. *Proceedings of the Royal Society, B: Biological Sciences 268*: 2531–5.

Jones, A.G., J.R. Arguello & S.J. Arnold (2002). Validation of Bateman's principles: a genetic study of sexual selection and mating patterns in the rough-skinned newt. *Proceedings of the Royal Society, B: Biological Sciences 269*: 2533–9.

Jones. G. & M.W. Holdereid (2007). Bat echolocation calls: Adaptation and convergent evolution. *Proceedings of the Royal Society, B: Biological Sciences 274*: 905–12.

Judson, O. (2002). *Dr. Tatiana's Sex Advice to All Creation.* London: Metropolitan Books.

Kaplan, H., K. Hill, J. Lancaster & A.M. Hurtado (2000). A theory of human life history evolution: Diet, intelligence and longevity. *Evolutionary Anthropology 9*: 156–85.

Kashi, Y. & D.G. King (2006). Simple sequence repeats as advantageous mutators in evolution. *Trends in Genetics 22*: 253–9.

Keller, M.C. & G.F. Miller (2006). Resolving the paradox of common, harmful heritable mental disorders: Which evolutionary models work best? *Behavioral and Brain Sciences 29*: 385–452.

Kempenaers, B., G.R. Verheyen & A.A. Dhondt (1997). Extra-pair paternity in the blue tit (*Parus caerulus*): Female choice, male characteristics and offspring quality. *Behavioral Ecology 8*: 481–92.

Kendler, K.S. & J.H. Baker (2007). Genetic influences on measures of the environment: A systematic review. *Psychological Medicine 37*: 615–26.

Kenward, R.E. (1978). Hawks and doves: Factors affecting success and selection in goshawk attacks on wood pigeons. *Journal of Animal Ecology 47*: 449–60.

Kimura, M. (1983). *The Neutral Theory of Molecular Evolution*. Cambridge: Cambridge University Press.

Kirkwood, T.B.L. (2008). Understanding ageing from an evolutionary perspective. *Journal of Internal Medicine 263*: 117–27.

Kitcher, P. (2007). *Living with Darwin: Evolution, Design, and the Future of Faith*. New York: Oxford University Press.

Koehler, N. & J.S. Chisholm (2007). Early psychosocial stress predicts extra-pair copulation. *Evolutionary Psychology 5*: 184–201.

Kokko, H. (2007). *Modelling for Field Biologists (and Other Interesting People)*. Cambridge: Cambridge University Press.

Kokko, H., R. Brooks, J.M. McNamara & A.I. Houston (2002). The sexual selection continuum. *Proceedings of the Royal Society, B: Biological Sciences 269*: 1331–40.

Kokko, H., R. Brooks, M.D. Jennions & J. Morley (2003). The evolution of mate choice and mating biases. *Proceedings of the Royal Society, B: Biological Sciences 270*: 653–64.

Krause, J., C. Lalueza-Fox, L. Orlando, *et al.* (2007). The derived *FOXP2* variant of modern humans was shared with neandertals. *Current Biology 17*: 1908–12.

Krebs, J.R. & N.B. Davies (1993). *An Introduction to Behavioural Ecology*. 3rd edn. Oxford: Blackwell.

Krings, M., A. Stone, R.W. Schmitz, H. Krainitzki, M. Stoneking & S. Paabo (1997). Neanderthal DNA sequences and the origins of modern humans. *Cell 90*: 19–30.

Kruuk, L.E.B., T.H. Clutton-Brock, L. Slate, *et al.* (2000). Heritability of fitness in a wild mammal population. *Proceedings of the National Academy of Sciences of the USA 97*: 698–703.

Lahdenperä, M., A.F. Russell & V. Lummaa (2007). Selection for long lifespan in man: Benefits of grandfathering? *Proceedings of the Royal Society of London, B 274*: 2437–44.

Laland, K.N. & G.R. Brown (2002). *Sense and Nonsense: Evolutionary Perspectives on Human Behaviour*. Cambridge: Cambridge University Press.

Langerhans, R.B. & T.J. DeWitt (2002). Plasticity constrained: Over-generalized induction cues cause maladaptive phenotypes. *Evolutionary Ecology Research 4*: 857–70.

Lawson, D. & R. Mace (2009). Trade-offs in modern parenting: A longitudinal study of sibling competition for parental care. *Evolution and Human Behavior* (in press).

Lear, J.T., B.B. Tan, A.G. Smith, *et al.* (1997). Risk factors for basal cell carcinoma in the UK: case-control study in 806 patients. *Journal of the Royal Society of Medicine 90*: 371–4.

Levy, S., G. Sutton, P.C. Ng, *et al.* (2007). The diploid genome sequence of an individual human. *PLOS Biology 5*, e254.

Lewin, R. & R.A. Foley (2003). *Principles of Human Evolution*. 2nd edn. Oxford: Blackwell.

Li, W.-H., M. Tanimura & P.M. Sharp (1987). An evaluation of the molecular clock hypothesis using mammalian DNA sequences. *Journal of Molecular Evolution 25*: 330–42.

Lieberman, D., J. Tooby & L. Cosmides (2003). Does morality have a biological basis? An empirical test of the factors governing moral sentiments relating to incest. *Proceedings of the Royal Society of London, B: Biological Sciences 270*: 819–26.

Lindblad-Toh, K., C.M. Wade & T.S. Mikkelsen, *et al.* (2005). Genome sequence, comparative analysis and haplotype structure of the domestic dog. *Nature 438*: 803–19.

Lively, C.M. (1987). Evidence from a New Zealand snail for the maintenance of sex by parasitism. *Nature 328*: 519–21.

Lively, C.M. & M.F. Dybdahl (2000). Parasite adaptation to locally common host genotypes. *Nature 405*: 679–81.

Lynch, M. & J.S. Connery (2003). The origins of genome complexity. *Science 302*: 1401–4.

Marlowe, F. (2000). Paternal investment and the human mating system. *Behavioural Processes 51*: 45–61.

Marmot, M.G., G.D. Smith, S. Stansfeld, *et al.* (1991). Health inequalities among British civil servants: The Whitehall II study. *The Lancet 337*: 1387–93.

Mathews, F., P.J. Johnson & A. Neil (2008). You are what your mother eats: Evidence for maternal pre-conception diet influencing foetal sex in humans. *Proceedings of the Royal Society of London, B: Biological Sciences 275*: 1661–8.

Maynard Smith, J. (1978). *The Evolution of Sex*. Cambridge: Cambridge University Press.

Maynard Smith, J. & E. Szathmáry (1995). *The Major Transitions in Evolution*. Oxford: Oxford University Press.

McBurney, D.H., J. Simon, S.J.C. Gaulin & A. Geliebter (2002). Matrilateral biases in the investment of aunts and uncles in a population presumed to have high paternity certainty. *Human Nature 13*: 391–402.

McElreath, R. & R. Boyd. (2007). *Mathematical Models of Social Evolution: A Guide for the Perplexed*. Chicago: University of Chicago Press.

Migliano, A.B., L. Vinicius & M.M. Lahr (2007). Life history trade-offs explain the evolution of human pygmies. *Proceedings of the National Academy of Sciences of the USA 104*: 20216–9.

Milinski, M., D. Semmann & H.-J. Krambeck (2002). Reputation helps solve the 'tragedy of the commons'. *Nature 415*: 424–6.

Milinski, M., D. Semmann, H.-J. Krambeck, & J. Marotzke (2006). Stabilizing the earth's climate is not a losing game: Supporting evidence from public goods experiments. *Proceedings of the National Academy of Sciences of the USA 103*: 3994–8.

Miller, J.D., E.C. Scott & S. Okamoto (2006). Public acceptance of evolution. *Science 313*: 765–6.

Møller, A.P. (1988). Female choice selects for male sexual tail ornaments in a swallow. *Nature 339*: 132–5.

Møller, A.P. (1990). Effects of a haematophagous mite on the barn swallow *Hirundo rustica*: A test of the Hamilton and Zuk hypothesis. *Evolution 44*: 771–84.

Mueller, U. & Mazur, A. 2001 Evidence of unconstrained directional selection for male tallness. *Behavioural Ecology and Sociobiology 50*: 302–311.

Munsinger, H. & A. Douglass (1976). The syntactic abilities of identical twins, fraternal twins, and their siblings. *Child Development 47*: 40–50.

Murdock, G.P. (1967). *Ethnographic Atlas*. Pittsburgh: University of Pittsburgh Press.

Nelson, L.D. & E.L. Morrison (2005). The symptoms of resource scarcity. *Psychological Science 16*: 167–73.

Nesby-O'Dell S., K.S. Scanlon, M.E. Cogswell, *et al.* (2002). Hypovitaminosis D prevalence and determinants among African American and white women of reproductive age: Third National Health and Nutrition Examination Survey, 1988–1994. *American Journal of Clinical Nutrition 76*: 187–92.

Nesse, R.M. & G.C. Williams (1995). *Evolution and Healing: The New Science of Darwinian Medicine*. London: Wiedenfeld & Nicholson. Original US edition was *Why We Get Sick*, New York: Random House, 1994.

Nettle, D. (2002). Women's height, reproductive success and the evolution of sexual dimorphism in modern humans. *Proceedings of the Royal Society, B: Biological Sciences 269*: 1919–23.

Nettle, D. (2006). The evolution of personality variation in humans and other animals. *American Psychologist 61*: 622–31.

Nettle, D. (2008). Why do some dads get more involved than others? Evidence from a large British cohort. *Evolution and Human Behavior 29*: 416–23.

Nettle, D. & T.V. Pollet (2008) Natural selection on male wealth in humans. *American Naturalist 172*: 658–66.

New, J., L. Cosmides & J. Tooby (2007). Category-specific attention for animals reflects ancestral priorities, not expertise. *Proceedings of the National Academy of Sciences of the USA 104*: 16593–603.

Nikaido, M., A.P. Rooney & S.R. Palumbi (1999). Phylogenetic relationships amongst cetartiodactyls based on insertions of long and short interspersed elements: Hippopotamuses are the closest extant relatives of whales. *Proceedings of the National Academy of Sciences of the USA 96*: 10261–6.

Nilsson, D.E. & S. Pelger (1994). A pessimistic estimate of the time required for an eye to evolve. *Proceedings of the Royal Society, B: Biological Sciences 256*: 53–8.

Nowak, R. (1994). Mining treasures from 'junk' DNA. *Science* 263: 608–10.

Okasha, S. (2006). *Evolution and the Levels of Selection.* Oxford: Oxford University Press.

Okoro, A.N. (1975). Albinism in Nigeria: A clinical and social study. *British Journal of Dermatology* 92: 485–492.

Orgel, L.E. & F.H. Crick (1980). Selfish DNA: The ultimate parasite. *Nature* 284: 604–7.

Packer, C., D. Scheel & A.E. Pusey (1990). Why lions form groups: Food is not enough. *American Naturalist* 136: 1–19.

Panchanathan, K. & R. Boyd (2004). Indirect reciprocity can stabilize cooperation without the second-order free rider problem. *Nature* 432: 499–501.

Panopoulou, G. & A.J. Poustka (2005). Timing and mechanism of ancient vertebrate genome duplications—the adventure of a hypothesis. *Trends in Genetics* 21: 559–67.

Parker, G.A. & J. Maynard-Smith (1990). Optimality theory in evolutionary biology. *Nature* 348: 27–33.

Pawlowski, B., Dunbar, R.I.M. & Lipowicz, A. 2000 Tall men have more reproductive success. *Nature* 403: 156.

Penke, L., J.A. Denissen & G.F. Miller (2007). The evolutionary genetics of personality. *European Journal of Personality* 21: 549–87.

Penny, D., L.R. Foulds & M.D. Hendy (1982). Testing the theory of evolution by comparing the phylogenetic trees constructed from five different protein sequences. *Nature* 297: 197–200.

Pesonen, A.-K., K. Raikonnen, K. Heinonen, E. Kajantie, T. Forsen & J.G. Eriksson (2008). Reproductive traits following a parent–child separation trauma during childhood: A natural experiment following World War II. *American Journal of Human Biology* 20: 345–51.

Platek, S.M., D.M. Raines, G.G. Gallup, *et al.* (2004). Reaction to children's faces: Males are more affected by resemblance than females are, and so are their brains. *Evolution and Human Behavior* 25: 394–405.

Plomin, R. & D. Daniels (1987). Why are children from the same family so different from each other? *Behavioral and Brain Sciences* 10: 1–16.

Plomin, R. & Y. Kovas (2005). Generalist genes and learning disabilities. *Psychological Bulletin* 131: 592–617.

Pope, T.R. (1998). Effects of demographic change on group kin structure and gene dynamics of populations in red howler monkeys. *Journal of Mammalogy* 79: 692–712.

Pollet, T.V., D. Nettle & M. Nelissen (2007). Maternal grandmothers do go the extra mile: Factoring distance and lineage into differential investment in grandchildren. *Evolutionary Psychology* 5: 832–43.

Pride, E.R. (2005). Optimal group size and seasonal stress in ring-tailed lemurs (*Lemur catta*). *Behavioral Ecology* 16: 550–60.

Promislow, D.E.L. & P.H. Harvey (1990). Living fast and dying young: A comparative analysis of life-history variation among mammals. *Journal of Zoology* 220: 417–37.

Raaum, R.L., K.N. Sterner, C.M. Noviello, C.B. Stewart & T.R. Disotell (2005). Catarrhine primate divergence dates estimated from complete mitochondrial genomes: concordance with fossil and nuclear DNA evidence. *Journal of Human Evolution* 48: 237–57.

Randerson, J.P. & L.D. Hurst (2001). The uncertain evolution of the sexes. *Trends in Ecology and Evolution* 16: 571–9.

Ratcliffe, J.M., M. Brock Fenton & B.G. Galef (2003). An exception to the rule: Common vampire bats do not learn taste aversions. *Animal Behaviour* 65: 385–9.

Redon, R., S. Ishikawa, K.R. Fitch, *et al.* (2006). Global variation in copy number in the human genome. *Nature* 444: 444–54.

Richardson, K. & S. Norgate (2005). The equal environments assumption of classical twin studies may not hold. *British Journal of Educational Psychology* 75: 339–50.

Richerson, P.J. & R. Boyd (2004). *Not by Genes Alone: How Cultured Transformed Human Evolution.* Chicago: University of Chicago Press.

Ridley, M. (1993). *The Red Queen: Sex and the Evolution of Human Nature.* London: Penguin.

Ridley, M. (1996). *Evolution.* 2nd edn. Cambridge, MA: Blackwell.

Roberts, G. (2005). Cooperation through inter-dependence. *Animal Behaviour 70*: 901–8.

Roberts, S.C. & L.M. Gosling (2003). Genetic similarity and quality interact in mate choice decisions by female mice. *Nature Genetics 35*: 103–6.

Rogers, A.R. (1988). Does biology constrain culture? *American Anthropologist 90*: 819–31.

Roff, A. (1992). *The Evolution of Life Histories*. London: Chapman & Hall.

Romero, T. & F. Aureli (2008). Reciprocity of support in coatis (*Nasua nasua*). *Journal of Comparative Psychology 122*: 19–25.

Rouquier, S., S. Taviaux, B.J. Trask, *et al.* (1998). Distribution of olfactory receptor genes in the human genome. *Nature Genetics 18*: 243–50.

Rozin, P. & J.W. Kalat (1971). Specific hungers and poison avoidance as adaptive specializations of learning. *Psychological Review 78*: 459–86.

Sapolsky, R.M. (2005). The influence of social hierarchy on primate health. *Science 348*: 648–52.

Schmitt. D.P. and 118 members of the International Sexuality Description Project (2003). Universal sex differences in the desire for sexual variety: Tests from 52 nations, 6 continents and 13 islands. *Journal of Personality and Social Psychology 85*: 85–104.

Sear, R. & R. Mace (2008). Who keeps kin alive? A review of the effects of kin on child survival. *Evolution and Human Behavior 29*: 1–18.

Sear, R., R. Mace, D. Shanley & I.A. McGregor (2001). The fitness of twin mothers: Evidence from rural Gambia. *Journal of Evolutionary Biology 14*: 433–43.

Sear, R., D. Lawson & T.E. Dickins (2007). Synthesis in the human evolutionary behavioural sciences. *Journal of Evolutionary Psychology 5*: 3–28.

Sella, G. & D.H. Ardell (2006). The coevolution of genes and genetic codes: Crick's frozen accident revisited. *Journal of Molecular Evolution 63*: 297–313.

Shanley, D.P., R. Sear, R. Mace & T.B.L. Kirkwood (2007). Testing evolutionary theories of menopause. *Proceedings of the Royal Society of London, B 274*: 2943–9.

Shepher, J. (1971). Mate selection amongst second generation kibbutz adolescents and adults: Incest avoidance and negative imprinting. *Archives of Sexual Behavior 1*: 293–307.

Sherman, P.W. & S.M. Flaxman (2002). Nausea and vomiting of pregnancy in an evolutionary perspective. *American Journal of Obstetrics and Gynecology 186*: S190–7.

Shtulman, A. (2006). Qualitative divergences between naïve and scientific theories of evolution. *Cognitive Psychology 52*: 170–94.

Shultz, S. & R.I.M. Dunbar (2007). The evolution of the social brain: Anthropoid primates contrast with other vertebrates. *Proceedings of the Royal Society of London, B 274*: 2429–36.

Sibly, R.M. & J.H. Brown (2007). Effects of body size and lifestyle on evolution of mammal life histories. *Proceedings of the National Academy of Sciences of the USA 104*: 17707–12.

Silk, J.B. (2007). The adaptive value of sociality in mammalian groups. *Proceedings of the Royal Society, B 362*: 539–59.

Simmons, J. (1996). *The Scientific 100: A Ranking of the Most Influential Scientists, Past and Present*. New York: Citadel Press.

Smith, E.A. (2000). Three styles in the evolutionary analysis of human behavior. In L. Cronk, N. Chagnon & W. Irons (eds), *Adaptation and Human Behavior: An Anthropological Perspective*, pp. 27–47. New York: Aldine de Gruyter.

Sommerfeld, R.D., H.-J. Krambeck, D. Semmann & M. Milinski (2007). Gossip as an alternative for direct observation in games of indirect reciprocity. *Proceedings of the National Academy of Sciences of the USA 104*: 17435–40.

Stearns, S.C. (1992). *The Evolution of Life Histories*. New York: Oxford University Press.

Stevenson, R.E., W.P. Allen, P. Shashidhar, *et al.* (2000). Decline in prevalence of neural tube defects in a high-risk region of the United States. *Pediatrics 106*: 677–83.

Strachan, T. & A.P. Read (2003). *Human Molecular Genetics 3*. New York: Garland.

Stromswold, K. (2001). The heritability of language: A review and meta-analysis of twin, adoption and linkage studies. *Language 77*: 647–723.

Sutter, N.B., C.D. Bustamente, K. Chase, *et al.* (2007). A single *IGF1* allele is a major determinant of small size in dogs. *Science* 316: 112–5.

Taft, R.J., M. Pheasant & J.S. Mattick (2007). The relationship between non-protein-coding DNA and eukaryotic complexity. *BioEssays* 29: 288–99.

Terkel, J. (1996). Cultural transmission of feeding behavior in the black rat (*Rattus rattus*). In C.M. Heyes & B.G. Galef (eds), *Social Learning in Animals: The Roots of Culture*, pp. 17–48. San Diego: Academic Press.

Thewissen J.G.M., E.M. Williams, L.J. Roe & S.T. Hussain (2001). Skeletons of terrestrial cetaceans and the relationship of whales to artiodactyls. *Nature* 413: 277–81.

Tinbergen, N. (1963). On aims and methods in ethology. *Zeitschrift für Tierpsychologie 20*: 410–33.

Tinbergen, N., G.H. Broekhuysen, F. Feekes, *et al.* (1962). Egg shell removal by the black-headed gull *Larus ridibundus*: A behaviour component of camouflage. *Behaviour 19*: 74–117.

Tooby, J. & L. Cosmides (1992). The psychological foundations of culture. In J.H. Barkow, L. Cosmides & J. Tooby (eds), *The Adapted Mind: Evolutionary Psychology and the Generation of Culture*, pp. 19–136. New York: Oxford University Press.

Tovée, M.J., V. Swami, A. Furnham & R. Mangalparsad (2006). Changing perceptions of attractiveness as observers are exposed to a different culture. *Evolution and Human Behavior 27*: 443–56.

Trivers, R.L. (1971). The evolution of reciprocal altruism. *Quarterly Review of Biology 46*: 35–57.

Trivers, R.L. (1972). Parental investment and sexual selection. In B. Campbell (ed.), *Sexual Selection and the Descent of Man*, pp. 139–79. Chicago: Aldine.

Trivers, R.L. (1974). Parent–offspring conflict. *American Zoologist 14*: 249–64.

Trivers, R.L. & D.E. Willard (1973). Natural selection of parental ability to vary the sex ratio of offspring. *Science 191*: 249–63.

Trut, L.N. (1999). Early canid domestication: The farm-fox experiment. *American Scientist 87*: 160–70.

Turkheimer, E., A. Haley, M. Waldron, B. Onofrio & I.I. Gottesman (2003). Socioeconomic status modifies the heritability of IQ in young children. *Psychological Science 14*: 623–8.

Uskul, A.K., S. Kitayama & R.E. Nisbett (2008). Eco-cultural basis of cognition: Farmers and fishermen are more holistic than herders. *Proceedings of the National Academy of Sciences of the USA 105*: 8552–6.

Vassilieva, L.L., A.M. Hook & M. Lynch (2000). The fitness effects of spontaneous mutations in *Caenorhabditis elegans*. *Evolution 54*: 1234–46.

Voight, B.F., S. Kudaravalli, X. Wen & J.K. Pritchard (2006). A map of recent positive selection in the human genome. *PLoS Biology 4*: e72.

Wang, E.T., G. Kodama, P. Baldi & R.K. Moyzis (2006). Global landscape of recent inferred Darwinian selection for *Homo sapiens*. *Proceedings of the National Academy of Sciences of the USA 103*: 135–140.

Wedekind, C. & S. Furi (1997). Body odour preferences in men and women: Do they aim for specific MHC combinations or simply heterozygosity? *Proceedings of the Royal Society, B: Biological Sciences 264*: 1471–9.

Wedekind, C. & M. Milinski (2000). Cooperation through image scoring in humans. *Science 288*: 850–2.

West, S.A., C.M. Lively & A.F. Read (1999). A pluralist approach to sex and recombination. *Journal of Evolutionary Biology 12*: 1003–12.

West, S.A., A.S. Griffin & A. Gardner (2006). Social semantics: Altruism, cooperation, mutualism, strong reciprocity and group selection. *Journal of Evolutionary Biology 82*: 327–48.

West, S.A., A.S. Griffin & A. Gardner (2007). Evolutionary explanations for cooperation. *Current Biology 17*: R661–72.

Wheeler, P.E. (1991). The thermoregulatory advantages of hominid bipedalism in open equatorial environments: The contribution of increased convective heat loss and cutaneous evaporative cooling. *Journal of Human Evolution 21*: 107–15.

White, S.A., S.E. Fisher, D.H. Geschwind, C. Scharff & T.E. Holy (2006). Singing mice, songbirds, and more: Models for FOXP2 function and dysfunction in human speech and language. *Journal of Neuroscience 26*: 10376–9.

White, T.D., B. Asfaw, D. DeGusta, *et al.* (2003). Pleistocene *Homo sapiens* from Middle Awash, Ethiopia. *Nature* 423: 742–7.

Wilson, D.S. (2007). *Evolution for Everyone: How Evolution Can Change the Way We Think About Our Lives.* New York: Delacorte Press.

Wilson, D.S. & E.O. Wilson (2007). Rethinking the theoretical foundation of socio-biology. *Quarterly Review of Biology* 82: 327–48.

Wilson, E.O. & B. Hölldobler (2005). Eusociality: Origin and consequences. *Proceedings of the National Academy of Sciences of the USA* 102: 13367–71.

Wilson, M. & M. Daly (1985). Competitiveness, risk-taking and violence: The young male syndrome. *Ethology and Sociobiology* 6: 59–73.

Wilson, M. & M. Daly (1997). Life expectancy, economic inequality, homicide, and reproductive timing in Chicago neighbourhoods. *British Medical Journal* 314: 1271–4.

Winterhalder, B. & E.A. Smith (2000). Analyzing adaptive strategies: Human behavioral ecology at twenty-five. *Evolutionary Anthropology* 9: 51–72.

Witte, K., U. Hirschler & E. Curio (2000). Sexual imprinting on a novel adornment influences mate preferences in the Javanese Mannikin *Lonchura leucogastroides.* *Ethology* 106: 349–63.

Wolf, A.P. (1995). *Sexual Attraction and Childhood Association: A Chinese Brief for Edward Westermarck.* Stanford: Stanford University Press.

Wolff, J.O. & D.W. Macdonald (2004). Promiscuous females protect their offspring. *Trends in Ecology and Evolution* 19: 127–34.

Wood, B. (2002). Hominid revelations from Chad. *Nature* 418: 133–5.

Wynne-Edwards, J.C. (1962). *Animal Dispersion in Relation to Social Behaviour.* Edinburgh: Oliver & Boyd.

Young, L.J., R. Nilsen, K.G. Waymire, *et al.* (1999). Increased affiliative response to vasopressin in mice expressing the V1a receptor from a monogamous vole. *Nature* 400: 766–8.

Zahavi, A. (1975). Mate selection—a selection for a handicap. *Journal of Theoretical Biology* 53: 205–14.

Zeifman, D.M. (2001). An ethological analysis of human infant crying: Answering Tinbergen's four questions. *Developmental Psychobiology* 39: 265–85.

Figure acknowledgements

Figure 1.2 Reproduced from Figure 14.11, p. 506, in L. Wolpert, T. Jessell, P. Lawrence, E. Meyerowitz, E. Robertson & J. Smith (2006). *Principles of Development*, 3rd edn. By permission of Oxford University Press.

Figure 1.6 Reproduced from Figure 2.8a, p. 46, in S. Freeman & J.C. Herron (2004). *Evolutionary Analysis*. © 2004 by Pearson Education, Inc. Reprinted by permission.

Figure 1.9a From P.T. Boag & P.R. Grant (1981). Intense natural selection in a population of Darwin's finches (Geospizinae) in the Galápagos. *Science* 214: 82–5. Reprinted with permission from AAAS.

Figure 1.9b Reproduced from P.R. Grant (1986). *Ecology and Evolution of Darwin's Finches*. © 1986 Princeton University Press. Reprinted by permission of Princeton University Press.

Figure 1.9c Reprinted by permission from Frontiers in Bioscience: P. Grant & B. Grant (2003). What Darwin's finches can teach us about the evolutionary origin and regulation of biodiversity. *BioScience* 53, 965–75.

Figure 1.10a Reproduced from Figure 2.8a, p. 46, in S. Freeman & J.C. Herron (2004). *Evolutionary Analysis*, 3rd edn. © 2004 by Pearson Education, Inc. Reprinted by permission.

Figure 1.10b Reproduced from P.D. Gingerich, B.H. Smith & E.L. Simons (1990). Hind limbs of Eocene *Basilosaurus*: Evidence of feet in whales. *Science* 249: 154–7. Reprinted with permission from AAAS.

Figure 1.10c Reproduced from J.G.M. Thewissen, S.T. Hussain & M. Arif (1994). Fossil evidence for the origin of aquatic locomotion in archaeocete whales. *Science* 263: 210–12. Reprinted with permission from AAAS.

Figure 1.11 Reproduced from P.W. Sherman & S.M. Flaxman (2002). Nausea and vomiting of pregnancy in an evolutionary perspective. *American Journal of Obstetrics and Gynecology* 186: S190–7.

Figure 1.12 Reproduced from P.W. Sherman, & S.M. Flaxman (2002). Nausea and vomiting of pregnancy in an evolutionary perspective. *American Journal of Obstetrics and Gynecology* 186: S190–7.

Figure 2.11 Reproduced from L.L. Vassilieva, A.M. Hook & M. Lynch (2000). The fitness effects of spontaneous mutations in *Caenorhabditis elegans*. *Evolution* 54: 1234–46.

Figure 2.14 Reproduced from N.B. Sutter, C.D. Bustamente, K. Chase, *et al.* (2007). A single IGF1 allele is a major determinant of small size in dogs. *Science* 316: 112–5.

Figure 2.15 Reprinted by permission from Macmillan Publishers Ltd: L.J. Young, R. Nilsen, K.G. Waymire, *et al.* (1999). Increased affiliative response to vasopressin in mice expressing the V1a receptor from a monogamous vole. *Nature* 400: 766–8. © 1999.

Figure 3.9 Reproduced from E. Turkheimer, A. Haley, M. Waldron, B. Onofrio & I.I. Gottesman (2003). Socioeconomic status modifies the heritability of IQ in young children. *Psychological Science* 14: 623–8.

Table 3.2 Adapted from T.J. Bouchard, M. McGue, Y.-M. Hur & J. M. Horn (1998). A genetic and environmental analysis of the California Psychological Inventory using adult twins reared apart and together. *European Journal of Personality* 12: 307–20.

Table 3.3 Adapted from H. Munsinger & A. Douglass (1976). The syntactic abilities of identical twins, fraternal twins, and their siblings. *Child Development* 47: 40–50.

Figure 4.8 Adapted from J.L. Hoogland (1983). Nepotism and alarm calling in the black-tailed prairie dog (*Cynomys ludovicianus*). *Animal Behaviour* 31: 472–9.

Figure 5.1 Adapted from D. Nettle (2002). Women's height, reproductive success and the evolution of sexual dimorphism in modern humans. *Proceedings of the Royal Society, B: Biological Sciences* 269: 1919–23.

Figure 5.2 Adapted from D. Nettle (2002). Women's height, reproductive success and the evolution of sexual dimorphism in modern humans. *Proceedings of the Royal Society, B: Biological Sciences* 269: 1919–23.

Figure 5.3 Reprinted from J.W. Dudley (2007). From means to QTL: The Illinois long-term selection experiment as a case study in quantitative genetics. *Crop Science* 47: S3, S20–31. With the kind permission of Professor John Dudley.

Figure 5.6 Reproduced from M.T. Gross (1991). Evolution of alternative reproductive strategies: Frequency-dependent sexual selection in male bluegill sunfish. *Philosophical Transactions of the Royal Society, B: Biological Sciences* 332: 59–66.

Figure 5.7 Adapted from N.J. Dingemanse, C. Both, P.J. Drent & J.M. Tinbergen (2004). Fitness consequences of avian personalities in a fluctuating environment. *Proceedings of the Royal Society, B: Biological Sciences* 271: 847–52.

Figure 5.8 Adapted from K. Foerster, T. Coulson, B.C. Sheldon, J.M. Pemberton, T.H. Clutton-Brock & L.E. Kruuk (2007). Sexually antagonistic genetic variation for fitness in red deer. *Nature* 447: 1107–10.

Figure 5.9 Adapted from D.E. Nilsson & S. Pelger (1994). A pessimistic estimate of the time required for an eye to evolve. *Proceedings of the Royal Society, B: Biological Sciences* 256: 53–8.

Figure 5.11 Adapted from R.E. Stevenson, W.P. Allen, P. Shashidhar, *et al.* (2000). Decline in prevalence of neural tube defects in a high-risk region of the United States. *Pediatrics* 106: 677–83; and S. Nesby-O'Dell, K.S. Scanlon, M.E. Cogswell, *et al.* (2002). Hypovitaminosis D prevalence and determinants among African American and white women of reproductive age: Third National Health and Nutrition Examination Survey, 1988–1994. *American Journal of Clinical Nutrition* 76: 187–92.

Figure 5.12 Adapted from N.G. Jablonski & G. Chaplin (2000). The evolution of human skin coloration. *Journal of Human Evolution* 39: 57–106.

Figure 6.3 Adapted from M.R. Goddard, H. Charles, J. Godfray & A. Burt (2005). Sex increases the efficacy of natural selection in experimental yeast populations. *Nature* 434: 636–40.

Figure 6.4 Reproduced from M. Pagel (ed.) *The Encyclopedia of Evolution*, p. 1027. Oxford: Oxford University Press. Adapted from C.M. Lively (1987). Evidence from a New Zealand snail for the maintenance of sex by parasitism. *Nature* 328: 519–21.

Figure 6.5 Adapted from S. Consuegra & C. Garcia de Leaniz (2008). MHC-mediated mate choice increases parasite resistance in salmon. *Proceedings of the Royal Society, B: Biological Sciences* 275: 1397–403.

Figure 6.7 Adapted from A.G. Jones, J.R. Arguello & S.J. Arnold (2002). Validation of Bateman's principles: A genetic study of sexual selection and mating patterns in the rough-skinned newt. *Proceedings of the Royal Society, B: Biological Sciences* 269: 2533–9.

Figure 6.8 Adapted from M. Borgerhoff Mulder (1988). Reproductive success in three Kipsigis cohorts. In T.H. Clutton- Brock (ed.), *Reproductive Success: Studies of Individual Variation in Contrasting Breeding Systems*, pp. 419–35. Chicago: University of Chicago Press.

Figure 6.9 Adapted from R.D. Alexander, J.L. Hoogland, R.D. Howard, K.M. Noonan & P.W. Sherman (1979). Sexual dimorphism and breeding systems in pinnipeds, ungulates, primates and humans. In N.A. Chagnon & W. Irons (eds), *Evolutionary Biology and Human Social Behavior: An Anthropological Perspective*, pp. 402–35. Scituate, MA: Duxbury Press.

Figure 6.10 Adapted from A.P. Møller (1988). Female choice selects for male sexual tail ornaments in a swallow. *Nature* 339: 132–5.

Figure 6.11 Reprinted by permission from Macmillan Publishers Ltd: T.C.M. Baker (1993). Positive genetic correlation between female preference and preferred male ornament in sticklebacks. *Nature* 363, 255–7. © 1993.

Figure 6.12 Adapted from Møller, A.P. (1990). Effects of a haematophagous mite on the barn swallow *Hirundo rustica*: A test of the Hamilton and Zuk hypothesis. *Evolution* 44: 771–84.

Figure 6.13 Adapted from A.G. Jones, D. Walker & J.C. Avise (2001). Genetic evidence for extreme polyandry and extraordinary sex-role reversal in a pipefish. *Proceedings of the Royal Society, B: Biological Sciences* 268: 2531–5.

Figure 6.14 Adapted from B. Kempenaers, G.R. Verheyen & A.A. Dhondt (1997). Extra-pair paternity in the blue tit (*Parus caerulus*): Female choice, male characteristics and offspring quality. *Behavioral Ecology* 8: 481–92.

Figure 6.15 Adapted from M. Wilson & M. Daly (1985). Competitiveness, risk-taking and violence: The young male syndrome. *Ethology and Sociobiology* 6: 59–73.

Figure 6.16 Adapted from D.P. Schmitt, L. Alcalay, J. Allik, *et al.* (2003). Universal sex differences in the desire for sexual variety: Tests from 52 nations, 6 continents and 13 islands. *Journal of Personality and Social Psychology* 85: 85–104.

Figure 6.17 Adapted from D.M. Buss (1989). Sex differences in human mate preferences: Evolutionary hypotheses tested in 37 cultures. *Behavioral and Brain Sciences* 12: 39–49.

Table 6.1 Adapted from M. Daly & M. Wilson (1983). *Sex, Evolution and Behavior*, 2nd edn. Belmont, CA: Wadsworth; M. Borgerhoff Mulder (1988). Reproductive success in three Kipsigis cohorts. In T.H. Clutton-Brock (ed.), *Reproductive Success: Studies of Individual Variation in Contrasting Breeding Systems*, pp. 419–35. Chicago: University of Chicago Press; and D. Nettle (2008). Why do some dads get more involved than others? Evidence from a large British cohort. *Evolution and Human Behavior* 29: 416–23.

Figure 7.3 Adapted from D.E.L. Promislow & P.H. Harvey (1990). Living fast and dying young: A comparative analysis of life-history variation among mammals. *Journal of Zoology* 220: 417–37.

Figure 7.4 Adapted from A.B. Migliano, L. Vinicius & M.M. Lahr (2007). Life history trade-offs explain the evolution of human pygmies. *Proceedings of the National Academy of Sciences of the USA* 104: 20216–9.

Figure 7.5 Adapted from M. Wilson & M. Daly (1997). Life expectancy, economic inequality, homicide, and reproductive timing in Chicago neighbourhoods. *British Medical Journal* 314: 1271–4.

Figure 7.7 © Paul Heasman/Fotolia.com. Adapted from S. Daan, C. Dijkstra & J.M. Tinbergen (1990). Family planning in the kestrel (*Falco tinninculus*): The ultimate control of covariation of laying date and clutch size. *Behaviour* 114: 83–116.

Figure 7.8 Adapted from R. Sear, R. Mace, D. Shanley & I.A. McGregor (2001). The fitness of twin mothers: Evidence from rural Gambia. *Journal of Evolutionary Biology* 14: 433–43.

Figure 7.10 © Martin McCarthy/istock.com. Adapted from T.H. Clutton-Brock, S.D. Albon & F.E. Guinness (1986). Great expectations: Dominance, breeding success and offspring sex ratios in red deer. *Animal Behaviour* 34: 460–71.

Figure 7.11 Adapted from M. Gibson & R. Mace (2003). Strong mothers bear more sons in rural Ethiopia. *Proceedings of the Royal Society of London (Biology Letters)* 270: S108–9.

Figure 7.14 Adapted from M. Daly & M. Wilson (1988). Evolutionary social psychology and family homicide. *Science* 242: 519–24.

Figure 8.2 Adapted from R.E. Kenward (1978). Hawks and doves: Factors affecting success and selection in goshawk attacks on wood pigeons. *Journal of Animal Ecology* 47: 449–60.

Figure 8.3 Adapted from C. Packer, D. Scheel & A.E. Pusey (1990). Why lions form groups: Food is not enough. *American Naturalist* 136: 1–19.

Figure 8.4 Adapted from K. Isvaran (2007). Intraspecific variation in group size in the blackbuck antelope: The roles of habitat structure and forage at different spatial scales. *Oecologia* 154: 435–44.

Figure 8.5 Adapted from K.B. Armitage & O.A. Schwartz (2000). Social enhancement of fitness in yellow-bellied marmots. *Proceedings of the National Academy of Sciences of the USA* 97: 12149–52.

Figure 8.6 Adapted from E.R. Pride (2005). Optimal group size and seasonal stress in ring-tailed lemurs (*Lemur catta*). *Behavioral Ecology* 16: 550–60.

Figure 8.8 Adapted from R.I.M. Dunbar & S. Shultz (2007). Understanding primate brain evolution. *Philosophical Transactions of the Royal Society, B* 362: 649–658.

Figure 8.9 Adapted from R.I.M. Dunbar & S. Shultz (2007). Evolution in the social brain. *Science* 317: 1344–7.

Figure 8.12 Reproduced with the permission of Royal Society Publishing from M. Bateson, D. Nettle & G. Roberts (2006). Cues of being watched enhance cooperation in a real-world setting. *Biology Letters* 12: 412–14.

Figure 8.13 Adapted from E. Fehr & S. Gächter (2002). Altruistic punishment in humans. *Nature* 415: 137–40.

Figure 9.1 Adapted from T.J. DeWitt (1998). Costs and limits of phenotypic plasticity: Tests with predator-induced morphology and life history in a freshwater snail. *Journal of Evolutionary Biology* 11: 465–80; and R.B. Langerhans & T.J. DeWitt (2002). Plasticity constrained: Over-generalized induction cues cause maladaptive phenotypes. *Evolutionary Ecology Research* 4: 857–70.

Figure 9.2 Adapted from L.S. Adair (2001). Size at birth predicts age at menarche. *Pediatrics* 107: e59–66.

Figure 9.4 Adapted from K. Witte, U. Hirschler & E. Curio (2000). Sexual imprinting on a novel adornment influences mate preferences in the Javanese Mannikin *Lonchura leucogastroides*. *Ethology* 106: 349–63.

Figure 9.5 Adapted from T. Bereczkei, P. Gyuris & G.E. Weisfeld (2004). Sexual imprinting in human mate choice. *Proceedings of the Royal Society, B: Biological Sciences* 271: 1129–34.

Figure 9.7 Adapted from B.G. Galef & D.J. White (1998). Mate choice copying in Japanese quail, *Coturnix coturnix japonica*. *Animal Behaviour* 55: 545–52.

Figure 9.8 Adapted from A.R. Rogers (1988). Does biology constrain culture? *American Anthropologist* 90: 819–31; and R. Boyd & P.J. Richerson (1995). Why does culture increase human adaptability? *Ethology and Sociobiology* 16: 125–43.

Figure 9.9a Reprinted from J. Terkel (1996). Cultural transmission of feeding behaviour in the black rat (*Rattus rattus*). In C.M. Hayes & B.G. Galef (eds), *Social Learning in Animals: The Roots of Culture*. With permission from Elsevier.

Figure 9.9b Reprinted by permission from Macmillan Publishers Ltd: F.B.M. de Waal (1999). Cultural primatology comes of age. *Nature* 399: 635–6. © 1999.

Figure 9.10 Reprinted from C. A. Caldwell & A.E. Millen (2008). Experimental models for testing hypotheses about cumulative cultural evolution. *Evolution and Human Behaviour* 29: 165–71. With the kind permission of Dr Christine Caldwell.

Figure 9.11 Adapted from C. A. Caldwell & A.E. Millen (2008). Experimental models for testing hypotheses about cumulative cultural evolution. *Evolution and Human Behavior* 29: 165–71.

Figure 10.4 Adapted from M. Nikaido, A.P. Rooney & S.R. Palumbi (1999). Phylogenetic relationships amongst cetartiodactyls based on insertions of long and short interspersed elements: Hippopotamuses are the closest extant relatives of whales. *Proceedings of the National Academy of Sciences of the USA* 96: 10261–6.

Figure 10.5 Reprinted by permission from Macmillan Publishers Ltd: K. Lindblad-Toh, C. Wade, T. Mikkelsen, *et al.* (2005). Genome sequence, comparative analysis and haplotype structure of the domestic dog. *Nature* 438: 803–19. © 2005.

Figure 10.6a Reprinted by permission from Macmillan Publishers Ltd: M. Friedman (2008). The evolutionary origin of flatfish asymmetry. *Nature* 454: 209–12. © 2008.

Figure 10.6b Reprinted by permission from Macmillan Publishers Ltd: P. Janvier (2008). Palaeontology: Squint of the fossil flatfish. *Nature* 454: 169–70. © 2008.

Figure 10.9 Adapted from R.L. Raaum, K.N. Sterner, C.M. Noviello, C.B. Stewart & T.R. Disotell (2005). Catarrhine primate divergence dates estimated from complete mitochondrial genomes: Concordance with fossil and nuclear DNA evidence. *Journal of Human Evolution* 48: 237–57.

Figure 10.10 Left: Reprinted by permission from Macmillan Publishers Ltd: B. Wood (2002). Palaeoanthropology: Hominid revelations from Chad. *Nature* 418: 133–5. © 2002. Right: Reprinted by permission from Macmillan Publishers Ltd: T. White, B. Asfaw, D. DeGusta, *et al.* (2003). Pleistocene *Homo sapiens* from Middle Awash, Ethiopia. *Nature* 423: 742–7. © 2003.

Figure 10.11 Reprinted by permission from Macmillan Publishers Ltd: M. Ingman, H. Kaessmann, S. Pääbo & U. Gyllensten (2000). Mitochondrial genome variation and the origin of modern humans. *Nature* 408: 708–13, © 2000.

Figure 10.13 Adapted from L.C. Aiello & P. Wheeler (1995). The expensive-tissue hypothesis: The brain and the digestive system in human and primate evolution. *Current Anthropology* 36: 199–221.

Figure 10.14 Adapted from H. Kaplan, K. Hill, J. Lancaster & A.M. Hurtado (2000). A theory of human life history evolution: Diet, intelligence and longevity. *Evolutionary Anthropology* 9: 156–85.

Figure 11.1 Adapted from D. Nettle & T.V. Pollet (2008). Natural selection on male wealth in humans. *American Naturalist* 172: 658–66.

Figure 11.2 Adapted from D.T.A. Eisenberg, B. Campbell, P.B. Gray & M.D. Sorenson (2008). Dopamine receptor genetic polymorphisms and body composition in undernourished pastoralists: An exploration of nutrition indices among nomadic and recently settled Ariaal men of northern Kenya. *BMC Evolutionary Biology* 8: 173.

Figure 11.3 Reprinted from A.K. Uskul, S. Kitayama & R.E. Nisbett (2008). Ecocultural basis of cognition: Farmers and fishermen are more holistic than herders. *Proceedings of the National Academy of Sciences of the USA* 105: 8552–6. © 2008 National Academy of Sciences of the USA.

Figure 11.4a Adapted from M.J. Tovée, V. Swami, A. Furnham & R. Mangalparsad (2006). Changing perceptions of attractiveness as observers are exposed to a different culture. *Evolution and Human Behavior* 27: 443–56.

Figure 11.4b Adapted from L.D. Nelson & E.L. Morrison (2005). The symptoms of resource scarcity. *Psychological Science* 16: 167–73.

Figure 11.5 Adapted from A. Drewnowski & S.E. Specter (2004). Poverty and obesity: The role of energy density and energy costs. *American Journal of Clinical Nutrition* 79: 6–16.

Figure 11.6 Adapted from N. Darmon, E.L. Ferguson & A. Briend (2002). A cost-constraint alone has adverse effects on food selection and nutrient density: An analysis of human diets by linear programming. *Journal of Nutrition* 132: 3764–71; and A. Drewnowski & S.E. Specter (2004). Poverty and obesity: The role of energy density and energy costs. *American Journal of Clinical Nutrition* 79: 6–16.

Table 11.1 Adapted from D.M. Zeifman (2001). An ethological analysis of human infant crying: Answering Tinbergen's four questions. *Developmental Psychobiology* 39: 265–85.

Index